Radio on the Road

The Traveler's Companion

(8th Edition)

William Hutchings

Robert Luce Publishing

Fairfield, CT

Radio on the Road – The Traveler's Companion. Copyright © 1995-2006 by William Hutchings. Printed and bound in the United States of America. All rights reserved. No part of this book may be reproduced in any form or by any electronic or mechanical means including information storage and retrieval systems without permission in writing from the publisher, except by a reviewer, who may quote brief passages in a review. Published by Robert Luce Publishing, 2490 Black Rock Turnpike # 342, Fairfield, CT 06432

Library of Congress Catalog Card Number 95-80287

Hutchings, William
radio on the road – the traveler's companion / William Hutchings

2nd Edition – June 1997
3rd Edition – August 1998
4th Edition – November 1999
5th Edition – November 2000
6th Edition – January 2002
7th Edition – June, 2003
8th Edition – February 2006

Publisher's Cataloging in Publication

Hutchings, William
 Radio on the road : the traveler's companion / William Hutchings.
 p. cm.
 Preassigned LCCN: 95-80287.
 ISBN 0-88331-221-2   UPC 859974001
  1.United States--Guidebooks. 2. Radio stations--United States--Directories.
I. Title.

E158.H88 1995   917.304'929
                           QBI95-20539

NPR is a registered trademark of National Public Radio, Inc.

Attention Corporations and Professional Organizations: Quantity discounts are available on bulk purchases of this book.

For David Sigurd Anderson

Friend and fellow flyfisherman

Introduction

In 1930, Paul Galvin coined the name *Motorola* for a new product, linking the concept of motion and radio. A small company was born. In 1930, a car radio was a rarity. Today, there are more than 145,000,000 cars and nearly 45,000,000 trucks on the road. Over 95 percent of these vehicles have a radio - a total of nearly 190,000,000 radios on the road. *Motorola* became an electronics giant.

In 1930, there were fewer than 800 radio stations in the entire United States. Sixty-five years later, *Radio on the Road* has listed 14,187 licensed AM, FM, and TV stations in the U.S. and Canada. Each station is listed by its program format.

The combination of so many stations and programming formats presents a frustrating problem for the listener without some form of guide. Not only must we know what stations are in what town, we need to know what formats these stations are transmitting.

We no longer have the simple formats of news, drama, weather, and sports. Today's broadcast programming is specialized, with continuity ranging from rap to religion. We have travel news, business news, world news, and sports news. We have opera, talk radio, radio for kids, and eclectic essays. All one has to do is *tune in*.

Easier said than done. Many times the "Seek" functions on the newer radios won't even pick up the signals of the weaker stations. If we know the frequency, we **can** tune in. But unless we have a guide to the station formats, *we still have a lot of dial-twisting*. With *Radio on the Road*, the traveler, the vacationer, the RV'er, the trucker, the salesman – even the freeway commuter – has a convenient, organized directory to help choose their program of choice without the frustration of constant channel surfing.

What's New - Today

This will be the 8th edition of *Radio on the Road*. For this edition we've added a few new things. Things our readers suggested.

National Public Radio is more popular than ever. More than eighteen million fans listen to NPR, and these fans are devoted. While we have always listed the primary NPR affiliates, we've also included most of their translator-repeaters. This brings a total of nearly 1,200 NPR locations which can be found in a separate section of the book.

Also new in this edition is an expanded section on **Talk Radio.** This element of radio has experienced a phenomenal growth, and National Hosts enjoy audiences that are almost rabid in following their favorites. Our new format includes not only the local station, but the local airtime. Our line-up of talk shows now includes:

Rush Limbaugh is heard on approximately 600 stations nationwide and enjoys radio's largest listening audience. Rush is the dominant force in Talk Radio today. He's earned that position. Rush is controversial, entertaining and very intelligent. Love him or hate him, but listen to his show. (www.rushlimbaugh.com)

Dr. Laura Schlessinger is heard in the U.S. and Canada - with many affiliates and a multitude of loyal fans. She's a best-selling author and deservedly so. Dr. Laura "calls them as she hears them" with a refreshing candor. (www.drlaura.com)

O'Reilly Factor – Bill O'Reilly has become one of Radio's most listened-to hosts. His popularity is growing at an incredible rate. He was banned until recently in Canada, but Bill's presence is now being felt by our neighbors to the north. Bill truly epitomizes "fair and balanced," and has become a true force in broadcasting. (www.billoreilly.com)

Clark Howard – America's popular consumer guru informs us on everything from travel to consumer products, and the scams to be avoided. If you haven't listened to Clark, you've missed a real treat. There's no telling how much money (and grief) he's saved his audience. Clark is great. (www.clarkhoward.com)

Dr. Dean Edell – is America's most listened-to physician. Warm and practical, Dr. Edell has helped millions of listeners with solid advice. His warm and compassionate delivery make it seem as though you are having a personal consultation. (www.drdeanedell.com)

Bob Brinker – Bob's "Money Talk" show is widely respected and followed by a passionate group of investors – new and experienced. His practical advice on investment is delivered in a down to earth approach that will make you a loyal fan, and probably, a wealthier fan. (www.bobbrinker.com)

Car Talk – heard on nearly 700 NPR stations, the zany but practical brothers, Tom and Ray Magliozzi, give us great advice and diagnoses regarding our cars – liberally sprinkled with a raucous humor. Prepare to laugh as you learn. They're like the Marx brothers with wrenches, but they really know cars and are accomplished mechanics. Their loyal listeners are totally addicted to the show and have been known to miss football games, cocktail parties, and Sunday dinner so they could listen up. (www.cartalk.com)

Laura Ingraham - Laura worked as a White House speechwriter in the final two years of the Reagan Administration, went on to graduate from the University of Virginia School of Law, and served as a law clerk to Supreme Court Justice Clarence Thomas. She is unabashedly conservative and very entertaining. Regardless of one's politics, Laura brings a wonderful energy to her listeners. (www.lauraingraham.com)

Prairie Home Companion – the ever-popular and talented Garrison Keillor delights us each weekend with his humor, music, and a wit that will make you laugh so hard your stomach will hurt - guaranteed. Don't miss this wonderful show whether you're traveling or at home in your living room. He tops everyone's list for variety and humor. (www.prairiehomecompanion.com)

Dennis Prager – is a man of great knowledge and common sense. Widely sought after by television shows for his opinions, he's appeared on Larry King Live, Hardball, Hannity & Colmes, CBS Evening News, The Today Show and many others. Listen to Dennis once and you won't want to miss him. (www.dennisprager.com)

Satellite Sisters – what a great group of gals! The five Dolan Sisters – Julie, Lian, Monica, Sheila, and Liz, launched their show in 2000. It has won three Gracie Awards for excellence from American Women in Radio and Television. Their show deals with books, health, humor, entertainment, and contemporary topics. Refreshing and fun! (www.satellitesisters.com)

We've included these hosts and their local stations as well as the local air times. The shows may also be heard on many stations which re-feed (taped) versions of their live broadcasts at different times.

These personalities enjoy an audience of *millions,* and we know you don't want to miss their shows as you travel. You will be able to find them on the radio dial wherever you may travel. Incidentally, all hosts are enjoying a lot of popularity, and popularity means change. By the time this book is published, each host will probably be heard on more affiliates than we've been able to include.

For sports fans, the affiliate stations for the home teams in **Major League Baseball,** the **National Football League,** have been included. If you're an "Angel" fan and you are traveling in another state, you will be able to find a local radio station carrying the "away" games.

Radio's Future

What's coming? The future is now. Select a station for a personalized traffic report – tailored to your route. Press another button and listen to your favorite Talk Show while you're thousands of miles away. What's happening? Digital radio. Satellite transmissions. Satellites are nothing new in communications, but until now, they have been used as transmitters to relay network programs to stations – except for satellite TV.

In the U.S., Lucent Technologies is cooperating with Sirius and XM Radio. The technology enables any car to receive satellite channels. The plan is priced at $9.95 per month – and there are no commercials on the music channels. Service is now available. This is just the beginning.

The FCC is grappling with the means of regulating the digital landscape so that the technology can be presented in an orderly fashion. It will be upon us before you know it, and the impact will be as great as that of Cable for television and the Internet for computers. (They soon may all be homogenous) Digital technology, the language of computers, will transform radio just as it has every other type of media.

Some networks are now "on-line" with the computer "servers" like America On Line, and MSN.   For example, ABC has started their ABC Radio Online. Microsoft has teamed up with NBC. A listener can access NASA broadcasts and listen to live, unedited feeds from the Congress, State Department, Pentagon, National Press Club, the White House, etc. This is a combination of radio and personal computers. Almost all of the "talk shows" can be tuned in over the internet.

While you won't (or shouldn't) be operating a computer from a moving vehicle, there are hundreds of thousands of RV'ers out there who log on when they stop at their favorite campground. In addition, for those travelers with computer modems, **www.broadcast.com** provides a wonderful array of programming on the Internet.

ABC has given e-mail access to many of the popular correspondents such as Paul Harvey and Joe Templeton. They are also providing direct forums with on-line discussions, as well as providing a Home Page on the Internet. National Public Radio also provides E-Mail and internet access. Listener participation and comments are invited just as do most of the major talk show hosts.

The trend is clear: a melding of computers, networks, and radio. Radio is getting stronger as a communications medium. And, the communication opportunities are becoming 2-way. In the meantime, in the military vernacular, "Listen up," and enjoy *Radio on the Road,* whether your preference is Music, News, Sports or Talk.

National Public Radio

Over 18 million listeners a week enjoy the variety of programming presented by National Public Radio, although many people may still be unfamiliar with NPR.

For twenty-five years this network has excelled with an eclectic and wide-ranging programming format. *Morning Edition, All Things Considered,* and *Weekend Edition* – just three of NPR's newsmagazines – have won millions of loyal listeners and every prestigious broadcast journalism award.

Performance Today is a daily two-hour classical music program, but NPR's commitment to a broad scope of musical selection includes *Jazz*

from Lincoln Center, The Composer's Voice, BluesStage, St. Louis Symphony, NPR World of Opera, The Thistle & Shamrock, and JazzSet with Winston Marsalis. These are only a few of their many offerings.

Drama was an early component in the history of radio, and NPR continues the tradition with literary as well as comedy selections on *NPR Playhouse.* And, children (as well as adults) will be enthralled with *Rabbit Ears Radio* hosted by Mel Gibson and Meg Ryan. Sir John Gielgud reading *The Emperor Has No Clothes* is typical fare. For car buffs, the zany show *Car Talk,* will hit home with a large audience. All in all, NPR presents a broad variety of unique programming.

NPR has nearly 700 member stations. Many stations enhance their range by employing repeater locations. We have included a large number of these repeaters in the NPR section. These are especially helpful in mountain locations, since FM broadcasts are line-of-sight.. Programming is distributed by satellite and each station designs its own format, combining local programming with the offerings from NPR. When you're in range of an NPR station, try it - you'll like it. Check the new, separate section for National Pubic Radio. And, when you return from your trip, tune in your local NPR station.

About Formats

Radio Program Directors are an ingenious lot. Their careers are based on choosing a format that will penetrate the market to an extent whereby they can lure advertisers into their web, and increase their rates over their competitors.

Someone coined the phrase "... if you don't like the weather, wait twenty minutes. It'll change." That person was undoubtedly a radio program director. We have done our best to keep *Radio on the Road* current with the latest changes, but there will be changes that have transpired between the writing and publishing.

Stations in *Radio on the Road* are organized and sorted:
   1. By State
   2. By City
   3. By Frequency within the above sorting procedure

The listed "**City**" is generally the locale where the broadcast or transmitter is located.

AM stations **never** contain a decimal, i.e., "640," "1270" etc.
FM stations **always contain a decimal**, i.e., "89.7," "105.3" etc. The number after the decimal is always an "odd" number, i.e., 1, 3, 5, 7, or 9.

The next page lists the format abbreviations we've used.

Format Abbreviations

These are the current definitions used in Radio on the Road.

AC	Adult Contemporary. No hard rock, adult oriented.
AC/H	Adult Contemporary - Hot
AC/S	Adult Contemporary - Soft
AD/A	Adult alternative. College popular. Mix of folk, jazz, rock, and blues
AH	Adult Hits - Music from the 70's forward
AS	Adult Standards – Soft popular music. Many vocals
BIZ	Business news
CH	Classic Hits – Rock and Pop from the 60's through the 80's
CLA	Classical music
CW	Country western – from traditional to classic
EZL	Easy Listening. Mostly instrumental, Some soft rock
JAZ	Jazz – from Dixieland to Progressive
N/Tlk	Primarily News and talk
N/A	Format not available at time of publication.
OLG	Popular music from the 50's, 60's, and 70's.
RK	Rock. Can be "Hard", "Soft, "New," etc.
RK/C	Classic rock
RL	Religion. Mostly educational. Can be any religion
RL/CC	Religion, contemporary Christian. Includes music
RL/G	Religion, gospel. Can be "Southern," "Black," or "Evangelical"
SIL	Silent. Station is currently "off the air." May resume at any time.
SP	Spanish Language & Music
SP/MEX	Primarily Mexican programming
SP/N	Spanish language news
SP/OM	Spanish – other music, i.e., CH, Tropicana, Ranchero
SP/RL	Spanish. Religion
SP/TK	Spanish, Talk
SP/TO	Spanish, Tejano music
SPTS	Sports programming & Talk
TLK+	Talk Plus – generally includes news & weather
TRAV	Travel information
URB	Urban. May be AC, Rhythm & Blues, etc.
VAR	Variety of formats. Many NPR stations are N/TLK, CLA, and Jazz
YTH	Children, Pre-teen, & Teen programming. Many are Disney stations.

Most stations utilize multiple formats. For example, most stations with News also have some Talk (N/TLK) programming. Conversely, (Talk+) stations also are heavy in News. Some Contemporary Hits (CH) stations also play Rock (RK). The formats listed in *Radio on the Road* are the station's primary formats as most recently reported. Talk Radio has become so popular that if a station has multiple formats, i.e., News and Talk, or Country Western and Talk, the format listed will be TLK+.

Table of Contents

Table of Contents

City	Station	Freq	Format	City	Station	Freq	Format
Abbeville	WIZB	94.3	RL/CC	Birmingham	WYSF	94.5	A/CS
Addison	WQAH	105.7	CW	Birmingham	WMJJ	96.5	AC
Alabaster	WQCR	1500	SP/MEX	Birmingham	WZRR	99.5	RK/CL
Albertville	WAVU	630	RL/G	Birmingham	WZZK	104.7	CW
Albertville	WWGC	1090	SP/MEX	Birmingham	WOBT	106.9	OLG
Albertville	WQSB	105.1	CW	Birmingham	WRAX	107.7	RK/NU
Alexander City	WRFS	1050	RL/G	Birmingham	WUHT	107.7	URB
Alexander City	WSTH	106.1	CW	Boaz	WBSA	1300	RL
Andalusia	WSTF	91.5	RL	Brantley	WAOQ	100.3	CW
Andalusia	WAAQ	103.7	CW	Brewton	WEBJ	1240	TLK+
Anniston	WHMA	1390	TLK+	Brewton	WELJ	90.9	RL/SG
Anniston	WDNG	1450	N/TLK	Brewton	WKNU	106.3	CW
Anniston	WANA	1490	RL/G	Bridgeport	WYMR	1480	TLK+
Anniston	WGRW	90.7	RL/CC	Brundidge	WTBF	94.7	A/CS
Arab	WRAB	1380	CW	Butler	WKZB	93.5	RK/CL
Arab	WAFN	92.7	OLG	Calera	WBYE	1370	RL/G
Ashland	Whma	95.5	CW	Camden	WCOX	1450	RL/G
Athens	WVNN	770	TLK+	Camden	WYVD	102.3	RL/G
Athens	WKAC	1080	SPN	Carrollton	WRAG	590	RL/G
Athens	WZYP	104.3	CHR	Carrollton	WZBQ	94.1	CHR
Atmore	WASG	550	RL	Carrville	WACQ	1130	OLG
Atmore	WYOK	104.1	CH	Centre	WEIS	990	CW
Atmore	WNSI	105.9	SPTS	Centre	WZTQ	1560	RL/G
Attalla	WKXX	102.9	AC/H	Centre	WRHY	105.9	CW
Auburn	waud	1230	AS	Centreville	WBIB	1110	CW
Auburn	WEGL	91.1	AD/A	Chickasaw	WDLT	98.3	UR/AC
Auburn	WKKR	97.7	CW	Citronelle	WQUA	102.1	YTH
Bay Minette	WBCA	1110	RL/G	Clanton	WKLF	980	RL/G
Bay Minette	WNSP	105.5	SPT	Columbia	WJJN	92.1	URB
Bessemer	WZGX	1450	SP/MEX	Columbiana	WQEM	101.5	CHR
Birmingham	WAGG	610	RL/G	Cordova	WXJC	92.5	RL/G
Birmingham	WJOX	690	SPT	Cordova	WFFN	95.3	SIL
Birmingham	WYDE	850	TLK+	Cullman	WFMH	1340	TLK+
Birmingham	WXIC	850	RL/G	Cullman	WMCJ	1460	RL
Birmingham	WATV	900	UR/RB	Cullman	WKUL	92.1	TLK+
Birmingham	WERC	960	TLK+	Cullman	WYDE	101.1	TLK+
Birmingham	WAPI	1070	NWS+	Dadevill	WKGA	100.3	CW
Birmingham	WAYE	1220	RL/G	Dadeville	WDLK	1450	RL/G
Birmingham	wyde	1260	TLK+	Dadeville	WELL	88.7	RL/G
Birmingham	WZZK	1320	CW	Daleville	WCMA	1560	SIL
Birmingham	WLJR	88.5	RL	Daphne	WAVH	106.5	OLG
Birmingham	WBFR	89.5	RL	Decatur	WHOS	800	NWS+
Birmingham	WBHM	90.3	N/TLK	Decatur	WWTM	1400	TLK+
Birmingham	WDYF	90.3	RL	Decatur	WAJF	1490	RL
Birmingham	WVSU	91.1	JAZ	Decatur	WYFD	91.7	RL
Birmingham	WBHM	91.1	RK/H	Decatur	WDRM	102.1	CW
Birmingham	WVSU	91.1	JAZ	Demopolis	WXAL	1400	RL/G
Birmingham	WGIB	91.9	RL	Demopolis	WZNJ	106.5	OLG
Birmingham	WGIB	91.9	RL	Dixons Mills	WMBV	91.9	RL
Birmingham	WDJC	93.7	RL/CC	Dora	WPYK	1010	CW

City	Station	Freq	Format	City	Station	Freq	Format
Dothan	WOOF	560	SPT	Glencoe	WGMZ	93.1	TOP
Dothan	WGZS	700	SP/MX	Glencoe	WGMX	93.1	POP
Dothan	WAGF	1320	RL/G	Goodwater	WZLM	97.5	A/C
Dothan	WWNT	1450	TLK+	Greensboro	WDGM	99.1	OLG
Dothan	WRWA	88.7	TLK+	Greenville	WGYV	1380	TLK+
Dothan	WGTF	89.5	RL	Greenville	WQZX	94.3	CW
Dothan	WVOB	91.3	RL/SG	Greenville	WKXN	95.9	UR/RB
Dothan	WTVY	95.5	CW	Grove Hill	WFOW	106.1	CW
Dothan	WOOF	99.7	AC	Guntersville	WGSV	1270	TLK+
Dothan	WAGF	101.3	UR/AC	Guntersville	WJIA	88.5	RL/CC
Dothan	WESP	102.5	OLG	Guntersville	WTWX	95.9	CW
Elba	WELB	1350	CW	Haleyville	WJBB	1230	RL/G
Elba	WZTZ	101.1	CW	Haleyville	WJBB	92.7	CW
Enterprise	WDJR	96.9	CW	Hamilton	WERH	970	RL/CW
Enterprise	WKMX	106.7	A/CH	Hamilton	WERH	92.1	CW
Eufaula	WULA	1240	SPTS	Hanceville	WXRP	1170	RK/CL
Eufaula	WKZJ	92.7	URB	Hartselle	WYAM	890	SP/MEX
Eufaula	WRVX	97.9	A/CH	Hartselle	WTAK	106.1	RK/CL
Eutaw	WQZZ	104.3	UR/RB	Harvest	WAYH	88.1	REL
Eva	WRJL	99.9	RL/G	Hazel Green	WBXR	1140	RL
Evergreen	WIJK	1470	SPTS	Headland	WBCD	105.3	URB
Evergreen	WPGG	93.3	CW	Heflin	WKINF	89.1	RL/G
Fairfield	WJLD	1400	RL/G	Heflin	WPIL	91.7	VAR
Fairhope	WDLT	660	RL/G	Hobson City	WHOG	1120	UR/RB
Fairhope	WABF	1220	POP	Holly Pond	WFMH	95.5	CW
Fairhope	WZEW	92.1	AP	Holly Pond	WRSA	96.9	A/CS
Fayette	WLDX	990	CW	Homewood	WBPT	106.9	TOP
Fayette	WTXT	98.1	CW	Huntsville	WDJL	1000	RL/G
Florala	WKWL	1230	RL	Huntsville	WBHP	1230	TLK+
Florence	WBCF	1240	TLK+	Huntsville	WTKI	1450	SPTs
Florence	WSBM	1340	UR/AC	Huntsville	WLOR	1550	URB
Florence	WFIX	91.3	RL/CC	Huntsville	WEUP	1600	RL/G
Florence	WXFL	96.1	CW	Huntsville	WEUV	1700	RL/G
Florence	WQLT	107.3	AC/G	Huntsville	WLRH	89.3	TLK+
Foley	WHEP	1310	AC	Huntsville	WOCG	90.1	RL
Fort Mitchell	WAGH	98.3	UR/AC	Huntsville	WOCG	90.1	RL
Fort Payne	WZOB	1250	CW	Huntsville	WJAB	90.9	JAZ
Fort Payne	WFPA	1400	OLG	Huntsville	WRTT	95.1	RK/C
Fort Rucker	WLDA	100.5	RL	Huntsville	WAHR	99.1	AC/G
Fruithurst	WCKS	102.7	AC/H	Irondale	WLPH	1480	RL/G
Gadsden	WAAX	570	TLK+	Jackson	WRJX	1230	RL/G
Gadsden	WMGJ	1240	UR/RB	Jackson	WCKS	94.5	CW
Gadsden	WGAD	1350	OLG	Jacksonville	WNSI	810	TLK+
Gadsden	WTBB	89.9	RL	Jacksonville	WLJA	91.9	TLK+
Gadsden	WSGN	91.5	RK/CL	Jasper	WLYJ	1240	RL/G
Gadsden	WQEN	103.7	CHR	Jasper	WZPQ	1360	RL/G
Gardendale	WNCB	97.3	CW	Jasper	WDXB	102.5	CW
Geneva	WGEA	1150	CW	Jemison	WEZZ	97.7	CW
Geneva	WRJM	93.7	TLK	Langdale	WEBT	91.5	RL
Georgiana	WFXX	107.7	AC	Level Plains	WIRB	1490	TLK+

City	Station	Freq	Format	City	Station	Freq	Format
Lexington	WJHX	620	SP/MX	Oneonta	WKLD	97.7	CW
Linden	WINL	98.5	CW	Opelika	WANI	1400	TLK+
Linden	WNPT	102.9	CW	Opelika	WTLM	1520	POP
Lineville	WZZX	780	CW	Opelika	WMXA	96.7	AC/H
Lisman	WPRN	107.7	CW	Opp	WAMI	860	CW
Littleville	WMXV	103.5	A/C	Opp	WOPP	1290	CW
Luverne	WHLW	104.3	RL/G	Opp	WJIF	91.9	RL/SG
Madison	WUMP	730	SPTS	Opp	WAMI	102.3	CW
Marion	WJUS	1310	RL/G	Orange Beach	WCSN	105.7	AC
Meridianville	WXQW	94.1	OLG	Orrville	WJAM	107.9	UR/AC
Millbrook	WWMG	97.1	UR/AC	Oxford	WOXR	1580	OLG
Mobile	WPMI	710	TLK+	Oxford	WTBJ	91.3	RL
Mobile	WBHY	840	RL	Oxford	WVOK	97.9	AC/H
Mobile	WGOK	900	RL/G	Ozark	WOZK	900	POP
Mobile	WMOB	1360	RL	Ozark	WQLS	1210	RL/G
Mobile	WLVV	1410	RL	Ozark	WAQG	91.7	RL/CC
Mobile	WABB	1480	NWS+	Ozark	WJRL	103.9	RK/C
Mobile	WBHY	88.5	RL/CC	Ozark	WOAB	104.9	CW
Mobile	WHIL	91.3	CLA	Parrish	WALN	89.3	RL
Mobile	WBLX	92.9	UR/RB	Pell City	WFHK	1430	CW
Mobile	WKSJ	94.9	TLK+	Pepperell	WZMG	910	CW
Mobile	WRKH	96.1	RK/C	Phenix City	WGSY	100.1	AC
Mobile	WABB	97.5	RL	Phenix City	WHAL	1460	RL/G
Mobile	WMXC	99.9	AC	Piedmont	WPID	1280	AC
Monroeville	WVMI	930	A/C	Piedmont	WJCK	88.3	RL
Monroeville	WMFC	1360	RL/G	Pine Hill	WKXK	96.7	UR/RB
Monroeville	WMFC	99.3	OLG	Prattville	WIQR	1410	SPTS
Montgomery	WMSP	740	SPTS	Prattville	WXFX	95.1	RK
Montgomery	WMGY	800	RL/SG	Priceville	WQAH	1310	RL/SG
Montgomery	WNZZ	950	POP	Prichard	WLPR	960	RL/CC
Montgomery	WACV	1170	TLK+	Pritchard	WIJD	1270	TLK+
Montgomery	WLWI	1440	TLK+	Rainbow City	WJBY	930	RL
Montgomery	WXVI	1600	RL/G	Rainsville	WVSM	1500	RL/SG
Montgomery	WLBF	89.1	RL	Reform	WBEI	101.7	AC/H
Montgomery	WVAS	90.7	JAZ	Repton	WFNU	101.1	A/C
Montgomery	WLWI	92.3	CW	Roanoke	WELR	1360	SPT
Montgomery	WQKS	96.1	RK/C	Roanoke	WELR	102.3	CW
Montgomery	WBAM	98.9	CW	Robertsdale	WNSI	1000	SPTS
Montgomery	WHHY	101.9	CHR	Rogersville	WYL	93.9	SPTS
Montgomery	WMXS	103.3	AC	Russellville	WGOL	920	CW
Montgomery	WTSU	89.9	N/TLK	Russellville	WKAX	1500	RL/SG
Moody	WURL	760	RL/SG	Scottsboro	WWIC	1050	CW
Moulton	WHIY	1190	RL/G	Scottsboro	WCZT	1330	RL/SG
Moulton	WEUP	103.1	URB	Scottsboro	WKEA	98.3	CW
Muscle Shls	WLAY	1450	OLG	Selma	WMRK	1340	OLG
Muscle Shls	WQPR	88.7	N/TLK	Selma	WHBB	1490	TLK+
Muscle Shls	WLAY	105.5	RK/C	Selma	WAPR	88.3	N/TLK
Northport	WSJL	88.1	N/TLK	Selma	WRNF	89.5	N/TLK
Northport	WRAX	100.5	RK/C	Selma	WAQU	91.1	RL/CC
Oneonta	WCRL	1570	SP/OM	Selma	WDXX	100.1	CW

City	Station	Freq	Format		City	Station	Freq	Format
Selma	WALX	100.9	AC/H					
Selma	WBFZ	105.3	UR/AC					
Sheffield	WBTG	1290	TLK+					
Sheffield	WAKD	89.9	RL/CC					
Sheffield	WBTG	106.3	RL/SG					
Smiths	WBFA	101.3	URB					
Stevenson	WMXN	101.7	AC					
Sumiton	WRSM	1540	URB					
Sylacauga	WYEA	1290	RL/SG					
Sylacauga	WFEB	1340	TLK/S					
Sylacauga	WAWV	98.3	AC/H					
Talladega	WNUZ	1230	RL/G					
Talladega	WTDR	92.7	CW					
Tallassee	WTLS	1300	SPTs					
Tallassee	WACQ	99.9	RK/NU					
Thomaston	WSMO	95.7	SIL					
Thomasville	WJDB	630	OLG					
Thomasville	WJDB	95.5	CW					
Trinity	WWXQ	92.5	OLG					
Troy	WTBF	970	TLK+					
Troy	WAXU	91.1	RL/CC					
Trussville	WENN	105.9	UR/AC					
Tuscaloosa	WTSK	790	RL/G					
Tuscaloosa	WSPZ	1150	TOP					
Tuscaloosa	WTBC	1230	TLK+					
Tuscaloosa	WWPG	1280	RL/G					
Tuscaloosa	WACT	1420	SPTS					
Tuscaloosa	WMFT	88.9	N/TLK					
Tuscaloosa	WVUA	90.7	RK/C					
Tuscaloosa	WTUG	92.9	UR/RB					
Tuscaloosa	WBHJ	95.7	UR/RB					
Tuscaloosa	WRTR	105.5	RK					
Tuscumbia	WZZA	1410	UR/RB					
Tuscumbia	WVNA	1590	TLK+					
Tuscumbia	WLAY	100.3	CW					
Tuskegee	WBIL	580	RL/G					
Tuskegee	WTGZ	95.9	RK/NU					
Union Springs	WQSI	94.1	CW					
Uniontown	WVFG	107.5	SIL					
Valley	WRLD	95.3	OLG					
Valley Head	WQRX	870	SP/RL					
Vernon	WVSA	1380	SPTs					
Vernon	WJEC	106.5	RL/G					
Warrior	WBHK	98.7	UR/AC					
Wetumpka	WAPZ	1250	RL/G					
Wetumpka	WJWZ	97.9	UR/AC					
Winfield	WKXM	1300	CW					
Winfield	WKXM	97.7	OLG					
York	WYLS	670	POP					
York	WSLY	104.9	UR/RB					

City	Station	Freq	Format	City	Station	Freq	Format
Anchorage	KTZN	550	SPTS	Ft. Yukon	KZPA	900	VAR
Anchorage	KHAR	590	POP	Galena	KIYU	910	VAR
Anchorage	KENI	650	TLK+	Girdwood	KEUL	88.9	VAR
Anchorage	KBYR	700	TLK+	Glennallen	KCAM	790	VAR
Anchorage	KBYR	700	TLK+	Glennallen	KXGA	90.5	N/TLK
Anchorage	KFQD	750	TLK+	Haines	KHNS	102.3	N/TLK
Anchorage	KASH	1080	TLK+	Homer	KGTL	620	POP
Anchorage	KRUA	88.1	AP	Homer	KBBI	890	N/TLK
Anchorage	KAKL	88.5	RL	Homer	KMJG	88.9	OLG
Anchorage	KATB	89.3	RL	Homer	KWVV	103.5	AC
Anchorage	KNBA	90.3	A/C	Houston	KJHA	88.7	RL
Anchorage	KSKA	91.1	TLK+	Houston	KQEZ	92.1	AC/S
Anchorage	KFAT	92.9	CHR	Houston	KADX	94.7	SIL
Anchorage	KAFC	93.7	RL/CC	Houston	KRPM	96.3	RK/C
Anchorage	KEAG	97.3	OLG	Juneau	KJNO	630	TLK+
Anchorage	KLEF	98.1	CLA	Juneau	KINY	800	AC
Anchorage	KYMG	98.9	AC	Juneau	KFMG	100.7	AC/H
Anchorage	KBFX	100.5	RK/CL	Juneau	KSRJ	102.7	AC/S
Anchorage	KGOT	101.3	CHR	Juneau	KTOO	104.3	N/TLk
Anchorage	KDBZ	102.1	AC	Juneau	KTKU	105.1	CW
Anchorage	KMXS	103.1	AC/H	Juneau	KSUP	106.3	RK
Anchorage	KBRJ	104.1	CW	Kasilof	KWVK	89.5	SIL
Anchorage	KNIK	105.3	JAZ	Kasilof	KWMD	90.5	SIL
Anchorage	KWHL	106.5	RK	Kasilof	KWJG	91.5	OLG
Anchorage	KASH	107.5	CW	Kenai	KDLL	91.9	N/TLK
Barrow	KBRW	680	VAR	Kenai	KWHQ	100.1	CW
Barrow	KBRW	91.9	VAR	Ketchikan	KTKN	930	TLK+
Bethel	KYUK	640	N/TLK	Ketchikan	KFMJ	99.9	OLG
Bethel	KYKD	100.1	RL	Ketchikan	KRBD	105.9	N/TLK
Big Lake	KAGV	1110	RL	Ketchikan	KGTW	106.7	CW
Chevak	KCUK	88.1	VAR	Kodiak	KVOK	560	CW
College	KUWL	103.9	OLG	Kodiak	KMLXT	100.1	N/TLK
Cordova	KLAM	1450	TOP	Kodiak	KRXX	101.1	N/TLK
Cordova	KCDV	100.9	AC/H	Kotzebue	KOTZ	720	VAR
Deadhorse	KCDS	88.1	VAR	Mccarthy	KXKM	89.7	N/TLK
Dillingham	KDLG	670	VAR	Mcgrath	KSKO	870	VAR
Dillingham	KRUP	99.1	TLK+	Naknek	KAKN	100.9	RL/G
Eagle River	KAXX	1020	SIL	Nenana	KIAM	630	RL/G
Fairbanks	KFAR	660	NWS+	Nikiski	KSBA	93.3	OLG
Fairbanks	KCBF	820	SPTS	Nome	KNOM	780	VAR
Fairbanks	KFBX	970	TLK/S	Nome	KICY	850	RL
Fairbanks	KUAC	89.9	N/TLK	Nome	KNOM	96.1	VAR
Fairbanks	KSUA	91.5	A/C	Nome	KICY	100.3	RL
Fairbanks	KXLR	95.9	RK/CL	North Pole	KJNP	1170	RL/G
Fairbanks	KXLR	95.9	RK/C	North Pole	KJNP	100.3	RL/G
Fairbanks	KYSC	96.9	A/C	Palmer	KJLP	88.9	N/TLK
Fairbanks	KWLF	98.1	RL	Petersburg	KRSA	580	VAR
Fairbanks	KAKQ	101.1	AC	Petersburg	KFSK	100.9	N/TLK
Fairbanks	KIAK	102.3	CW	Sand Point	KSDP	830	VAR
Fairbanks	KKED	104.7	RK	Seward	KSWD	950	CW

21

City	Station	Freq	Format	City	Station	Freq	Format
Seward	KPFN	105.9	A/C				
Sitka	KIFW	1230	AC/H				
Sitka	KSBZ	103.1	RK				
Sitka	KCAW	104.7	N/TLK				
Soldotna	KSRM	920	TLK+				
Soldotna	KSLD	1140	RK/CL				
Soldotna	KKIS	96.5	AC/H				
Soldotna	KPEN	101.7	CW				
St. Paul	KUHB	91.9	N/TLK				
Sterling	KRAW	90.1	SIL				
Talkeetna	KTNA	88.5	N/TLK				
Tok	KUDU	91.9	RL/CC				
Unalakleet	KNSA	930	VAR				
Unalaska	KIAL	1450	VAR				
Valdez	KCHU	770	N/TLK				
Valdez	KVAK	1230	CW				
Valdez	KVAK	93.3	AC/H				
Wasilla	KMBQ	99.7	AC				
Wrangell	KSTK	101.7	N/TLK				
Seward	KPFN	105.9	A/C				

City	Station	Freq	Format	City	Station	Freq	Format
Apache Jct	KVVA	107.1	SP/MX	Glendale	KPXQ	1360	RL
Arizona City	KKMR	106.5	SP/MX	Glendale	KKFR	92.3	UR/RB
Bagdad	KRCI	103.1	TOP	Glendale	KLNZ	103.5	SP/MX
Benson	KAVV	97.7	CW	Globe	KJAA	1240	RL
Bisbee	KRMB	90.1	SP/MEX	Globe	KLKA	88.5	N/TLK
Bisbee	KWRB	90.9	RL	Globe	KVJC	91.9	RL
Bisbee	KWCD	92.3	CW	Globe	KRDE	94.1	CW
Black Cyn	KMIA	710	SP/MX	Globe	KRXS	97.3	OLG
Buckeye	KDVA	106.9	SP/MX	Globe	KDDJ	100.3	SP/MEX
Bullhead Cty	KFLG	1000	AS	Grand Cyn	KNAG	90.3	N/TLK
Bullhead Cty	KZZZ	1490	N/TLK	Green Vly	KGVY	1080	AS
Casa Grand	KLVA	105.5	RL	Green Vly	KFMA	92.1	RK/NU
Cave Creek	KFNX	1100	TLK+	Green Vly	KTZR	97.1	SP/MEX
Chandler	KMLE	107.9	CW	Holbrook	KDJI	1270	N/TLK
Chinle	KFXR	107.3	CW	Holbrook	KBMH	90.3	RL
Chino Vly	KFPB	94.3	A/CH	Holbrook	KZUA	92.1	CW
Claypool	KIKO	106.1	AC/S	Hotevilla	KUYI	88.1	VAR
Clifton	KCUZ	1490	CW	Kachina	KFLX	105.1	AP
Clifton	KWRQ	102.3	AC	Kearny	KZLZ	105.3	SP/MEX
Colorado Cy	KMXM	107.3	EZL	Kingman	KAAA	1230	NWS+
Coolidge	KCKY	1150	SP/MEX	Kingman	KFLG	94.7	CW
Coolidge	KCOO	89.9	RL	Kingman	KGMN	99.9	CW
Cortaro	KEVT	1030	SP/MEX	Lk Havasu	KNTR	980	TLK+
Cottonwood	KYBC	1600	AS	Lk Havasu	KNLB	91.1	RL
Cottonwood	KZGL	95.9	RK	Lk Havasu	KNLB	91.1	RL
Cottonwood	KVRD	105.7	CW	Lk Havasu	KJJJ	92.7	CW
Dewey-Hldt	KZLB	97.5	SP/MX	Lk Havasu	KBBC	96.7	AS
Dolan Spgs	KRRN	92.7	SP/MX	Lk Havasu	KRCY	96.7	OLG
Dolan Sprs	KOAS	105.7	JAZ	Lk Havasu	KRRK	101.1	RK/CL
Douglas	KAPR	930	RL	Lk Havasu	KJJJ	102.3	CW
Douglas	KDAP	1450	SP/MEX	Lk Havasu	KZUL	104.5	AC
Douglas	KRMC	91.7	SP/MEX	Mammoth	KLTU	88.1	N/TLK
Douglas	KCDQ	95.3	A/C	Marana	KSAZ	580	AS
Douglas	KDAP	96.5	CW	Marana	KOHT	98.3	UR
Drake	KJZA	89.5	JAZ	Mesa	KXAM	1310	TLK+
Duncan	KJIK	100.7	A/CS	Mesa	KFNN	1510	BUS
Eagar	KTHQ	92.5	CW	Mesa	KDKB	93.3	RK
Flagstaff	KVNA	600	TLK+	Mesa	KZZP	104.7	CHR
Flagstaff	KAFF	930	CW	Miami	KIKO	1340	OLG
Flagstaff	KNAU	88.7	N/TLK	Miami	KQSS	98.3	CW
Flagstaff	KJTA	89.9	RL	Nogales	KOFH	91.1	SP/MEX
Flagstaff	KJTA	89.9	RL	Nogales	KRDX	98.3	SP/MEX
Flagstaff	KPUB	91.7	N/TLK	Nogales	KOFH	99.1	SP/RL
Flagstaff	KAFF	92.9	CW	Oracle	KGMG	106.3	OLG
Flagstaff	KAFF	92.9	CW	Oro Valley	KSZR	97.5	AS
Flagstaff	KMGN	93.9	RK/CL	Oro Valley	KCMT	101.9	RK/CL
Flagstaff	KNVA	100.1	A/C	Oro Valley	KCMT	101.9	CW
Florence	KCDX	103.1	TOP	Oro Valley	KCMT	102.1	SP/MEX
Font'n Hills	KLVK	89.1	RL	Page	KPGE	1340	OLG
Gilbert	KEDJ	103.9	RK/NU	Page	KNAD	91.7	N/TLK

City	Station	Freq	Format	City	Station	Freq	Format
Page	KXAZ	93.3	AC/S	Red Mesa	KRMH	89.7	ET/O
Paradise Vly	KHOT	105.9	SP/MEX	Safford	KATO	1230	TLK+
Parker	KLPZ	1380	CW	Safford	KXKQ	94.3	CW
Parker	KWFH	90.1	RL	Sahuarita	KQTL	1210	SP/MEX
Parker	KRIT	93.9	SP/MEX	Scottsdale	kazg	1440	OLG
Payson	KMOG	1420	CW	Scottsdale	KSLX	100.7	RK/CL
Payson	KSXX	99.3	TOP	Sedona	KAZM	780	N/TLK
Payson	KNRJ	101.1	CH	Sedona	KQST	102.9	CHR
Payson	KAJM	104.3	UR/RB	Sedona	KSED	107.5	CW
Phoenix	KFYI	550	TLK+	Seligman	KZKE	103.3	OLG
Phoenix	KTAR	620	TLK+	Sells	KOHN	91.9	ALT
Phoenix	KIDR	740	SP/MEX	Show Low	KVWM	970	N/TLK
Phoenix	KMVP	860	SPTS	Show Low	KVSL	1450	AS
Phoenix	KMVP	860	SPTS	Show Low	KNAA	90.7	N/TLK
Phoenix	KGME	910	SPTS	Show Low	KSNX	93.5	OLG
Phoenix	KKNT	960	TLK+	Show Low	KRFM	96.5	AC
Phoenix	KOY	1230	AS	Sierra Vista	KTAN	1420	TLK+
Phoenix	KXEG	1280	RL	Sierra Vista	KNXN	1470	RL
Phoenix	KSUN	1400	SP/MEX	Sierra Vista	KZMK	100.9	AC/H
Phoenix	KPHX	1480	TLK+	Sierra Vista	KKYZ	101.7	OLG
Phoenix	KASA	1540	SP/MEX	South Tucson	KJLL	1330	N/TLK
Phoenix	KBAQ	89.5	CLA	South Tucson	KXEW	1600	SP/TO
Phoenix	KFLR	90.3	RL	Springerville	KRVZ	1400	OLG
Phoenix	KJZZ	91.5	N/TLK	Springerville	KQAZ	101.7	A/CS
Phoenix	KOOL	94.5	OLG	St. Johns	KWKM	95.7	A/CH
Phoenix	KYOT	95.5	JAZ	Sun City	KOMR	106.3	SP/MEX
Phoenix	KMXP	96.9	AC/H	Sun City West	KFMR	95.1	URB
Phoenix	KKLT	98.7	AC/S	Tempe	KDUS	1060	SPT
Phoenix	KESZ	99.9	AC	Tempe	KMIK	1580	YTH
Phoenix	KZON	101.5	AC	Tempe	KUPD	97.9	RK
Phoenix	KNIX	102.5	CW	Texarkana	KWLL	89.3	N/TLK
Phoenix	KTWC	103.5	AS	Thatcher	KFMM	99.1	CW
Phoenix	KBZR	103.9	RK/NU	Tolleson	KXEG	1010	N/TLK
Phoenix	KZZP	104.7	AC/H	Tolleson	KMYL	1190	N/TLK
Phoenix	KRDS	105.3	RL/CC	Tuba City	KTBA	1050	RL
Phoenix	KLVA	105.5	RL/CC	Tuba City	KGHR	91.5	VAR
Phoenix	KEDJ	106.3	RK/NU	Tucson	KVOI	690	N/TLK
Phoenix	KMJK	106.9	AC/S	Tucson	KNST	790	TLK+
Phoenix	KMLE	107.9	CW	Tucson	KFLT	830	RL/CC
Phoenix	KPHF	88.3	RL	Tucson	KGMS	940	RL
Prescott	KNOT	1450	AS	Tucson	KTKT	990	SP/OM
Prescott	KYCA	1490	TLK+	Tucson	KQTL	1210	SP/MEX
Prescott	KNAQ	89.3	N/TLK	Tucson	KCUB	1290	SPTS
Prescott	KGCB	90.9	RL/CC	Tucson	KTUC	1400	AS
Prescott	KTMG	99.1	A/CS	Tucson	KWFM	1450	OLG
Prescott	KAHM	102.1	EZL	Tucson	KFFN	1490	SPT
Prescott Vly	KQNA	1130	N/TLK	Tucson	KUAZ	1550	N/TLK
Prescott Vly	KKLD	98.3	OLG	Tucson	KFLT	88.5	N/TLK
Prescott Vly	KPPV	106.7	AC	Tucson	KJZZ	88.9	VAR
Prescott Vly	KBUX	94.3	VAR	Tucson	KUAZ	89.1	N/TLK

City	Station	Freq	Format		City	Station	Freq	Format
Safford	KATO	1230	TLK+					
Safford	KXKQ	94.3	CW					
Sahuarita	KQTL	1210	SP/MEX					
Tucson	KCUB	1290	SPTS					
Tucson	KTUC	1400	AS					
Tucson	KWFM	1450	OLG					
Tucson	KFFN	1490	SPT					
Tucson	KUAZ	1550	N/TLK					
Tucson	KFLT	88.5	N/TLK					
Tucson	KJZZ	88.9	VAR					
Tucson	KUAZ	89.1	N/TLK					
Tucson	KUAT	90.5	CLA					
Tucson	KXCI	91.3	VAR					
Tucson	KEKO	92.1	AP					
Tucson	KWFM	92.9	CW					
Tucson	KWMT	92.9	ALT					
Tucson	KRQQ	93.7	CHR					
Tucson	KMXZ	94.9	AC					
Tucson	KLPX	96.1	RK/CL					
Tucson	KIIM	99.5	CW					
Tucson	KZPT	104.1	AC/H					
Tucson	KHYT	107.5	CH					
Tucson	KVET	1030	SP/MEX					
Tucson	KGVY	1080	AS					
Tucson	KRKR	1110	N/A					
Tucson	KJLL	1330	TLK+					
Tusayan	KSGC	92.1	AC					
Van Buren	KLFS	90.3	RL					
Waldo	KWDO	99.1	A/CH					
Wellton	KCEC	104.5	SP/MEX					
Whiteriver	KNNB	88.1	VAR					
Wickenburg	KBSZ	1250	CW					
Wickenburg	KSWG	96.3	SIL					
Wickenburg	KHOV	105.3	SP/MEX					
Willcox	KHIL	1250	CW					
Willcox	KWCX	104.9	AC/S					
Williams	KYET	1180	SIL					
Williams	KWMX	96.7	OLG					
Window Rk	KTNN	660	CW					
Window Rk	KWRK	96.1	VAR					
Window Rk	KWIM	104.9	RL/CC					
Winslow	KINO	1230	CW					
Yuma	KBLU	560	TLK/S					
Yuma	KAWC	1320	N/TLK					
Yuma	KJOK	1400	OLG					
Yuma	KCFY	88.1	RL/CC					
Yuma	KYRM	91.9	SP/MEX					
Yuma	KLJZ	93.1	CHR					
Yuma	KTTI	95.1	CW					
Yuma	KQSR	100.9	A/C					
Yuma	KAWC	88.9	N/TLK					

Radio on the Road   Arkansas

City	Station	Freq	Format	City	Station	Freq	Format
Arkadelphia	KVRC	1240	TOP	Conway	KCON	1230	AC
Arkadelphia	KSWH	91.1	VAR	Conway	KTOD	1330	SILNT
Arkadelphia	KDEL	100.9	AC	Conway	KUCA	91.3	AC/H
Ashdown	KMJI	93.3	AC	Conway	KASR	92.7	SPT
Ashdown	KHSP	103.9	CW	Conway	KHDX	93.1	VAR
Atkins	KVLD	99.3	OLG	Conway	KMJX	105.1	RK/CL
Augusta	KJSM	97.7	RL	Corning	KCCB	1260	AC
Bald Knob	KAPZ	710	TLK+	Corning	KBKG	93.5	AC
Bald Knob	KKSY	107.1	A/CH	Crossett	KAGH	800	RL/G
Barling	KOLX	94.5	N/TLK	Crossett	KAGH	104.9	CW
Batesville	KAAB	1130	SP/MEX	Danville	KYEL	105.5	CW
Batesville	KBTA	1340	SPTS	Dardanelle	KCAB	980	N/TLK
Batesville	KZLE	93.1	RK	Dardanelle	KWXT	1490	RL/SG
Batesville	KBTA	99.5	AC/S	Dardanelle	KCJC	102.3	CW
Beebe	KBGR	101.5	RL	De Queen	KDQN	1390	SPN
Bella Vista	KBVA	106.5	POP	De Queen	KBPU	88.7	RL
Bellefonte	KNWA	1600	SPTS	De Queen	KDQN	92.1	CW
Benton	KEWI	690	CW	De Witt	KDEW	97.3	CW
Benton	KHKN	106.7	CW	Dermott	KXSA	103.1	CW
Bentonville	KAPG	88.1	RL/CC	Dermott	KRKD	105.7	RL/CC
Bentonville	KSEC	95.7	OLG	Des Arc	KBDO	91.7	RL/CC
Bentonville	KFAY	98.3	CW	Des Arc	KFLI	104.7	OLG
Bentonville	KREB	1190	SPTS	Dumas	KDDA	1560	SIL
Berryville	KTHS	1480	CW	Dumas	KFLI	104.7	OLG
Berryville	KTHS	107.1	CW	Earle	KCJF	103.9	RK/C
Blytheville	KLCN	910	CW	E. Camden	KCXY	95.3	CW
Blytheville	KBCM	88.3	RL/CC	El Dorado	KDMS	1290	N/TLK
Blytheville	KOUX	91.5	CW	El Dorado	KELD	1400	SP/MEX
Blytheville	KHLS	96.3	CW	El Dorado	KKDU	88.9	N/TLK
Booneville	KBHN	89.7	RL	El Dorado	KBSA	90.9	N/TLK
Booneville	KRBK	104.7	OLG	El Dorado	KAGL	93.3	RK/CL
Brinkley	KBRI	1570	RL/SG	El Dorado	KMRX	96.1	A/CH
Brinkley	KBRI	1570	RL	El Dorado	KLBQ	98.7	A/CH
Bryant	KKZR	93.3	RK	El Dorado	KMLK	101.5	URB
Cabot	KZTD	1350	SP/MEX	El Dorado	KIXB	103.3	CW
Cabot	KPZK	102.5	RL/G	England	KVDW	1530	RL
Calico Rock	KEZG	97.1	N/TLK	England	KHTE	96.5	RK/NU
Camden	KNHD	1450	RL/SG	Eudora	KAVH	101.5	RL/G
Camden	KCAC	89.5	AP	Eureka S.	KTCN	100.9	RL/SG
Camden	KCXY	95.3	CW	Fairfield B.	KFFB	106.1	POP
Camden	KAMD	97.1	AC	Farmington	KFAY	1030	N/TLK
Camden	KMGC	104.5	UR/RB	Fayetteville	KOFC	1250	RL/SG
Cave City	KZIG	89.9	VAR	Fayetteville	KZUA	88.3	ALT
Cherokee V.	KFCM	100.9	OLG	Fayetteville	KAYH	89.3	RL
Clarksville	KLYR	1360	CW	Fayetteville	KBNV	90.1	RL/CC
Clarksville	KLYR	92.7	CW	Fayetteville	KUAF	91.3	N/Tlk
Clarksville	KXIO	106.9	CW	Fayetteville	KKEG	92.1	RK
Clinton	KGFL	1110	OLG	Fayetteville	KKIX	103.9	CW
Clinton	KHPQ	92.1	CW	Fayetteville	KEZA	107.9	AC
Colt	KTRQ	102.3	OLG	Fordyce	KBJT	1570	TLK+

ARKANSAS

Radio on the Road

City	Station	Freq	Format	City	Station	Freq	Format
Fordyce	KQEW	102.3	CW	Hot Spr Vly	KVRE	92.9	POP
Forrest City	KXJK	950	CH	Hoxie	KJLV	105.3	RL/CC
Forrest City	KARH	88.1	RL/CC	Humnoke	KVLO	101.7	RL/G
Forrest City	KBFC	93.5	CW	Huntsville	KAKA	99.5	SP/MEX
Fort Smith	KFSA	950	RL/CC	Jacksonville	KDJE	100.3	RK
Fort Smith	KFPW	1230	POP	Jonesboro	KNEA	970	RL
Fort Smith	KYHN	1320	NWS+	Jonesboro	KBTM	1230	NWS+
Fort Smith	KTCS	1410	RL/SG	Jonesboro	KAOG	90.5	RL/CC
Fort Smith	KWHN	1650	N/tLK	Jonesboro	KASU	91.9	N/TLK
Fort Smith	KAOW	88.9	RL/CC	Jonesboro	KDEZ	100.5	RK/CL
Fort Smith	KISR	93.7	CHR	Jonesboro	KIYS	101.9	CHR
Fort Smith	KMAG	99.1	CW	Jonesboro	KFIN	107.9	CW
Fort Smith	KTCS	99.9	CW	Lake City	KDXY	104.9	CW
Fort Smith	KBBQ	100.7	OLG	Lk Village	KUUZ	95.9	RL
Fouke	KLMZ	104.3	RL	Lk Village	KZYQ	103.5	AC
Glenwood	KWXI	670	RL	Lakeview	KKTZ	93.5	AC
Glenwood	KWXE	104.5	CW	Little Rock	KARN	920	NWS+
Gosnell	KAMJ	93.9	UR/RB	Little Rock	KJBN	1050	RL
Gould	KAPN	102.5	CW	Little Rock	KAAY	1090	RL
Greenwood	KZKZ	106.3	RL/CC	Little Rock	KZPK	1250	RL/G
Gurdon	KYXK	106.9	CW	Little Rock	KITA	1440	RL/G
Hamburg	KHMB	99.5	AC/S	Little Rock	KABF	88.3	VAR
Hampton	KBPW	88.1	RL/CC	Little Rock	KAUA	89.1	N/TLK
Hampton	KELD	106.5	N/TLK	Little Rock	KLRE	90.5	CLA
Hardy	KSRB	1570	OLG	Little Rock	KKPT	94.1	CH
Hardy	KOOU	104.7	CH	Little Rock	KSSN	95.7	CW
Harrisburg	KWHG	95.9	CW	Little Rock	KURB	98.5	AC/H
Harrison	KHOZ	900	N/TLK	Little Rock	KDIS	99.5	YTH
Harrison	KBPB	91.9	RL/SG	Little Rock	KABZ	103.7	TLK+
Harrison	KCWD	96.1	RK/CL	Little Rock	KMJX	105.1	RK/H
Harrison	KHOZ	102.9	CW	Little Rock	KDDK	106.7	CW
Hatfield	KILX	104.1	A/CH	Little Rock	KYTN	107.7	RL/CC
Heber Spr	KAWW	1370	TLK+	Lonoke	KOLL	106.3	OLG
Heber Spr	KBMJ	89.5	RL/CC	Lowell	KMXF	101.9	CHR
Heber Spr	KAWW	100.7	RK/CL	Magnolia	KVMA	630	CW
Helena	KFFA	1360	CW	Malvern	KBOK	1310	CW
Helena	KJIW	94.5	RL/G	Malvern	KISI	101.5	OLG
Helena	KFFA	103.1	CW	Mammoth S	KAMS	95.1	CW
Hope	KXAR	1490	N/TLK	Marianna	KAKJ	105.3	UR/RB
Hope	KBYB	101.7	TOP	Marion	KXHT	107.1	UR/RB
Hope	KHPA	104.9	CW	Marked Tree	KJBR	93.7	RL/CC
Horseshoe B	KKIK	106.5	CH	Marshall	KCGS	960	RL/SG
Hot Springs	KPZA	590	SP/MEX	Marshall	KBCN	104.3	CW
Hot Springs	KZNG	1340	TLK+	Marvell	KVRN	90.7	RL/CC
Hot Springs	KXOW	1420	CW	Maumelle	KOLL	94.9	ALT
Hot Springs	KLRO	90.1	RL/CC	Maumelle	KWLR	96.9	RL/CC
Hot Springs	KALR	91.5	RL/CC	McGehee	KVSA	1220	CW
Hot Springs	KLXQ	96.7	A/CH	Melbourne	KAEN	90.3	N/TLK
Hot Springs	KQUS	97.5	CW	MENA	KENA	1450	RL/SG
Hot Springs	KLAZ	105.9	CHR	MENA	KTTG	96.3	SPTS

28

CITY	STATION	FREQ	FORMAT	CITY	STATION	FREQ	FORMAT
Mena	KENA	102.1	CW	Russellville	KXRJ	91.9	VAR
Mena	KQOR	105.3	OLG	Russellville	KWKK	100.9	AC/S
Monticello	KHBM	1430	AS	Salem	KHOM	100.9	CW
Monticello	KHBM	93.7	RK/C	Searcy	KWCK	1300	TLK+
Monticello	KGPQ	99.9	AC	Searcy	KWCK	99.9	CW
Morrilton	KVOM	800	NWS+	Sheridan	KGHT	880	RL/SG
Morrilton	KVOM	101.7	CW	Sheridan	KANX	91.1	RL/CC
Mountain Hm	KTLO	1240	CW	Sheridan	KARN	102.9	N/TLK
Mountain Hm	KCMH	91.5	RL/G	Sherwood	KMTL	760	RL
Mountain Hm	KTLO	97.9	POP	Sherwood	KOKY	102.1	UR/AC
Mountain Hm	KPFM	105.5	CW	Siloam Spr	KUOA	1290	N/TLK
Mountain Hm	KOMT	107.5	POP	Siloam Spr	KLRC	101.1	RL/CC
Mountain Pin	KZBR	101.9	AC/H	Siloam Spr	KMCK	105.7	CHR
Mountain Vu	KWOZ	103.3	CW	Springdale	KZRA	1590	SPN
Murfreesboro	KMTB	99.5	CW	Springdale	KXNA	104.9	RK/NU
Nashville	KBHC	1260	SP/MEX	Stamps	KZHE	100.5	CW
Nashville	KSSW	96.9	AC	Stuttgart	KWAK	1240	SPTS
Nashville	KNAS	105.5	OLG	Stuttgart	KWAK	105.5	OLG
Newark	KLLN	90.9	RL/SG	Texarkana	KOSY	790	A/S
Newport	KNBY	1280	N/TLK	Texarkana	KTOY	104.7	UR/AC
Newport	KOKR	96.7	CW	Texarkana	KYGL	106.3	RK/CL
N. Crossett	KWLT	102.7	RK/C	Texarkana	KTWN	107.1	CW
N. Little Rock	KDXE	1380	SPTS	Trumann	KJBX	106.7	AC
N. Little Rock	KWBF	101.1	ALT	Turrell	KKLV	94.7	RL/CC
Ola	KARV	101.3	N/TLK	Van Buren	KAYR	1060	SPN
Osceola	KQDD	107.3	RK	Van Buren	KFDF	1580	N/TLK
Ozark	KDYN	1540	CW	Van Buren	KLSZ	102.7	RK/CL
Ozark	KDYN	96.7	CW	Waldron	KRWA	103.1	SPTS
Pangburn	KSMD	99.1	N/TLK	Walnut R	KRLW	1320	OLG
Paragould	KDRS	1490	SPTS	Walnut R	KRLW	106.3	OLG
Paragould	KDRS	107.1	A/CH	Warren	KWRF	860	CW
Paris	KERX	95.3	RK/C	Warren	KWRF	105.5	CW
Piggott	KBOA	105.5	POP	W. Helena	KCLT	104.9	UR/RB
Pine Bluff	KPBA	1270	SIL	W.Memphis	KSUD	730	SPTS
Pine Bluff	KCAT	1340	RL/G	White Hall	KTRN	104.5	AC/S
Pine Bluff	KCLA	1400	RL/G	Wilson	KOSE	860	RL/SG
Pine Bluff	KOTN	1490	SPTS	Wrightsville	KLAL	107.7	CHR
Pine Bluff	KUAP	89.7	JAZ	Wynne	KWYN	1400	CW
Pine Bluff	KIPR	92.3	UR/RB	Wynne	KWYN	92.5	CW
Pine Bluff	KZYP	99.3	UR/AC	Yellville	KCTT	101.7	OLG
Pine Bluff	KPBQ	101.3	CW				
Pocahontas	KPOC	1420	SPTS				
Pocahontas	KPOC	104.1	AC				
Prairie Grove	KYNF	94.9	A/C				
Prescott	KTPA	1370	RL/SG				
Rogers	KURM	790	N/TLK				
Rogers	KREB	1390	N/SPT				
Rogers	KAMO	94.3	CW				
Russellville	KARV	610	N/TLK				
Russellville	KMTC	91.1	RL/CC				

City	Station	Freq	Format	City	Station	Freq	Format
Alameda	KXJO	92.7	CH	Bakersfield	KUZZ	107.9	CW
Alturas	KKFJ	570	AS	Banning	KMET	1490	JAZ
Alturas	KCNO	94.5	CW	Barstow	KSZL	1230	N/TLK
Alturas	KALT	106.5	RJ/C	Barstow	KIQQ	1310	SP/MEX
Anaheim	KXMX	1190	N/A	Barstow	KDUC	94.3	N/TLK
Anaheim	KFSH	95.9	RL	Barstow	KXXZ	95.9	SP/MEX
Anderson	KEWB	94.7	CH/D	Bayside	KZPN	91.5	N/TLK
Angwin	KNDL	89.9	RL/CC	Beaumont	KAEH	100.9	SPN
Apple Valley	KIXW	960	TLK+	Berkeley	KVTO	1400	ET/C
Apple Valley	KWRN	1550	SPN	Berkeley	KPFB	89.3	VAR
Apple Valley	KZXY	102.3	AC/H	Berkeley	KALX	90.7	VAR
Arcadia	KSSE	107.1	SPN	Berkeley	KPFA	94.1	VAR
Arcata	KATA	1340	SPTS	Berkeley	KBLX	102.9	URB
Arcata	KHSU	90.5	N/VAR	Bev Hills	KGIL	1260	OLG
Arcata	KXGO	93.1	RK/CL	B.Bear City	KBHR	93.3	AC/S
Arnold	KBYN	95.9	SPN	B. Bear Lk	KXSB	101.7	SPN
Arnold	KCFA	106.1	SP/MEX	Big Pine	KRHV	93.3	AC/H
Arroyo Grde	KLFF	890	RL	Bishop	KBOV	1230	OLG
Arroyo Grde	KXTK	1280	SPTS	Bishop	KWTW	88.5	RL
Arvin	KMYX	92.5	SP/MEX	Bishop	KIBS	100.7	CW
Atascadero	KIQO	104.5	OLG	Blue Lake	KCIK	1450	N/TLK
Atherton	KCEA	89.1	AS	Blythe	KERU	88.5	SPN
Atwater	KBRE	92.5	URB	Blythe	KJMB	100.3	AC
Auberry	KLBN	105.1	SPN	Brawley	KROP	1300	CW
Auburn	KAHI	950	OLG	Brawley	kseh	94.5	SPN
Auburn	KSMH	1620	RL	Brawley	KSIQ	96.1	CHR
Auburn	KHYL	101.1	OLG	Buena Pk	KBPK	90.1	AC/H
Avalon	KBRT	740	RL/TLK	Burney	KNCA	89.7	TLK/S
Avalon	KISL	88.7	VAR	Burney	KIBC	90.5	RL
Avenal	KAAX	95.1	SIL	Burney	KRRX	106.1	RK
Baker	KHRQ	94.9	RK	Calexico	KICO	1490	SPN
Baker	KIXF	101.5	CW	Calexico	KUBO	88.7	SPN
Bakersfield	KUZZ	550	CW	Calexico	KQVO	97.7	N/TLK
Bakersfield	KDFO	800	SPT	California C.	KCEL	106.9	SP/MEX
Bakersfield	KZTK	970	TLK+	Calipatria	KSSB	100.9	RK/C
Bakersfield	KAFY	1100	SPN	Calistoga	KXST	100.9	SPN
Bakersfield	KGEO	1230	TLK/S	Camarillo	KMRO	90.3	SPN
Bakersfield	KBID	1350	OLG	Camarillo	KOCP	95.9	RK/C
Bakersfield	KERN	1410	TLK+	Canyon C.	KHTS	1220	N/TLK
Bakersfield	KWAC	1490	SPN	Carlsbad	KUSS	95.7	CW
Bakersfield	KNZR	1560	TLK+	Carmel	KRML	1410	JAZ
Bakersfield	KPRX	89.1	N/TLK	Carmel	KBOQ	95.5	CLA
Bakersfield	KTQX	90.1	SPN	Carmel	KBTU	101.7	A/CH
Bakersfield	KFRB	91.3	RL/CC	Carmel Vly	KXME	540	SP/MEX
Bakersfield	KPSL	92.1	SPN	Carmichael	KFIA	710	RL
Bakersfield	KISV	94.1	CHR	Carnelian B.	KODS	103.7	CH
Bakersfield	KBKO	96.5	CW	Carpinteria	KSBL	101.7	AC
Bakersfield	KKBB	99.3	OLG	Cartago	KWTY	102.9	RK/C
Bakersfield	KGFM	101.5	AC/S	Cathedral Cty	KWXY	1340	EZL
Bakersfield	KCWR	107.1	CW	Cathedral Cty	KWXY	98.5	ESL

California   Radio on the Road

City	Station	Freq	Format	City	Station	Freq	Format
Ceres	KVIN	920	AS	Earlimart	KNAC	93.5	A/C
Ceres	KBES	89.5	N/AO	East L A	KLAX	97.9	SP/MX
Chester	KBNF	98.9	A/C	E.Porterville	KMQA	100.5	SP/Mx
Chico	KPAY	1290	TLK+	E. Sonora	KARQ	89.5	RL
Chico	KHAP	89.1	RL/CC	El Cajon	KECR	910	RL/CC
Chico	KZFR	90.1	VAR	El Cajon	KHTS	93.3	CHR
Chico	KCHO	91.7	N/TLK	El Centro	KXO	1230	OLG
Chico	KLRS	92.7	CHR	El Centro	KWST	1430	CW
Chico	KFMF	93.9	RK	El Centro	KXO	107.5	AC
Chico	KMXI	95.1	AC	El Cerrito	KECG	88.1	JAZ
China Lake	KFRJ	91.1	N/TLK	El Rio	KMLA	103.7	SP/MX
China Lake	KSSI	102.7	CH	Ellwood	KSPE	94.5	SPN
Chowchilla	KNTO	93.3	SPN	Encinitas	KPRI	102.1	ALT
Chualar	KHDC	90.9	SPN	Escondido	KFSD	1450	CLA
Claremont	KSPC	88.7	VAR	Escondido	KSOQ	92.1	CW
Cloverdale	KSRT	107.1	CHR	Esparto	KTTA	97.9	SP/MX
Clovis	KOOR	790	SPN	Essex	KHWY	98.9	AC
Clovis	KOND	92.1	SP/MEX	Eureka	KWSW	790	N/TLK
Coachella	KNWZ	970	N/TLK	Eureka	KINS	980	TLK+
Coachella	KBXO	90.3	RL	Eureka	KGOE	1480	TLK/S
Coachella	KPSH	90.9	RL	Eureka	KMUE	88.3	VAR
Coachella	KCLB	93.7	RK/CL	Eureka	KRED	92.3	CW
Coalinga	KDKL	88.3	RL	Eureka	KFMI	96.3	AC/H
Coalinga	KDKL	88.3	RL/CC	Eureka	KEKA	101.5	CW
Coalinga	KNGS	100.1	SIL	Eureka	KKHB	105.5	OLG
Columbia	KCVR	98.9	SPN	Fair Oaks	KSSJ	94.7	JAZ
Colusa	KKCY	103.1	CW	Fairfield	KASK	91.5	N/TLK
Colusa	KQPT	107.5	A/C	Fairmead	KLVY	91.1	RL/CC
Compton	KJLH	102.3	URB	Fallbrook	KSSD	107.1	SPN
Concord	KABN	1480	SIL	Felton	KTEE	93.7	OLG
Concord	KVHS	90.5	AP	Ferndale	KAJK	99.1	A/CH
Copperopolis	KRVR	105.5	JAZ	Firebaugh	KAJP	94.7	AS
Corcoran	KkXQX	102.3	SP/MX	Ford City	KZPE	102.1	N/TLK
Corning	KTHU	100.7	RK/C	Fort Bragg	KDAC	1230	SP/MX
Corona	KWRM	1370	SPN	Fort Bragg	KOZT	95.3	ALT
Crescent	KPOD	1240	AS	Fort Bragg	KSAY	98.5	AC
Crescent	KFVR	1310	SP/MEX	Fortuna	KNCR	1090	SP/MX
Crescent	KHSR	91.9	N/TLK	Fortuna	KWPT	100.3	AS
Crescent	KCRE	94.3	AC	Fountain V.	KLIt	92.7	AS
Crescent	KPOD	97.9	CW	Fowler	KQEQ	1220	SPN
Culver City	KIEV	1500	N/TLK	Fowler	KEZL	96.7	JAZ
Cupertino	KKUP	91.5	VAR	Frazier Pk	KJPG	1050	RL
Davis	KDVS	90.3	VAR	Freedom	KPlg	107.5	AS
Davis	KHZZ	104.3	SP/MEX	Fremont	KOHL	89.3	CHR
Delano	KCHJ	1010	SP/OM	Fresno	KMJ	580	TLK+
Delano	KDFO	98.5	AS	Fresno	KOQO	790	SPN
Delano	KKDJ	105.3	AC	Fresno	KBIF	900	RL
Dinuba	KRDU	1130	RL/G	Fresno	KEYQ	980	SPN
Dinuba	KSOF	98.9	AC/S	Fresno	KYNO	1300	SP/RL
Dunsmuir	KZRO	100.1	OLG	Fresno	KCBL	1340	SPT

City	Station	Freq	Format	City	Station	Freq	Format
Fresno	KFIG	1430	SPTS	Hemet	KXRS	105.7	SPN
Fresno	KIRV	1510	RL	Hesperia	KRAK	910	AS
Fresno	KXEX	1550	SP/MX	Hollister	KMPG	1520	SPN
Fresno	KGST	1600	SPN	Hollister	KHRI	90.7	RL/CC
Fresno	KFCF	88.1	VAR	Hollister	KOTR	93.5	AC
Fresno	KFNO	90.3	RL/CC	Holtville	KGBA	100.1	RL/CC
Fresno	KFSR	90.7	JAZ	Hoopa	KIDE	91.3	VAR
Fresno	KSJV	91.5	SPN	Hydesville	KSLG	94.1	RK/NU
Fresno	KSKS	93.7	CW	Idyllwild	KATY	101.3	AC
Fresno	KJFX	95.7	RK/CL	Imperial	KMXX	99.3	SPN
Fresno	KMGV	97.9	URB	Independence	KSRW	92.5	AC
Fresno	KJWL	99.3	AS	Indio	KESQ	1400	SPN
Fresno	KWYE	101.1	CHR	Indio	KCRI	89.3	N/TLK
Fresno	KOQO	101.9	SPN	Indio	KKUU	92.7	CH/D
Fresno	KALZ	102.7	AC	Indio	KJJZ	102.3	JAZ
Fresno	KRNC	105.9	SPN	Inglewood	KTYM	1460	RL
Garberville	KMUD	91.1	VAR	Inglewood	KRCD	103.9	SP/MX
Garberville	KLVG	103.7	RL/CC	Inyokern	KZLU	88.7	N/TLK
Garberville	KHUM	104.7	AP	Irvine	KUCI	88.9	VAR
Garden Grove	LEBM	94.3	SP/MX	Jackson	KSFS	94.3	RL
George	KATJ	100.7	CW	Johannesburg	KRAJ	100.9	OLG
Gilroy	KAZA	1290	SPN	Johannesburg	KEDD	103.9	SP/MX
Gilroy	KBAY	94.5	AC/S	Joshua Tree	KQCM	92.1	CHR
Glendale	KRLA	870	TLK+	Julian	KSDG	890	N/TLK
Goleta	KMGQ	106.3	JAZ	Julian	KLVJ	100.1	RL/CC
Gonzales	KKMC	880	RL	June Lake	KWTM	90.9	RL
Gonzales	KMBY	104.3	RK	Kerman	KOKO	94.3	OLG
Grass Valley	KNCO	830	TLK+	Kerman	KBHH	95.3	SPN
Grass Valley	KNCO	94.1	A/C	Kernville	KCNQ	102.5	CW
Grass Valley	KLVS	99.3	RL	King City	KRKC	1490	CW
Grass Valley	KCEE	103.3	N/TLK	King City	KDRY	91.3	RL
Green Acres	KAXL	88.3	RL/CC	King City	KHDV	93.9	SPN
Green Acres	KRAB	106.1	RK	King City	KRKC	102.1	A/CH
Greenfield	KLOK	99.5	SPN	Kings Beach	KSRN	107.7	SP/MX
Greenfield	KSEA	107.9	SPN	Kingsburg	KFYE	106.3	RL/CC
Greenville	KPJP	89.3	RL	La Quinta	KUNA	96.7	SP/MX
Gridley	KMJE	101.5	A/C	Lk Arrowhead	KCXX	103.9	RK
Groveland	KXSR	91.7	CLA	Lake Isabella	KQAB	1140	TLK+
Grover Beach	KQJZ	107.3	RK	Lake Isabella	KVLI	104.5	OLG
Guadalupe	KIDI	105.5	SPN	Lakeport	KXBX	1270	AS
Gualala	KTDE	100.5	A/C	Lakeport	KXBX	98.3	AC
Hamilton	KRER	101.7	N/TLK	Lakeport	KNTI	99.5	AC
Hanford	KIGS	620	SPN	Lancaster	KAVL	610	SPT
Hanford	KGEN	94.5	SPN	Lancaster	KWJL	1380	SP/MX
Hanford	KRZR	103.7	RK/CL	Lancaster	KTLW	88.9	RL
Hanford	KMPH	107.5	URB	Lancaster	KGMX	106.3	AC
Hayward	KCRH	89.9	VAR	Le Grand	KEFR	89.9	RL/CC
Healdsburg	KFGY	92.9	CW	Lemoore	KJOP	1240	RL
Healdsburg	KRSH	95.9	A/C	Lenwood	KHDR	96.9	RK
Hemet	KSDT	1320	SPN	Lenwood	KBTW	104.5	SP/MX

California

Radio on the Road

City	Station	Freq	Format	City	Station	Freq	Format
Lenwood	KIXW	107.3	CW	Los Angeles	KOST	103.5	AC
Lincoln	KXCL	103.9	TOP	Los Angeles	KBIG	104.3	AC
Lindsay	KZPO	103.3	AS	Los Angeles	KMZT	105.1	CLA
Livermore	KKIQ	101.7	AC	Los Angeles	KPWR	105.9	URB
Livingston	KLVN	88.3	RL	Los Angeles	KLVE	107.5	SPN
Livingston	KCJH	89.1	RL	Los Banos	KLBS	1330	ETH
Livingston	KSKD	95.9	SPN	Los Banos	KQLB	106.9	ET/O
Lodi	KCVR	1570	SPN	Los Gatos	KRTY	95.3	CW
Lodi	KWIN	97.7	CHR	Los Molinos	KCEZ	102.1	OLG
Loma Linda	KCAA	1050	N/TLK	Los Osos-	KSTT	101.3	A/C
Lompoc	KTME	1410	TLK+	Lucerne Valley	KIXA	106.5	RK
Lompoc	KLWG	88.1	RL	Ludlow	KHWZ	100.1	RK
Lompoc	KRQZ	91.5	RL	Ludlow	KDUQ	102.5	CHR
Lompoc	KRQK	100.3	SPN	Madera	KHOT	1250	RL
Lompoc	KBOX	104.1	AC	Madera	KMMM	107.1	SPN
Lompoc	KWSZ	105.1	AP	Magalia	KLVC	88.3	RL/CC
Lompoc	KSMY	106.7	SP/MX	Mammoth Lks	KMMT	106.5	AC/H
Long Beach	KFRN	1280	RL/CC	Manteca	KMRQ	96.7	RK
Long Beach	KLTX	1390	RL	Marina	KTOM	92.7	CW
Long Beach	KKJZ	88.1	JAZ	Mariposa	KUBB	96.3	CW
Long Beach	KBUE	105.5	SPN	Mariposa	KDJK	103.9	CH
Los Altos	KFJC	89.7	VAR	Marysville	KMYC	1410	TLK+
Los Altos	KFFG	97.7	AP	Marysville	KRCX	99.9	SPN
Los Angeles	KLAC	570	SPTS	Mcfarland	KIWI	102.9	SPN
Los Angeles	KFI*	640	TLK+	Mecca	KRCK	97.7	TOP
Los Angeles	KSPN	710	SPTS	Mendocino	KPMO	1300	TLK+
Los Angeles	KABC	790	TLK+	Mendocino	KAKX	89.3	AS
Los Angeles	KHJ	930	SPN	Mendocino	KMFB	92.7	OLG
Los Angeles	KFWB	980	NWS+	Mendota	KMEN	100.5	SP/MEX
Los Angeles	KTNQ	1020	SPN	Merced	KYOS	1480	TLK+
Los Angeles	KNX*	1070	NWS+	Merced	KTIQ	1660	SP/MX
Los Angeles	KTLK	1150	SPT	Merced	KBKY	94.1	SPTS
Los Angeles	KYPA	1230	N/AK	Merced	KABX	97.5	OLG
Los Angeles	KWKW	1330	SPN	Merced	KAMB	101.5	RL/CC
Los Angeles	KMPC	1540	SPT	Merced	KHPO	106.3	CH
Los Angeles	KXLU	88.9	AP	Middletown	KSXY	98.7	CHR
Los Angeles	KPFK	90.7	VAR	Mission Viejo	KSBR	88.5	JAZ
Los Angeles	KUSC	91.5	CLA	Modesto	KPMP	840	N/TLK
Los Angeles	KHHT	92.3	TOP	Modesto	KTRB	860	NWS+
Los Angeles	KCBS	93.1	TOP	Modesto	KESP	970	SPT
Los Angeles	KZLA	93.9	CW	Modesto	KFIV	1360	TLK+
Los Angeles	KTWV	94.7	JAZ	Modesto	KMPO	88.7	SPN
Los Angeles	KLOS	95.5	RK/CL	Modesto	KADV	90.5	RL/CC
Los Angeles	KXOL	96.3	SPN	Modesto	KBBU	93.9	SP/MX
Los Angeles	KLSX	97.1	TLK+	Modesto	KJSN	102.3	AC
Los Angeles	KYSR	98.7	AC	Modesto	KATM	103.3	CW
Los Angeles	KKLA	99.5	RL	Modesto	KHKK	104.1	RK/C
Los Angeles	KKBT	100.3	URB	Mojave	KAVC	1340	AS
Los Angeles	KRTH	101.1	OLG	Mojave	KCRY	88.1	N/TLK
Los Angeles	KIIS	102.7	CHR	Mojave	KVVS	97.7	CHR

34

City	Station	Freq	Format	City	Station	Freq	Format
Monte Rio	KVRV	97.7	RK	Pacific Grove	KAZU	90.3	N/TLK
Montecito	KJEE	92.9	RK	Pacific Grove	KCAQ	104.7	CHR
Monterey	KIDD	630	AS	Palm Desert	KHCS	91.7	RL
Monterey	KNRY	1240	TLK+	Palm Springs	KPSI	920	TLK+
Monterey	KWAV	96.9	AC	Palm Springs	KNWQ	1140	TLK+
Moraga	KSMC	89.5	VAR	Palm Springs	KGAM	1450	TLK+
Moreno Valley	KHPY	1530	SP/MX	Palm Springs	KPSI	100.5	A/CH
Morgan Hill	KSQQ	96.1	SPN	Palm Springs	KDES	104.7	OLG
Morro Bay	KLMM	94.1	SPN	Palm Springs	KPLM	106.1	CW
Morro Bay	KXTY	99.7	SPTS	Palmdale	KUTY	1470	N/TLK
Moss Beach	KLSI	89.3	N/TLK	Palo Alto	KNTS	1220	SPTS
Mount Shasta	KMJC	620	TLK+	Paradise	KKXX	930	RL
Mount Shasta	KMJC	107.9	AP	Paradise	KZAP	96.7	RK/CL
Mountain Pass	KHYZ	99.5	AC	Paradise	KHSL	103.5	CW
Mountain View	KSFH	87.9	RK	Pasadena	KDIS	1110	TLK+
Mt. Bullion	KCIV	99.9	RL	Pasadena	KAZN	1300	ET/K
Mt. Shasta	KNSQ	88.1	N/TLK	Pasadena	KPCC	89.3	N/TLK
Napa	KVON	1440	TLK+	Pasadena	KROQ	106.7	RK/NU
Needles	KTOX	1340	TLK+	Paso Robles	KPRL	1230	TLK+
Needles	KNKK	97.9	AC	Paso Robles	KKAL	92.5	CW
Needles	KLUK	97.9	RK	Paso Robles	KLUN	103.1	SPN
Nevada City	KVMR	89.5	VAR	Patterson	KOSO	93.1	A/CH
Newberry Spr	KIQQ	103.7	SP/MX	Patterson	KTSE	97.1	SPN
Newport Beach	KDLE	103.1	RK	Pebble Beach	KSPB	91.9	VAR
N Highlands	KQEI	89.3	RL/CC	Pescadero	KPDO	89.3	N/TLK
N Highlands	KQEI	89.9	N/TLK	Petaluma	KTOB	1490	SPN
Northridge	KCSN	88.5	CLA	Philo	KZYX	90.7	N/TLK
Oakdale	KHOP	95.1	CHR	Piedmont	KMZT	1510	OLG
Oakhurst	KTNS	1090	A/CS	Pismo Beach	KXTZ	95.3	POP
Oakhurst	KAAT	103.1	SP/MX	Pittsburg	KATD	990	SPN
Oakland	KNEW	910	TLK+	Pittsburg	KATD	990	SP/MX
Oakland	KQKE	960	AS	Placerville	KZSA	92.1	SPN
Oakland	KMKY	1310	Youth	Pomona	KWKU	1220	SPN
Oceanside	KKSM	1320	ALT	Pomona	KAHZ	1600	N/A
Ohai	KFYV	105.5	CHR	Port Hueneme	KVTA	1520	TLK+
Oildale	KLLY	95.3	AC	Porterville	KTIP	1450	TLK+
Ojai	KLFH	105.5	RL	Porterville	KIOO	99.7	RK
Ontario	KSPA	1510	AS	Prunedale	KLVM	89.7	RL/CC
Ontario	KDAI	93.5	ETH/K	Quincy	KPCO	1370	POP
Orange	KMXE	830	SP/MX	Quincy	KQNC	88.1	N/TLK
Orange Cove	KMAK	100.3	SP/MX	Quincy	KJCQ	88.5	N/TLK
Orcutt	KGDP	660	RL	Quincy	KNLF	95.9	RL
Orcutt	KPAT	95.7	RL/SG	Quincy	KSPY	100.3	BUS
Orland	KRQR	106.7	RK	Ro Cordova	KSTE	650	N/TLK
Oroville	KEWE	1340	SPTS	Ro Mirage	KMRJ	99.5	RK
Oroville	KHHZ	97.7	SPN	Randsburg	KGBM	89.7	RL
Oxnard	KOXR	910	SPN	Red Bluff	KBLF	1490	AC/S
Oxnard	KCRU	89.1	N/TLK	Red Bluff	KALF	95.7	N/TLK
Oxnard	KXLM	102.9		Red Bluff	KVLB	102.7	RK/CL
Oxnard	KCAQ	104.7		Redding	KVIP	540	RL

City	Station	Freq	Format	City	Station	Freq	Format
Redding	KLXR	1230	AS	Sacramento	KZZO	100.5	AC
Redding	KQMS	1400	TLK+	Sacramento	KBMB	103.5	CHR
Redding	KNRO	1670	SPT	Sacramento	KNCI	105.1	CW
Redding	KKRO	91.5	RL	Sacramento	KWOD	106.5	ALT
Redding	KFPR	88.9	N/TALK	Sacramento	KDND	107.9	CHR
Redding	KNCQ	97.3	CW	Salinas	KDBV	980	SP/RL
Redding	KVIP	98.1	RL/CC	Salinas	KZFX	1380	SIL
Redding	KSHA	104.3	AC/S	Salinas	KABL	1460	AS
Redlands	KCAL	1410	SP/MX	Salinas	KTGE	1570	SPN
Redlands	KUOR	89.1	JAZ	Salinas	KEBV	97.9	OLG
Redlands	KCAL	96.7	RK	Salinas	KPRC	100.7	SP
Redondo Bch	KDAY	93.5	CHR	Salinas	KDON	102.5	CHR
Redwood V.	KAIS	88.7	N/TLK	Salinas	KRAY	103.5	SPN
Ridgecrest	KLOA	1240	AS	San Ardo	KBDH	91.7	N/TLK
Ridgecrest	KWDJ	1360	CW	San Bernardino	KTIE	590	NTLK
Ridgecrest	KWTD	91.9	N/TLK	San Bernardino	KEZY	1240	SP/RL
Ridgecrest	KZIQ	92.7	AC/S	San Bernardino	KKDD	1290	YTH
Ridgecrest	KLOA	104.9	CW	San Bernardino	KTDD	1350	CW
Rio Dell	KNHT	107.3	N/TLK	San Bernardino	KVCR	91.9	N/CLA
Rio Vista	KRVH	101.5	SIL	San Bernardino	KFRG	95.1	CW
Riverbank	KCBC	770	RL	San Bernardino	KOLA	99.9	OLG
Riverside	KDIF	1440	SPN	San Clemente	KWVE	107.9	RL/G
Riverside	KPRO	1570	RL	San Diego	KOGO	600	TLK+
Riverside	KUCR	88.3	A/C	San Diego	KFMB	760	TLK+
Riverside	KSGN	89.7	RL	San Diego	XEMO	860	SPN
Riverside	KELT	92.7	AC/S	San Diego	KURS	1040	RL/G
Riverside	KLYY	97.5	SPN	San Diego	KSDO	1130	SP/RL
Riverside	KGGI	99.1	CH/D	San Diego	KCBQ	1170	TLK+
Rocklin	KEBR	1210	RL/CC	San Diego	KPRZ	1210	RL/TLK
Rohnert Park	KRPQ	104.9	CW	San Diego	KSON	1240	YTH
Rosamond	KLKX	92.5	CLA	San Diego	KLSD	1360	N/TLK
Rosamond	KOSS	105.5	AC	San Diego	KSDS	88.3	JAZ
Rosedale	KOGR	88.9	N/TLK	San Diego	KPBS	89.5	N/TLK
Roseville	KLIB	1110	SPN	San Diego	KMYI	94.1	A/CH
Roseville	KFSG	1690	SPN	San Diego	KBZT	94.9	RK
Roseville	KHWD	93.7	RK	San Diego	KYXY	96.5	AC/S
Sacramento	KHTK	1140	SPT	San Diego	KSON	97.3	CW
Sacramento	KSAC	1240	N/TLK	San Diego	KIFM	98.1	JAZ
Sacramento	KCTC	1320	AS	San Diego	KFMB	100.7	AC/H
Sacramento	KTKZ	1380	N/TLK	San Diego	KGB	101.5	RK/CL
Sacramento	KJAY	1430	ETH	San Diego	KLQV	102.9	SPN
Sacramento	KIID	1470	YTH	San Diego	KPLN	103.7	RK/C
Sacramento	KFBK	1530	TLK+	San Diego	KIOZ	105.3	RK
Sacramento	KEDR	88.1	RL	San Diego	KLNV	106.5	SPN
Sacramento	KXJZ	88.9	JAZ	San Fernando	KUBA	94.3	SPN
Sacramento	KXPR	90.9	CLA	San Francisco	KSFO	560	TLK+
Sacramento	KGBY	92.5	A/C	San Francisco	KFRC	610	RL
Sacramento	KYMX	96.1	AC/S	San Francisco	KNBR	680	SPTS
Sacramento	KSEG	96.9	RK/CL	San Francisco	KCBS	740	AM
Sacramento	KRXQ	98.5	RK	San Francisco	KGO	810	AM

City	Station	Freq	Format	City	Station	Freq	Format
San Francisco	KIQI	1010	RL	San Martin	KPRZ	1210	ETH
San Francisco	KFAX	1100	AC/S	San Mateo	KTCT	1050	SPT
San Francisco	KOIT	1260	ETH	San Mateo	KCSM	91.1	JAZ
San Francisco	KEST	1450	TLK+	San Mateo	KSAN	107.7	RK/CL
San Francisco	KYCY	1550	N/TLK	San Rafael	KSRH	88.1	VA
San Francisco	KQED	88.5	VA	San Rafael	KJQI	100.7	SP
San Francisco	KPOO	89.5	AP	Santa Ana	KVNR	1480	ETH
San Francisco	KUSF	90.3	N/TLK	Santa Ana	KWIZ	96.7	SPN
San Francisco	KALW	91.7	SP	Santa Ana	KALI	106.3	ETH
San Francisco	KRZZ	93.3	CHR	Santa Barbara	KTMS	990	TLK+
San Francisco	KYLD	94.9	POP	Santa Barbara	KZER	1250	SP
San Francisco	KZBR	95.7	AC/S	Santa Barbara	KZBN	1290	POP
San Francisco	KOIT	96.5	AC	Santa Barbara	KIST	1340	N/TLK
San Francisco	KLLC	97.3	UR/RB	Santa Barbara	KSPE	1490	SPN
San Francisco	KISQ	98.1	SPN	Santa Barbara	KQSC	88.7	CLA
San Francisco	KSOL	98.9	OLG	Santa Barbara	KSBX	89.5	N/CLA
San Francisco	KFRC	99.7	AC/H	Santa Barbara	KCSB	91.9	VAR
San Francisco	KIOI	101.3	CLA	Santa Barbara	KDB	93.7	CLA
San Francisco	KDFC	102.1	JAZ	Santa Barbara	KRUZ	97.5	AC
San Francisco	KKSF	103.7	AP	Santa Barbara	KTYD	99.9	RK
San Francisco	KFOG	104.5	RK/NU	Santa Barbara	KVYB	103.3	SP
San Francisco	KITS	105.3	UR/RB	Santa Barbara	KIST	107.7	RK
San Francisco	KMEL	106.1	RL/CC	Santa Clara	KVVN	1430	SPN
San Francisco	KEAR	106.9	ET/C	Santa Clara	KSCU	103.3	VAR
San Gabriel	KMRB	1430	CHR	Santa Clara	KARA	105.7	SP/RL
San Jacinto	KWIE	96.1	CW	Santa Cruz	KSCO	1080	TLK+
San Joaquin	KUUS	105.5	SPN	Santa Cruz	KZSC	88.1	VAR
San Jose	KLOK	1170	SPN	Santa Cruz	KUSP	89.9	N/VAR
San Jose	KZSF	1370	ETH/V	Santa Cruz	KSRI	90.7	RL/CC
San Jose	KSJX	1500	NWS+	Santa Cruz	KSQL	99.1	SPN
San Jose	KLIV	1590	VAR	Sta Margarita	KWWV	106.1	CHR
San Jose	KMTG	89.3	VA	Santa Maria	KSMA	1240	TLK+
San Jose	KSJS	90.5	SP	Santa Maria	KUHL	1440	TLK+
San Jose	KSJO	92.3	RK/CL	Santa Maria	KSBQ	1480	SPN
San Jose	KUFX	98.5	SPN	Santa Maria	KTAP	1600	SPN
San Jose	KBRG	100.3	AC/H	Santa Maria	KHFR	89.7	N/A
San Jose	KEZR	106.5	TLK+	Santa Maria	KGDP	90.7	SP/RL
S L Obispo	KVEC	920	RL/SP	Santa Maria	KXFM	99.1	RK/CL
S L Obispo	KJDJ	1030	TLK+	Santa Maria	KSNI	102.5	CW
S L Obispo	KYNS	1340	POP	Santa Monica	KBLA	1580	SPN
S L Obispo	KKJL	1400	RL/CC	Santa Monica	KCRW	89.9	N/VAR
S L Obispo	KLVH	88.5	RL	Santa Monica	KACD	103.1	RK
S L Obispo	KLFF	89.3	N/JZ	Santa Paula	KUNX	1400	SP
S L Obispo	KCBX	90.1	ALT	Santa Paula	KLJR	96.7	SPN
S L Obispo	KCPR	91.3	RK/CL	Santa Rosa	KSRO	1350	TLK+
S L Obispo	KZOZ	93.3	CHR	Santa Rosa	KRRS	1460	SPN
S L Obispo	KSLY	96.1	SPN	Santa Rosa	KBBF	89.1	SPN
S L Obispo	KLRM	97.1	RL	Santa Rosa	KRCB	91.1	VAR
S L Obispo	KKJG	98.1	AC/S	San Martin	KPRZ	1210	ETH
San Marcos	KPRZ	1210	RL	San Mateo	KTCT	1050	SPT

City	Station	Freq	Format	City	Station	Freq	Format
San Mateo	KCSM	91.1	JAZ	Seaside	KMBY	103.9	RK
San Mateo	KSAN	107.7	RK/CL	Seaside	KSES	107.1	SPN
San Rafael	KSRH	88.1	VA	Sebastopol	KJZY	93.7	JAZ
San Rafael	KJQI	100.7	SP	Selma	KQKL	88.5	RL
Santa Ana	KVNR	1480	ETH	Shafter	KGLV	89.5	N/A
Santa Ana	KWIZ	96.7	SPN	Shafter	KGZO	90.9	SP/RL
Santa Ana	KALI	106.3	N/A	Shafter	KKXX	93.1	POP
Santa Barbara	KTMS	990	TLK+	Shafter	KSMJ	97.7	JAZ
Santa Barbara	KZER	1250	SP	Shasta	KCNR	1460	SIL
Santa Barbara	KZBN	1290	POP	Shasta Lake	KJPR	1330	N/A
Santa Barbara	KIST	1340	N/TLK	Shasta Lake	KNNN	99.3	CHR
Santa Barbara	KSPE	1490	SPN	Shasta Lake	KESR	107.1	ALT
Santa Barbara	KQSC	88.7	CLA	Shingle Springs	KCCL	101.9	OLG
Santa Barbara	KSBX	89.5	N/CLA	Shingletown	KKXS	96.1	CW
Santa Barbara	KCSB	91.9	VAR	Shingletown	KRDG	105.3	OLG
Santa Barbara	KDB	93.7	CLA	Simi Valley	KIRN	670	ET/O
Santa Barbara	KRUZ	97.5	AC	Soledad	KMGX	700	SPN
Santa Barbara	KTYD	99.9	RK	Soledad	KFRS	89.9	RL
Santa Barbara	KVYB	103.3	SP	Soledad	KLUE	106.3	AC/H
Santa Barbara	KIST	107.7	RK	Solvang	KSYV	96.7	AC
Santa Clara	KVVN	1430	SPN	Sonoma	KSVY	91.3	VAR
Santa Clara	KSCU	103.3	VAR	Sonora	KVML	1450	TLK+
Santa Clara	KARA	105.7	SP/RL	Sonora	KZSQ	92.7	AC
Santa Cruz	KSCO	1080	TLK+	Soquel	KYAA	1200	OLG
Santa Cruz	KZSC	88.1	VAR	S Lake Tahoe	KTHO	590	TLK+
Santa Cruz	KUSP	89.9	N/VAR	S Lake Tahoe	KOWL	1490	TLK+
Santa Cruz	KSRI	90.7	RL/CC	S Lake Tahoe	KRLT	93.9	AC
Santa Cruz	KSQL	99.1	SPN	S Lake Tahoe	KWYL	102.9	CHR
Sta Margarita	KWWV	106.1	CHR	South Oroville	KYIX	104.9	RL/CC
Santa Maria	KSMA	1240	TLK+	Stanford	KZSU	90.1	VA
Santa Maria	KUHL	1440	TLK+	Stockton	KWG	1230	RL
Santa Maria	KSBQ	1480	SPN	Stockton	KUYL	1280	RL
Santa Maria	KTAP	1600	SPN	Stockton	KSTN	1420	OLG
Santa Maria	KHFR	89.7	N/A	Stockton	KYCC	90.1	RL/G
Santa Maria	KGDP	90.7	SP/RL	Stockton	KUOP	91.3	N/JAZ
Santa Maria	KXFM	99.1	RK/CL	Stockton	KJOY	99.3	AC
Santa Maria	KSNI	102.5	CW	Stockton	KQOD	100.1	UR/RB
Santa Monica	KBLA	1580	SPN	Stockton	KSTN	107.3	SPN
Santa Monica	KCRW	89.9	N/VAR	Sun City	KXFG	92.9	CW
Santa Monica	KACD	103.1	RK	Sunnyvale	KCNL	104.9	RK
Santa Paula	KUNX	1400	SP	Susanville	KSUE	1240	TLK+
Santa Paula	KLJR	96.7	SPN	Susanville	KFLL	88.1	N/A
Santa Rosa	KSRO	1350	TLK+	Susanville	KHJQ	92.3	AC/H
Santa Rosa	KRRS	1460	SPN	Susanville	KJDX	93.3	CW
Santa Rosa	KBBF	89.1	SPN	Sutter	KXJS	88.7	N/JAZ
Santa Rosa	KRCB	91.1	VAR	Taft	KBDS	103.9	CHR
Santa Rosa	KLVR	91.9	RL/CC	Tahoe City	KKTO	90.5	N/JAZ
Santa Rosa	KZST	100.1	AC	Tahoe City	KLCA	96.5	AC/S
Santa Rosa	KXFX	101.7	RK	Tehachapi	KYLU	88.7	N/A
Santa Ynez	KRAZ	105.9	CW	Tehachapi	KKZQ	100.1	CW

Radio on the Road   California

City	Station	Freq	Format	City	Station	Freq	Format
Tehachapi	KTPI	103.1	CW	Visalia	KDUV	88.9	RL/CC
Temecula	KRTM	88.9	RL	Visalia	KARM	89.7	RL/CC
Temecula	KMYT	94.5	JAZ	Visalia	KFSO	92.9	SP
Temecula	KTMQ	103.3	RK/CL	Visalia	KSLK	96.1	SP
Templeton	KXDZ	100.5	POP	Visalia	KSEQ	97.1	CHR
Thousand Oaks	KCLU	88.3	N/JAZ	Vista	KCEO	1000	BTLS
Thousand Oaks	KDSC	91.1	CLA	Walnut	KSAK	90.1	VAR
Thousand Oaks	KMLT	92.7	POP	Walnut Creek	KABL	92.1	POP
Thousand Plms	KXPS	1010	SPTS	Wasco	KFHL	91.7	N/A
Thousand Plms	KXPS	1270	TLK/S	Wasco-Gr Acres	KERI	1180	RL
Thousand Plms	KLOB	94.7	SP	Watsonville	KOMY	1340	TLK+
Tipton	KCRZ	104.9	POP	Weaveerville	KWCA	101.1	RK
Torrance	KFOX	1650	ETH	Weaverville	KHRD	103.1	RK
Tracy	KYKL	90.7	RL	Weed	KNTK	102.3	N/TLK
Tracy	KMIX	100.9	SPN	West Covina	KALI	900	SPN
Truckee	KTKE	101.5	ALT	West Covina	KALI	900	SP/RL
Tulare	KJUG	1270	CW	West Covina	KRCV	98.3	SP
Tulare	KGEN	1370	SPN	West Covina	KRCV	98.3	SP
Tulare	KBOS	94.9	CHR	Westwood	KTOR	99.7	RK
Tulare	KJUG	106.7	CW	Williams	KARA	99.1	RL
Tule Lake	KFLS	96.5	CW	Willits	KLLK	1250	SP
Turlock	KLOC	1390	SP	Willits	KZYZ	91.5	N/VAR
Turlock	KBDG	90.9	SPN	Willits	KMKX	93.5	RK
Turlock	KCSS	91.9	VA	Willows	KIQS	1560	SP
Turlock	KWNN	98.3	CHR	Willows	KKFS	105.5	RL
Twain Harte	KKBN	93.5	CW	Willows	KCHC	106.3	N/A
29 Palms	KQYN	1250	TLK	Windsor	KMHX	104.1	AC
29 Palms	KDHI	95.7	CW	Winton	KLOQ	98.7	SP
29 Palms	KXCM	96.3	CW	Woodlake	KUFW	90.5	SPN
29 Palms	KCDZ	107.7	AC	Woodlake	KFRR	104.1	RK/H
Ukiah	KUKI	1400	SP	Woodland	KSFM	102.5	CH/D
Ukiah	KPRA	89.5	RL/CC	Yermo	KNYR	91.3	N/CLA
Ukiah	KWNE	94.5	AC	Yermo	KRXV	98.1	AC
Ukiah	KULV	97.1	RL	Yermo	KRSX	105.3	OLG
Ukiah	KUKI	103.3	CW	Yreka	KSYC	1490	TLK+
Ukiah	KUKI	103.3	CW	Yreka	KSYC	103.9	CW
Ukiah	KQPM	105.9	CW	Yuba City	KOBO	1450	SPN
Vacaville	KUIC	95.3	AC	Yuba City	KUBA	1600	CW
Vallejo	KDYA	1190	AC	Yucaipa	KLRD	90.1	RL/CC
Vallejo	KDIA	1640	RL	Yucca Valley	KDGL	106.9	POP
Ventura	KVEN	1450	OLG				
Ventura	KKZZ	1590	POP				
Ventura	KBBY	95.1	AC				
Ventura	KHAY	100.7	CW				
Ventura	KSSC	107.1	SPN				
Victorville	KATJ	1590	SP				
Victorville	KHMS	88.5	RL/CC				
Victorville	KXRD	89.5	RL/CC				
Victorville	KVVQ	103.1	CW				
Visalia	KVBL	1400	SPT				

Radio on the Road   Colorado

City	Station	Freq	Format	City	Station	Freq	Format
Alamosa	KGIW	1450	OLG	Colorado Spr	KKCS	101.9	CW
Alamosa	KRZA	88.7	N/VAR	Colorado Spr	KAFA	104.5	RK
Alamosa	KASF	90.9	AP	Commerce	KLTT	670	RL
Alamosa	KALQ	93.5	CW	Cortez	KVFC	740	OLG
Arvada	KDDZ	1550	Youth	Cortez	KSJD	91.5	VAR
Aspen	KAJX	91.5	YTH	Cortez	KISZ	97.9	CW
Aspen	KSPN	103.1	ALT	Cortez	KRTZ	98.7	AC
Aspen	KPVW	107.1	SP	Craig	KRAI	550	CW
Aurora	KMXA	1090	SP	Craig	KPYR	88.3	N/A
Aurora	KEZW	1430	POP	Craig	KRAI	93.7	CHR
Avon	KZYR	97.7	ALT	Craig	KQZR	102.5	RK
Basalt	KNFO	106.1	TLK+	Crested B.	KBUT	90.3	N/ALT
Bayfield	KLJH	107.1	RL	Delta	KDTA	1400	CW
Bennet	KSIR	107.1	SIL	Delta	KKNN	95.1	RK
Boulder	KVCU	1190	RK	Delta	KPRU	103.3	N/CLA
Boulder	KCFC	1490	N/TLK	Denver	KLZ	560	SPTS
Boulder	KGNU	88.5	VAR	Denver	KHOW	630	TLK+
Boulder	KRKS	94.7	RL	Denver	KNUS	710	NWS+
Boulder	KBCO	97.3	AP	Denver	KOA	850	TLK+
Breckenridge	KSMT	102.3	RK/NU	Denver	KPOF	910	RL
Brighton	KLDC	800	RL/G	Denver	KKFN	950	SPT
Broomfield	KDJM	92.5	OLG	Denver	KRKS	990	RL
Brush	KSIR	1010	TLK+	Denver	KLVZ	1220	SP/RL
Brush	KBWA	89.5	N/A	Denver	KBNO	1280	SP
Brush	KPRB	106.3	A/C	Denver	KCFR	1340	N/TLK
Buena Vista	KSKE	1450	TLK+	Denver	KGNU	1390	VAR
Buena Vista	KBVC	104.1	CW	Denver	KBJD	1650	N/TLK
Burlington	KNAB	1140	POP	Denver	KUVO	89.3	VA
Burlington	KNAB	104.1	CW	Denver	KVOD	90.1	CLA
Canon City	KRLN	1400	TLK+	Denver	KMGG	95.7	SP
Canon City	KTLC	89.1	RL/G	Denver	KYGO	98.5	CW
Canon City	KSTY	104.5	CW	Denver	KQMT	99.5	RK/CL
Carbondale	KCJX	88.9	N/JAZ	Denver	KIMN	100.3	AC
Carbondale	KVOV	90.5	CLA	Denver	KOSI	101.1	AC/S
Castle Rock	KJMN	92.1	SPN	Denver	KRFX	103.5	RK/S
Colona	KAVP	1450	AC	Denver	KXKL	105.1	OLG
Colona	KTMH	89.9	RL	Denver	KALC	105.9	AC/H
Colorado Spr	KVOR	740	TLK+	Denver	KBPI	106.7	RK/NU
Colorado Spr	KRDO	1240	SPT	Denver	KWMX	107.5	CH
Colorado Spr	KKML	1300	SPTSX	Dolores	KTCF	89.5	RL
Colorado Spr	KCMN	1530	POP	Dolores	KKDC	93.3	ALT
Colorado Spr	KWYD	1580	N/TLK	Durango	KIUP	930	POP
Colorado Spr	KEPC	89.7	VAR	Durango	KDGO	1240	TLK+
Colorado Spr	KTLF	90.5	RL/G	Durango	KDUR	91.9	VA
Colorado Spr	KRCC	91.5	N/VAR	Durango	KPTE	99.7	AC
Colorado Spr	KSPZ	92.9	OLG	Durango	KRSJ	100.5	CW
Colorado Spr	KILO	94.3	RK	Durango	KIQX	101.3	AC
Colorado Spr	KRDO	95.1	AC/S	Eagle	KTUN	101.5	RK
Colorado Spr	KKFM	96.1	RL/CC	Eaton	KLCQ	88.9	RL
Colorado Spr	KKFM	98.1	RK/CL	Englewood	KNRC	1150	SIL

City	Station	Freq	Format	City	Station	Freq	Format
Estes Park	KEZZ	1470	A/CS	Johnstown	KHNC	1360	TLK+
Estes Park	KXDC	102.1	CW	Julesburg	KJBL	96.5	CW
Evergreen	KXPK	96.5	SP/RL	Kremmling	KKHI	106.3	OLG
Fort Collins	KIIX	1410	AS	La Junta	KBLJ	1400	TLK+
Fort Collins	KCSU	90.5	AP	La Junta	KRLJ	89.1	N/VAR
Fort Collins	KTCL	93.3	RK	La Junta	KBLJ	92.1	OLG
Fort Collins	KPAW	107.9	CH	La Junta	KFVR	106.5	CHR
Fort Morgan	KFTM	1400	CW	Lakewood	KCKK	1600	CW
Fort Morgan	KBRU	101.7	SIL	Lakewood	KFDN	88.1	RL
Fountain	KIBT	96.1	RL/G	Lakewood	KQKS	107.5	CHR
Frisco	KYSL	93.9	AC	Lamar	KLMR	920	CW
Fruita	KEKB	99.9	CW	Lamar	KLMR	93.3	RK
Ft. Collins	KLHV	88.3	N/A	Lamar	KVAY	105.7	CW
Ft. Collins	KFRC	88.9	ALT	Limon	KIIQ	93.7	POP
Glenwood Spr	KGLN	980	OLG	Limon	KAVD	103.1	CW
Glenwood Spr	KDNK	88.1	VAR	Littleton	KCUV	1510	UR/AC
Glenwood Spr	KDRH	91.9	RL/CC	Longmont	KRCN	1060	N/TLK
Glenwood Spr	KKCH	92.7	AC/H	Longmont	KGUD	90.7	EZL
Glenwood Spr	KRVG	95.5	RK	Longmont	KJCD	104.3	JAZ
Glenwood Spr	KMTS	99.1	CW	Loveland	KSXT	1570	TLK
Granby	KRKY	930	CW	Loveland	KXWA	89.7	RL
Grand Junct.	KRDY	620	RL	Loveland	KTRR	102.5	AC
Grand Junct.	KNZZ	1100	TLK+	Manitou Spr	KXRE	1490	SPN
Grand Junct.	KEXO	1230	SPN	Manitou Spr	KCME	88.7	CLA
Grand Junct.	KTMM	1340	SPT	Manitou Spr	KBIQ	102.7	RL/CC
Grand Junct.	KAFM	88.1	VAR	Meeker	KAYW	98.1	CW
Grand Junct.	KCIC	88.5	RL/G	Monte Vista	KSLV	1240	CW
Grand Junct.	KPRN	89.5	N/TLK	Monte Vista	KSLV	95.3	AC/S
Grand Junct.	KJOL	90.3	RL/CC	Montrose	KUBC	580	POP
Grand Junct.	KMSA	91.3	VA	Montrose	KPRH	88.3	N/TLK
Grand Junct.	KJYE	92.3	AC/S	Montrose	KVMT	89.1	VAR
Grand Junct.	KMGJ	93.1	AC	Montrose	KKXK	94.1	CW
Grand Junct.	KSNJ	100.7	CW	Montrose	KSTR	96.1	OLG
Grand Junct.	KMXY	104.3	CHR	Monument	KCBR	1040	RL/G
Grand Junct.	KBKL	107.9	OLG	Morrison	KLDV	91.1	RL/CC
Greeley	KFKA	1310	TLK+	New Castle	KCUV	94.5	N/A
Greeley	KGRE	1450	SPN	Norwood	KRYD	104.9	CW
Greeley	KUNC	91.5	N/VAR	Oak Creek	KFMU	104.1	ALT
Greeley	KSME	96.1	CHR	Otis	KATR	98.3	CW
Gunnison	KPKE	1490	CW	Ouray	KWGL	105.7	AC
Gunnison	KWSB	91.1	VA	Pagosa Spr	KPAG	1400	CW
Gunnison	KEJJ	98.3	OLG	Pagosa Spr	KTPS	89.7	RL
Gunnison	KVLE	102.3	RK	Pagosa Spr	KWUF	106.3	OLG
Gypsum	KLRY	91.3	N/A	Paonia	KVNF	90.9	N/ALT
Hayden	KHCO	90.1	N/A	Pierce	KJMP	870	POP
Hayden	KIDN	95.9	RK/NU	Placerville	KTEI	90.7	RL
Hayden	KRMR	107.3	N/TLK	Pueblo	KCSJ	590	TLK/S
Holyoke	KSTH	92.3	AC	Pueblo	KRMX	690	SPN
Ignacio	KUTE	90.1	VAR	Pueblo	KFEL	970	RL/CC
Ignacio	KSUT	91.3	N/ETH	Pueblo	KKPC	1230	N/TLK

CITY	Station	Freq	Format		CITY	Station	Freq	Format
Pueblo	KGHF	1350	SPTS		Widefield	KKLI	106.3	AC/S
Pueblo	KAVA	1480	SPN		Windsor	KVVS	1170	SPN
Pueblo	KTPL	88.3	RL		Windsor	KUAD	99.1	CW
Pueblo	KTSC	89.5	AP		Wray	KRDZ	1440	POP
Pueblo	KFRY	89.9	N/A		Wray	KATR	98.3	CW
Pueblo	KCFP	91.9	N/CLA		Yuma	KNEC	100.9	POP
Pueblo	KCCY	96.9	CW					
Pueblo	KKMG	98.9	CHR					
Pueblo	KVUU	99.9	AC/H					
Pueblo	KGFT	100.7	RL/G					
Pueblo	KNKN	106.9	SPN					
Pueblo	KDZA	107.9	OLG					
Pueblo West	KYZX	103.9	RK					
Ridgway	KBNG	103.7	A/CH					
Rifle	KRGS	690	SPTS					
Rifle	KZKS	105.3	CW					
Rocky Ford	KPHT	95.5	A/CH					
Rye	KRYI	89.7	N/A					
Rye	KXWY	90.9	N/A					
Rye	KRYE	104.9	N/A					
Salida	KVRH	1340	OLG					
Salida	KVRH	92.1	A/CH					
Salida	KSBV	93.7	RK					
Security	KLIM	1120	A/CH					
Security	KSKX	105.5	CW					
Snowmass Vil	KSNO	103.9	AS					
Steamboat Sp	KBCR	1230	OLG					
Steamboat Sp	KTAH	88.5	N/A					
Steamboat Sp	KLBV	89.3	N/A					
Steamboat Sp	KBCR	96.9	CW					
Steamboat Sp	KRMR	107.3	TLK+					
Sterling	KSTC	1230	OLG					
Sterling	KDRE	90.7	N/A					
Sterling	KLZV	91.3	RL					
Sterling	KNNG	104.7	CW					
Sterling	KPMX	105.7	AC					
Strasburg	KJEB	102.3	CW					
Telluride	KOTO	91.7	N/VAR					
Thornton	KKZN	760	TLK					
Timnath	KJAC	105.5	POP					
Trimble	KTDU	88.5	RL					
Trinidad	KCRT	1240	CW					
Trinidad	KCRT	92.5	RK/CL					
Vail	KSKE	610	TLK/S					
Vail	KPRE	89.9	N/CLA					
Vail	KSKE	104.7	CW					
Walsenburg	KSPK	102.3	CW					
Wellington	KCOL	600	TLK					
Wellington	KKQZ	94.3	RK					

City	Station	Freq	Format	City	Station	Freq	Format
Ansonia	WADS	690	SPN	Milford	WFIF	1500	RL
Berlin	WERB	97.3	VAR	Monroe	WMNR	88.1	CLA
Bloomfield	WDZK	1550	YTH	Naugatuck	WFNW	1380	SPN
Bridgeport	WICC	600	TLK+	New Britain	WYRM	840	SPN
Bridgeport	WCUM	1450	SPN	New Britain	WNEZ	910	SPN
Bridgeport	WDJZ	1530	SPN	New Britain	WRCH	100.5	AC/S
Bridgeport	WPKN	89.5	VAR	New Britain	WFCS	107.7	AD/A
Bridgeport	WEZN	99.9	AC	New Canaan	WSLX	91.9	CLA
Bristol	WPRX	1120	SPN	New Haven	WELI	960	TLK+
Brookfield	WINE	940	CW	New Haven	WAVZ	1300	AS
Brookfield	WRKI	95.1	RK/H	New Haven	WYBC	1340	URB
Danbury	WLAD	800	TLK+	New Haven	WYBC	94.3	URB
Danbury	WXCI	91.7	AD/A	New Haven	WPLR	99.1	RK/H
Danbury	WFAR	93.3	RL	New London	WCNI	91.1	VAR
Danbury	WDAQ	98.3	AC/H	New London	WTYD	100.9	AC/S
East Lyme	WNLZ	98.7	AS	Norwalk	WNLK	1350	TLK/S
Enfield	WPKX	97.9	CW	Norwalk	WEFX	95.9	RK
Fairfield	WVOF	88.5	VAR	Norwich	WICH	1310	TLK+
Greenwich	WGCH	1490	TLK+	Norwich	WCTY	97.7	CW
Groton	WSUB	980	TLK+	Old Saybrook	WLIS	1420	AC/S
Guilford	WGRS	91.5	CLA	Pawcatuck	WKCD	107.7	AC
Hamden	WQUN	1220	AS	Pomfret	wbvc	91.1	VAR
Hamden	WQAQ	88.3	AD/A	Putnam	WINY	1350	AC/S
Hamden	WKCI	101.3	CHR	Ridgefield	WREF	850	OLG
Hartford	WRYM	840	SPN	Salisbury	WKZE	98.1	AD/A
Hartford	WNEZ	910	MOT	Sharon	WKZE	1020	RK/C
Hartford	WTIC	1080	TLK+	Sharon	WQQQ	103.3	CLA
Hartford	WCCC	1290	SPT	Shelton	WRXC	90.1	CLA
Hartford	WDRC	1360	AS	Somers	WDJW	89.7	VAR
Hartford	WAS	1410	SPT	South Kent	WGSK	90.1	CLA
Hartford	WMMW	1470	AS	Southington	WNTY	990	SPN
Hartford	WKND	1480	URB	Stamford	WSTC	1400	TLK+
Hartford	WJMJ	88.9	EZL	Stamford	WKHL	96.7	OLG
Hartford	WRTC	89.3	VAR	Stonington	WAXK	102.3	RK
Hartford	WQTQ	89.9	URB	Storrs	WHUS	91.7	VAR
Hartford	WWYZ	92.5	CW	Torrington	WAD/AJ	89.9	VAR
Hartford	WZMX	93.7	URB	Vernon	WCTF	1170	RL
Hartford	WTIC	96.5	CH	Wallingford	WWEB	89.9	VAR
Hartford	WRCH	100.5	AC/S	Waterbury	WWCO	1240	TLK+
Hartford	WDRC	102.9	OLG	Waterbury	WATR	1320	TLK+
Hartford	WHCN	105.9	RK	Waterbury	WWYZ	92.5	CW
Hartford	WCCC	106.9	RK	Waterbury	WMRQ	104.1	RK/C
Ledyard	WBMW	106.5	AC	Wst Hartford	WCCC	1290	RK
Litchfield	WZBG	97.3	AC/S	Wst Hartford	WWUH	91.3	VAR
Manchester	WLAT	1230	SPN	West Haven	WHNU	88.7	AD/A
Meriden	WMMW	1470	AS	Westport	WWPT	90.3	AD/A
Meriden	WKSS	95.7	CH	Westport	WEBE	107.9	AC/G
Middletown	WMRD	1150	CHR	Willimantic	WILI	1400	AC
Middletown	WESU	88.1	VAR	Willimantic	WILI	98.3	CHR
Middletown	WIHS	104.9	RL/CC	Windsor	WKND	1480	URB

Radio on the Road   Delaware

City	Station	Freq	Format	City	Station	Freq	Format
Bethany Bch	WOSC	95.9	RK				
Bethany Bch	WJNE	103.5	AC				
Christiana	WXHL	89.1	RL/CC				
Dover	WDOV	1410	TLK+				
Dover	WQVL	1600	RL/G				
Dover	WRDX	94.7	RK/CL				
Fenwick Island	WLBW	92.1	OLG				
Georgetown	WJWL	900	POP				
Georgetown	WZBH	93.5	RK/CL				
Laurel	WJNE	95.3	RK/NU				
Lewes	WXJN	105.9	CW				
Milford	WYUS	930	SPN				
Milford	WAFL	97.7	CHR				
Milford	WXPZ	101.3	RL/CC				
Newark	WNRK	1260	SILNT				
Newark	WVUD	91.3	AP				
Ocean View	WZEB	101.7	CHR				
Rehoboth B.	WGMD	92.7	TLK+				
Seaford	WJWK	1280	POP				
Seaford	WGBG	98.3	CH				
Seaford	WGBG	98.5	CH				
Selbyville	WOCM	98.1	AP				
Smyrna	WDSD	92.9	CW				
Wilmington	WDEL	1150	TLK+				
Wilmington	WJBR	1290	POP				
Wilmington	WTMC	1380	Traf				
Wilmington	WILM	1450	TLK+				
Wilmington	WMPH	91.7	CHR				
Wilmington	WSTW	93.7	CH/40				
Wilmington	WJBR	99.5	AC				
Wilmington	WNNN	101.7	RL				
Wilmington	WXCY	103.7	CW				

47

Radio on the Road

District of Columbia

City	Station	Freq	Format		City	Station	Freq	Format
Washington	WMAL	630	TLK+					
Washington	WTEM	980	SPT					
Washington	WUST	1120	N/A					
Washington	WGAY	1260	BIZ					
Washington	WYCB	1340	RLG					
Washington	WOL	1450	TLK+					
Washington	WTOP	1500	NWS+					
Washington	WPFW	89.3	JAZ+					
Washington	WCSP	90.1	TLK+					
Washington	WKYS	93.9	URB					
Washington	WHUR	96.3	URB					
Washington	WASH	97.1	AC					
Washington	WMZQ	98.7	CW					
Washington	WJMO	99.5	CHR					
Washington	WBIG	100.3	OLG					
Washington	WGMS	103.5	CLA					
Washington	WRQX	107.3	AC/H					
Washington	WWDC	101.1	RK					

49

Florida

City	Station	Freq	Format	City	Station	Freq	Format
Alachua	WNDT	92.5	RK	Clearwater	WXTB	97.9	RK
Apalachicola	WOYS	100.5	AC	Clermont	WWFL	1340	TLK+
Apalahicola	WFCT	105.5	AS	Clermont	WLAZ	88.7	SPN
Apopka	WHIM	1520	RL	Clewiston	WAFC	590	SPN
Apopka	WPYO	95.3	CHR	Clewiston	WAFC	106.3	CW
Arcadia	WZTK	1480	TLK+	Cocoa	WRFB	860	EZL
Atlantic Beach	WQOP	1600	RL	Cocoa	WWBC	1510	RL
Atlantic Beach	WFYV	104.5	RK	Cocoa	WMIE	91.5	RL/CC
Auburndale	WTWB	1570	RL/G	Cocoa	WLRQ	99.3	AC
Avon Park	WAVP	1390	SILNT	Cocoa Beach	WXXU	1300	RL/G
Avon Park	WWOJ	99.1	CW	Cocoa Beach	WJRR	101.1	RK
Baker	WTJT	90.1	RL/G	Cocoa Beach	WTKS	104.1	TLK+
Baldwin	WXQL	105.7	SILNT	Coral Cove	WSRZ	107.9	OLG
Bartow	WWBF	1130	OLG	Coral Gables	WVCG	1080	VAR
Bartow	WBAR	1460	CW	Coral Gables	WRHC	1560	SPN
Belle Glade	WSWN	900	RL/G	Coral Gables	WVUM	90.5	AD/A
Belle Glade	WBGF	93.5	CW	Coral Gables	WHQT	105.1	URB
Belleview	WWKO	91.3	RL/G	Crawfordsville	WAKU	94.1	RL/CC
Beverly Hills	WGUL	106.3	AS	Crestview	WJSB	1050	CW
Big Pine Key	WWUS	104.1	CLA	Crestview	WAAZ	104.7	CW
Bithlo	WNTF	1580	SILNT	Cross City	WDFL	1240	CW
Blountstown	WYBT	1000	RL/G	Cross City	WKZY	106.9	AC
Blountstown	WPHK	102.3	CW	Crystal River	WAQV	90.9	RL/CC
Boca Raton	WSBR	740	TLK+	Crystal River	WXJC	91.9	RL/CC
Boca Raton	WKIS	99.9	CW	Crystal River	WKTK	98.5	AC
Bonifay	WYYX	97.7	RK	Cypress Gard	WHNR	1360	URB
Bonita Springs	WRXX	96.1	RK	Dade City	WDCF	1350	TLK+
Boynton Beach	WJNA	1040	TLK/S	Dade City	WMGG	96.1	SPN
Boynton Beach	WJNO	1290	TLK+	Davie	WAVS	1170	SPN
Boynton Beach	WRMB	89.3	RL	Daytona Bch	WNDB	1150	TLK+
Bradenton	WWPR	1490	AS	Daytona Bch	WROD	1340	AS
Bradenton	WJIS	88.1	RL/CC	Daytona Bch	WMFJ	1450	RL
Brandon	WLCC	760	SPN	Daytona Bch	WCFB	94.5	URB
Brooksville	WWJB	1450	TLK+	De Funiak Sp	WAKJ	91.3	SILNT
Bushnell	WKFL	1170	RL/G	De Funiak Sp	WGTX	1280	OLG
Callahan	WELX	1160	SPN	De Funiak Sp	WZEP	1460	CW
Callahan	WPLA	93.3	RK	De Funiak Sp	WMXZ	103.1	AC/H
Callaway	WMXP	103.5	AC/S	De Land	WYND	1310	RL/G
Cape Coral	WXKB	103.9	CHR	De Land	WXVQ	1490	TLK+
Carrabelle	WOCY	106.5	CW	De Land	WOCL	105.9	URB
Cedar Creek	WKSG	89.5	RL/CC	Delray Beach	WPBI	1420	NWS+
Cedar Key	WRGO	102.7	OLG	Destin	WBZR	1120	SILNT
Century	WPFL	105.1	OLG	Destin	WMMK	92.1	OLG
Chattahoochee	WTCL	1580	RL/G	Dogwood Lk	WJED	91.1	RL/G
Chiefland	WLQH	940	CW	Dunedin	WGUL	860	AS
Chiefland	WTBH	91.5	RL/G	Dunedin	WLVU	1470	BIZ
Chiefland	WLQH	107.9	CW	Dunnellon	WTRS	102.3	CW
Chipley	WBGC	1240	OLG	Eatonville	WRLZ	1270	SPN
Clearwater	WTAN	1340	TLK+	Edgewater	WKTO	88.7	RL/CC
Clearwater	WSSR	95.7	AC	Edgewater	WKRO	93.1	RK

51

CITY	STATION	FREQ	FORMAT	CITY	STATION	FREQ	FORMAT
Englewood	WENG	1530	TLK+	Gifford	WAVW	94.7	CW
Englewood	WSEB	91.3	RL	Goulds	WRTO	98.3	SPN
Englewood	WYNF	105.9	TLK+	Graceville	WYDA	101.7	CHR
Eustis	WKIQ	1240	TLK+	Green Cove	WJBT	92.7	URB
Fernandina B	WGSR	1570	RL/G	Gretna	WGWD	93.3	CW
Fernandina B	WNLE	91.7	RL/G	Gulf Breeze	WRNE	980	URB
Five Points	WCJX	106.5	RK/C	Haines City	WLVF	930	RL/G
Flagler Beach	WJLH	90.3	RL/CC	Haines City	WLVF	90.3	RL/G
Florida City	WMFL	88.5	RL/CC	Havana	WHTF	104.9	CHR
Ft Lauderdale	WFTL	1400	TLK+	Hernando	WRZN	720	AS
Ft Lauderdale	WSRF	1580	TLK+	Hialeah	WACC	830	SPN
Ft Lauderdale	WAFG	90.3	RL/CC	Hialeah	WCMQ	92.3	SPN
Ft Lauderdale	WHYI	100.7	CHR	High Springs	WXJZ	104.9	JAZ
Ft Lauderdale	WMGE	103.5	URB	Holiday	WSUN	97.1	OLG
Ft Lauderdale	WBGG	105.9	CH	Holly Hill	WAPN	91.5	RL/CC
Ft Lauderdale	WRMA	106.7	SPN	Holly Hill	WVYB	103.3	CHR
Fort Myers	WINK	1200	TLK+	Hollywood	WLQY	1320	SPN
Fort Myers	WCRM	1350	SPN	Holmes Bch	WLLB	98.7	URB
Fort Myers	WMYR	1410	CW	Homestead	WOIR	1430	SPN
Fort Myers	WAYJ	88.7	RL/CC	Homestead	WRGP	88.1	VAR
Fort Myers	WSRX	89.5	RL/CC	Homosassa	WXCV	95.3	AC/H
Fort Myers	WSOR	90.9	RL	Immokalee	WAFZ	1490	SPN
Fort Myers	WJYO	91.5	RL/CC	Immokalee	WGCQ	92.1	RL/G
Fort Myers	WTLT	93.5	AC/S	Indian Rock B	WXYB	1520	SPN
Fort Myers	WARO	94.5	CH	Indiantown	WPBZ	103.1	RK/H
Fort Myers	WOLZ	95.3	OLG	Inverness	WINV	1560	SILNT
Fort Myers	WINK	96.9	AC	Jacksonville	WBWL	600	SPT
Fort Myers	WWGR	101.9	CW	Jacksonville	WOKV	690	TLK+
Fort Myers	WSGL	104.7	AC/H	Jacksonville	WFXJ	930	SPT
Fort Myers	WQNU	105.5	CW	Jacksonville	WVOJ	970	TLK/S
Fort Pierce	WJNX	1330	TLK+	Jacksonville	WROS	1050	RL
Fort Pierce	WIRA	1400	TLK+	Jacksonville	WJAX	1220	EZL
Fort Pierce	WLDI	95.5	CHR	Jacksonville	WSVE	1280	RL/G
Fort Pierce	WKGR	98.7	RK	Jacksonville	WJGR	1320	TLK+
Fort Walton B	WFTW	1260	TLK+	Jacksonville	WCGL	1360	RL/G
Fort Walton B	WFAV	1400	AS	Jacksonville	WZAZ	1400	RL/G
Fort Walton B	WPSM	91.1	RL/CC	Jacksonville	WZNZ	1460	BIZ
Fort Walton B	WZNS	96.5	CHR	Jacksonville	WOBS	1530	RL/G
Fort Walton B	WKSM	99.5	RK/C	Jacksonville	WNCM	88.1	RL/CC
Ft. Myers	WJBX	99.3	RK/H	Jacksonville	WJFR	88.7	RL
Gainesville	WRUF	850	NWS+	Jacksonville	WKTZ	90.9	EZL
Gainesville	WLUS	980	AS	Jacksonville	WAPE	95.1	CHR
Gainesville	WGGG	1230	SPT	Jacksonville	WEJZ	96.1	AC/S
Gainesville	WAJD	1390	RK	Jacksonville	WKQL	96.9	OLG
Gainesville	WTMN	1430	SPT	Jacksonville	WQIK	99.1	CW
Gainesville	WYFB	90.5	RL/G	Jacksonville	WMXQ	102.9	AC/H
Gainesville	WJLF	91.7	RL/CC	Jacksonville	WROO	107.3	CW
Gainesville	WYGC	100.9	CW	Jacksonville	WIOJ	1010	RL/CC
Gainesville	WRUF	103.7	RK	Jensen Bch	WMBX	102.3	CHR
Gainesville	WYKS	105.3	CHR	Jupiter	WDBE	1000	NWS+

Radio on the Road   Florida

City	Station	Freq	Format	City	Station	Freq	Format
Jupiter	WJBW	99.5	AC	Marathon	WAVK	105.5	AC
Kendall	WRHB	1020	ET/O	Marco	WAVV	101.1	EZL
Key Colony B	WKYZ	101.3	RK/C	Marco Island	WODX	1480	AS
Key Largo	WMKL	91.7	RL/CC	Marco Island	WMIB	1660	NWS+
Key Largo	WZMQ	106.3	SPN	Marco Island	WGUF	98.9	CH
Key West	WKIZ	1500	SPN	Marianna	WTOT	980	TLK+
Key West	WKWF	1600	TLK/S	Marianna	WTYS	1340	CW
Key West	WJIR	90.9	RL/CC	Marianna	WJNF	91.1	RL/CC
Key West	WEOW	92.5	CHR	Marianna	WTYS	94.1	RL/G
Key West	WKEY	93.5	AC	Marianna	WJAQ	100.9	CW
Key West	WCNK	98.7	JAZ	Mary Esther	WYZB	105.5	CW
Key West	WAIL	99.5	RK/C	Mayo	WGSG	89.5	RL/G
Key West	WIIS	107.1	RK	Melbourne/Titus.	WMEL	920	TLK+
Key West	WVMQ	107.9	SPN	Melbourne/Titus.	WAMT	1060	AS
Kissimmee	WFIV	1080	AS	Melbourne/Titus.	WMMB	1240	AS
Kissimmee	WOTS	1220	SPN	Melbourne/Titus.	WTMS	1560	BIZ
Kissimmee	WWKQ	89.1	SPN	Melbourne/Titus.	WBVD	95.1	RK/C
La Belle	WBIY	88.3	RL/CC	Melbourne/Titus.	WGNE	98.1	CW
La Belle	WWWD	92.5	CW	Melbourne/Titus.	WCIF	106.3	RL/G
La Crosse	WBXY	99.5	AC/H	Melbourne/Titus.	WAOA	107.1	CHR
Lafayette	WWFO	99.9	RK	Mexico Beach	WPBH	99.3	OLG
Lake City	WGRO	960	AS	Miami	WINZ	940	TLK/S
Lake City	WDSR	1340	SPT	Miami Beach	WMBM	1490	RL/G
Lake City	WOLR	91.3	RL/CC	Miami Beach	WLVE	93.9	JAZ
Lake City	WNFB	94.3	AC	Miami Springs	WAFN	1700	RL/SP
Lake Wales	WIPC	1280	RL/SP	Miami/Ft Laud.	WQAM	560	SPT
Lake Worth	WLVS	1380	RL/SP	Miami/Ft Laud.	WIOD	610	SPT
Lakeland	WONN	1230	AS	Miami/Ft Laud.	WIOD	610	TLK/S
Lakeland	WWAB	1330	URB	Miami/Ft Laud.	WWFE	670	SPN
Lakeland	WLKF	1430	TLK+	Miami/Ft Laud.	WAQI	710	SPN
Lakeland	WKES	91.1	RL	Miami/Ft Laud.	WQBA	1140	SPN
Lakeland	WYFO	91.9	RL	Miami/Ft Laud.	WSUA	1260	SPN
Lakeland	WSJT	94.1	JAZ	Miami/Ft Laud.	WOCN	1450	SPN
Lantana	WPBR	1340	TLK+	Miami/Ft Laud.	WDNA	88.9	JAZ
Largo	WZTM	820	SPN	Miami/Ft Laud.	WMCU	89.7	RL/CC
Lecanto	WLMS	88.3	RL	Miami/Ft Laud.	WTMI	93.1	FNA
Leesburg	WLBE	790	TLK+	Miami/Ft Laud.	WPOW	96.5	CHR
Leesburg	WQBQ	1410	TLK+	Miami/Ft Laud.	WFLC	97.3	AC/G
Lehigh Acres	WWCL	1440	SPN	Miami/Ft Laud.	WEDR	99.1	URB
Lehigh Acres	WCKT	107.1	CW	Miami/Ft Laud.	WLYF	101.5	AC/S
Live Oak	WQHL	1250	TLK+	Miami/Ft Laud.	WAMR	107.5	SPN
Live Oak	WQHL	98.1	CW	Miami/Ft Laud.	WFTL	1400	TLK+
Live Oak	WLVO	106.1	OLG	Miami/Ft Laud.	WHYI	100.7	CHR
MacClenny	WJXR	92.1	TLK+	Miami/Ft Laud.	WMGE	103.5	URB
Madison	WMAF	1230	SILNT	Miami/Ft Laud.	WRMA	106.7	SPN
Madison	WIMV	102.7	URB	Micanopy	WSKY	97.3	TLK+
Marathon	VOA	1180	SPN	Midway	WOKL	100.7	OLG
Marathon	WFFG	1300	TLK+	Milton	WEBY	1330	TLK+
Marathon	WGMX	94.3	AC/S	Milton	WECM	1490	RL/G
Marathon	WWWK	97.7	CHR	Milton	WTGF	90.5	RL/G

Florida

CITY	STATION	FREQ	FORMAT	CITY	STATION	FREQ	FORMAT
Milton	WEGS	91.7	RL	Oviedo	WONQ	1030	SPN
Milton	WXBM	102.7	CW	Palatka	WPLK	800	CW
Mims	WPGS	840	TLK+	Palatka	WIYD	1260	CW
Miramar Beach	WSBZ	106.3	JAZ	Palatka	WFKS	99.9	AC/H
Monticello	WVHT	105.7	URB	Palm Bay	WWIA	88.5	RL
Mount Dora	WMGF	107.7	AC/S	Palm Bay	WEJF	90.3	RL/CC
Murdock	WHHD	98.9	AC/H	Palm Beach	WRMF	97.9	AC/H
Naples	WNOG	1270	TLK+	Palm City	WCNO	89.9	RL
Naples	WSRX	89.5	RL/CC	Panama City	WGNE	590	AS
Naples	WSOR	90.9	RL	Panama City	WLTG	1430	TLK+
Naples	WTLT	93.5	AC/S	Panama City	WPAP	92.5	CW
Naples	WARO	94.5	CLA	Panama City	WFSY	98.5	AC
Naples	WSGL	104.7	AC/H	Panama City	WQIM	100.1	AC
Naples Park	WQNU	105.5	CW	Panama City	WILN	105.9	CHR
Navarre	WGCX	95.7	RL/CC	Panama City	WLHR	107.9	CHR
New Pt Richey	WPSO	1500	SPN	Panama City	WPCF	1290	RL/CC
New Pt Richey	WLPJ	91.5	RL/CC	Panama City	WAKT	105.1	CW
New Pt Richey	WDUV	105.5	AS	Parker	WPPT	94.5	RK
New Smyrna B	WSBB	1230	AS	Pennsuco	WIRP	88.3	RL/SP
New Smyrna B	WJLU	89.7	RL/CC	Pensacola	WVTJ	610	RL/G
Newberry	WHHZ	100.5	CHR	Pensacola	WSWL	790	NWS+
Niceville	WNCV	100.3	AC/S	Pensacola	WTKX	1230	RL/G
Nocatee	WZSP	105.3	RL/SP	Pensacola	WCOA	1370	TLK+
No Fort Myers	WWCN	770	TLK/S	Pensacola	WBSR	1450	OLG
No Miami	WKAT	1360	SPN	Pensacola	WPCS	89.5	RL
Ocala	WMOP	900	SPT	Pensacola	WMEZ	94.1	AC/S
Ocala	WCFI	1290	SILNT	Pensacola	WWRO	100.7	CH
Ocala	WOCA	1370	TLK+	Pensacola	WTKX	101.5	RK
Ocala	WHIJ	88.1	RL/CC	Pensacola	WYCL	107.3	RK/H
Ocala	WMFQ	92.9	AC/S	Perry	WPRY	1400	CW
Ocala	WOGK	93.7	CW	Perry	WNFK	92.1	CW
Ocoee	WUNA	1480	SPN	Pine Castle	WAJL	1190	RL/G
Okeechobee	WOKC	1570	CW	Pine Island	WINK	1200	TLK+
Okeechobee	WWFR	91.7	RL	Pinellas Park	WTBN	570	RL
Orange Park	WAYR	550	RL/G	Pinellas Park	WWBA	1040	TLK/R
Orlando	WDBO	580	TLK+	Plant City	WTWD	910	RL
Orlando	WQTM	740	TLK/S	Plantation Key	WCTH	100.3	CW
Orlando	WTLN	950	RL	Plantation Key	WFKZ	103.1	CH/40
Orlando	WHOO	990	YTH	Pompano Beach	WHSR	980	SPN
Orlando	WRMQ	1140	SPN	Pompano Beach	WRBD	1470	TLK+
Orlando	WMFE	90.7	RK	Pompano Beach	WMXJ	102.7	OLG
Orlando	WWKA	92.3	CW	Port Charlotte	WVIJ	91.7	RL
Orlando	WHTQ	96.5	RK/C	Port Charlotte	WFSN	100.1	AC/S
Orlando	WMMO	98.9	CH	Port Charlotte	WKFF	100.1	TLK+
Orlando	WMMV	1350	TLK+	Port St. Joe	WEBZ	93.5	URB
Orlando	WFLF	540	TLK+	Port St. Lucie	WPSL	1590	TLK+
Orlando MLA	WSHE	100.3	OLG	Port St. Lucie	WHLG	101.3	AC
Orlando MLA	WOMX	105.1	AC/H	Punta Gorda	WCCF	1580	TLK+
Ormond Beach	WELE	1380	TLK+	Punta Gorda	WIKX	92.7	CW
Ormond/ Sea	WHOG	95.7	CH	Punta Rassa	WCCL	97.7	CLA

City	Station	Freq	Format	City	Station	Freq	Format
Quincy	WWSD	1230	URB	Stuart	WZZR	92.7	TLK+
Quincy	WXSR	101.5	RK/H	Summerland Key	WPIK	102.5	CW
Riviera Beach	WMNE	1600	YTH	Sunrise	WKPX	88.5	CH/40
Riviera Beach	WWLV	94.3	JAZ	Tallahassee	WFRF	1070	RL/CC
Rock Harbor	WKLG	102.1	AC	Tallahassee	WNLS	1270	TLK/S
Rockledge	WHKR	102.7	CW	Tallahassee	WCVC	1330	RL
Royal Palm B	WLVJ	640	RL	Tallahassee	WHBT	1410	CW
Royal Palm B	WPSP	1190	SPN	Tallahassee	WTAL	1450	TLK+
Safety Harbor	WYUU	92.5	OLG	Tallahassee	WVFS	89.7	AD/A
Santa Rosa B	WWAV	102.1	SPN	Tallahassee	WANM	90.5	URB
Sarasota	WUGL	930	AS	Tallahassee	WTNT	94.9	CW
Sarasota	WQSA	1220	SILNT	Tallahassee	WHBX	96.1	URB
Sarasota	WTMY	1280	BIZ	Tallahassee	WBZE	98.9	AC
Sarasota	WSPB	1450	TLK+	Tallahassee	WAIB	103.1	CW
Sarasota	WSMR	89.1	RL/CC	Tallahassee	WGLF	104.1	RK
Sarasota	WHPT	102.5	RK/C	Tallahassee	WTAL	105.7	TLK+
Sarasota	WKZM	104.3	RL/CC	Tallahassee	WWLD	106.1	URB
Sarasota	WCTQ	105.5	CW	Tampa/St Pete	WFLA	570	TLK+
Sebastion	WBKM	95.9	OLG	Tampa/St Pete	WRMD	680	SPN
Sebring	WJCM	960	OLG	Tampa/St Pete	WTIS	1110	RL
Sebring	WITS	1340	AS	Tampa/St Pete	WHNZ	1250	NWS+
Sebring	WWLL	105.7	RK/C	Tampa/St Pete	WWMI	1380	YTH
Sebring/Placid	WWTK	730	SPN	Tampa/St Pete	WAMA	1550	SPN
Seffner	WQYK	1010	SPT	Tampa/St Pete	WBVM	90.5	RL
Silver Springs	WNDD	95.5	RK	Tampa/St Pete	WFLZ	93.3	CHR
Solana	WKII	1070	EZL	Tampa/St Pete	WWRM	94.9	AC/S
Solana	WCVU	104.9	EZL	Tampa/St Pete	WMTX	100.7	AC/H
South Daytona	WPUL	1590	URB	Tampa/St Pete	WFJO	101.5	URB
South Miami	WAXY	790	VA	Tampa/St Pete	WRBQ	104.7	CW
Springfield	WRBA	95.9	RK/C	Tampa/St Pete	WFLA	970	TLK+
Springfield	WYOO	101.1	TLK+	Tampa/St Pete	WQYK	99.5	CW
St. Augustine	WFOY	1240	TLK+	Tampa/St Pete	WBBY	107.3	JAZ
St. Augustine	WAOC	1420	TLK+	Tarpon Springs	WYFE	88.9	RL/G
St. Augustine	WFCF	88.5	VA	Tavares	WXXL	106.7	CH/40
St. Augustine	WAYL	91.9	RL/CC	Tavernier	WIFL	96.9	AC/S
St. Augustine	WSOS	94.1	AC	Temple Terrace	WQBN	1300	SPN
St. Augustine	WFSJ	97.9	JAZ	Titusville	WAMT	1060	AS
St. Augustine	WKLN	1170	AS	Titusville	WPIO	89.3	RL
St. Augustine	WJQR	105.5	URB	Titusville	WNUE	98.1	SPN
St. Petersburg	WRMD	680	SPN	Trenton	WDJY	101.7	CW
St. Petersburg	WWMI	1380	YTH	Union Park	WPOZ	88.1	RL/CC
St. Petersburg	WFTI	91.7	RL	Valparaiso	WFSH	1340	AS
St. Petersburg	WQYK	99.5	CW	Venice	WAMR	1320	SPT
St. Petersburg	WFJO	101.5	OLG	Venice	WDDV	92.1	AS
St. Petersburg	WBBY	107.3	CH	Vero Beach	WAXE	1370	TLK+
St. Petersburg	WRXB	1590	URB	Vero Beach	WTTB	1490	TLK+
Starke	WEAG	1490	CW	Vero Beach	WSCF	91.9	RL/CC
Starke	WTLG	88.3	RL	Vero Beach	WGYL	93.7	AC
Starke	WEAG	106.3	CW	Vero Beach	WGNX	99.7	OLG
Stuart	WSTU	1450	OLG	Vero Beach	WCZR	101.7	TLK+

Florida

CITY	STATION	FREQ	FORMAT	CITY	STATION	FREQ	FORMAT
Vero Beach	WQOL	103.7	OLG				
Watertown	WQLC	102.1	CW				
Wauchula	WAUC	1310	SPN				
W Palm Beach	WAQI	710	SPN				
W Palm Beach	WBSR	740	TLK+				
W Palm Beach	WAXY	790	VA				
W Palm Beach	WDJA	850	TLK/S				
W Palm Beach	WSWN	900	URB				
W Palm Beach	WWNN	980	TLK+				
W Palm Beach	WJNA	1040	AS				
W Palm Beach	WBZT	1230	SPN				
W Palm Beach	WJNA	1230	AS				
W Palm Beach	WDBF	1420	AS				
W Palm Beach	WRBD	1470	RL/G				
W Palm Beach	WPOM	1600	RL/G				
W Palm Beach	WAYF	88.1	RL/CC				
W Palm Beach	WRLX	92.1	RK/H				
W Palm Beach	WEAT	104.3	AC/S				
W Palm Beach	WTPX	105.5	OLG				
W Palm Beach	WIRK	107.9	CW				
White City	WFLM	104.7	CH/40				
Wildwood	WHOF	640	RL/G				
Williston	WTMG	101.3	URB				
Wilton Manors	WEXY	1520	RL/G				
Winter Garden	WOKB	1600	RL/G				
Winter Garden	WTIR	1680	Traf				
Winter Haven	WSIR	1490	RL/G				
Winter Haven	WPCV	97.5	CW				
Winter Park	WPRD	1440	SPN				
Winter Park	WPRK	91.5	FNA				
Winter Park	WLOQ	103.1	JAZ				
Yankeetown	WBKX	96.3	CW				
Zephyrhills	WZHR	1400	AS				
Zolfo Springs	WZZS	106.9	CW				

CITY	STATION	FREQ	FORMAT	CITY	STATION	FREQ	FORMAT
Adel	WBIT	1470	RL/SG	Augusta	WTHB	1550	RL/G
Adel	WDDQ	92.1	AC/S	Augusta	WYRU	1630	TLK/S
Albany	WJYZ	960	RL/G	Augusta	WLPE	91.7	RL
Albany	WANL	1250	RL/CC	Augusta	WEKL	102.3	RK/CL
Albany	WGPC	1450	POP	Augusta	WFXA	103.1	UR/RB
Albany	WALG	1590	TLK+	Augusta	WBBQ	104.3	AC/H
Albany	WJIZ	96.3	UR/RB	Augusta	WZNY	105.7	CHR
Albany	WNUQ	101.7	CHR	Austell	WAOS	1600	SPN
Albany	WKAK	104.5	CW	Bainbridge	WXFX	97.3	RK
Alma	WAJQ	1400	RL/SG	Bainbridge	WJHW	101.9	CW
Alma	WAJQ	104.3	CW	Bainbridge	WMGR	930	TLK+
Alpharetta	WVNF	1400	RL/G	Bainbridge	WMGR	930	AS
Americus	WISK	1390	CW	Barnesville	WBAF	1090	CW
Americus	WDEC	94.3	AC/H	Baxley	WUFE	1260	RL/SG
Americus	WISK	98.7	CW	Baxley	WBYZ	94.5	CW
Ashburn	WFFM	105.7	UR	Blackshear	WGIA	1350	SPT
Athens	WRFC	960	TLK+	Blackshear	WKUB	105.1	CW
Athens	WGAU	1340	TLK+	Blakely	WBBK	1260	CW
Athens	WXAG	1470	SILNT	Blakely	WBBK	93.1	CW
Athens	WMSL	88.9	RL/G	Blue Ridge	WPPL	103.9	CW
Athens	WUOG	90.5	AP	Boston	WTUF	106.3	CW
Athens	WBTS	95.5	CHR	Bostwick	WMOQ	92.3	CW
Athens	WALR	104.7	UR/RB	Bremen	WGMI	1440	RL/SG
Atlanta	WDWD	590	Youth	Broxton	WULS	103.7	RL/SG
Atlanta	WGST	640	TLK+	Brunswick	WSFN	790	TLK/S
Atlanta	WCNN	680	SPT	Brunswick	WGIG	1440	TLK+
Atlanta	WSB	750	TLK+	Brunswick	WMOG	1490	POP
Atlanta	WQXI	790	TLK/S	Brunswick	WWRR	100.7	CH
Atlanta	WAEC	860	RL	Brunswick	WSOL	101.5	UR/AC
Atlanta	WAFS	920	RL/CC	Brunswick	WSEG	104.1	UR/RB
Atlanta	WNIV	970	RL/G	Buford	WXEM	1460	SPN
Atlanta	WGKA	1190	CLA	Buford	WLKQ	102.3	OLG
Atlanta	WALR	1340	TLK/S	Byron	WPWB	90.5	RL/G
Atlanta	WAOK	1380	RL/G	Cairo	WGRA	790	RL/SG
Atlanta	WYZE	1480	RL/G	Cairo	WSLE	102.3	AC
Atlanta	WRAS	88.5	VA	Calhoun	WJTH	900	CW
Atlanta	WREK	91.1	VAR	Calhoun	WEBS	1110	OLG
Atlanta	WZGC	92.9	RK/CL	Camilla	WQVE	105.5	UR/AC
Atlanta	WPCH	94.9	AC/S	Canton	WCHK	1290	CW
Atlanta	WKLS	96.1	RK	Canton	WMXV	105.7	TLK+
Atlanta	wmnw	96.3	RK/CL	Carrollton	WBTR	1330	TLK+
Atlanta	WSB	98.5	AC	Carrollton	WBTR	92.1	CW
Atlanta	WNNX	99.7	RK/NU	Cartersville	WYXC	1270	CW
Atlanta	WVEE	103.3	UR/RB	Cartersville	WBHF	1450	TLK+
Atlanta	WGUN	1010	RL	Cartersville	WCCV	91.7	RL/G
Augusta	WGAC	580	TLK+	Cedartown	WGAA	1340	CW
Augusta	WFAM	1050	RL	Chatsworth	WQMT	98.9	CW
Augusta	WKIM	1230	TLK+	Chauncey	WQIL	101.3	RL/SG
Augusta	WBBQ	1340	Youth	Clarkesville	WCHM	1490	RL/SG
Augusta	WRDW	1480	SPT	Clarkesville	WMJE	102.9	AC

CITY	STATION	FREQ	FORMAT	CITY	STATION	FREQ	FORMAT
Claxton	WCLA	1470	RL/G	Dublin	WXLI	1230	CW
Claxton	WCLA	107.3	UR/RB	Dublin	WMLT	1330	RL/G
Clayton	WGHC	1370	POP	Dublin	WKKZ	92.7	CHR
Clayton	WRBN	104.1	AC/S	Dublin	WQZY	95.9	CW
Cleveland	WRWH	1350	CW	East Point	WKGE	1160	POP
Cleveland	WAZX	101.9	SPN	East Point	WTJH	1260	RL/G
Cochran	WVMG	1440	TLK+	Eastman	WUFF	710	CW
Cochran	WVMG	96.7	CW	Eastman	WUFF	97.5	CW
College Park	WWWQ	100.5	CHR	Eatonton	WKVQ	1520	POP
Columbus	WDAK	540	SPT	Elberton	WSGC	1400	OLG
Columbus	WMLF	1270	SPT	Elberton	WWRK	92.1	CW
Columbus	WOKS	1340	UR/RB	Ellijay	WPGY	1560	CW
Columbus	WRCG	1420	TLK+	Ellijay	WLJA	93.5	CW
Columbus	WEAM	1580	RL	Evans	WAEG	92.3	CHR
Columbus	WYFK	89.5	RL	Fayetteville	WEGF	97.5	RL/G
Columbus	WFRC	90.5	RL/G	Fitzgerald	WBHB	1240	RL/SG
Columbus	WVRK	102.9	RK	Fitzgerald	WRDO	96.9	AC/S
Columbus	WFXE	104.9	UR/RB	Folkston	WOKF	92.5	CW
Columbus	WCGQ	107.3	AC	Forsyth	WFXM	100.1	RL/G
Commerce	WJJC	1270	CW	Fort Valley	WXKO	1150	RL/G
Conyers	WPBS	1050	RL/SG	Fort Valley	WJTG	91.3	RL/G
Cordele	WAEF	90.3	RL/CC	Fort Valley	WIBB	97.9	UR/RB
Cordele	WKKN	98.3	CW	Fort Valley	WQBZ	106.3	RK
Cornelia	WCON	1450	RL/SG	Gainesville	WDUN	550	TLK+
Cornelia	WCON	99.3	CW	Gainesville	WLBA	1130	SPN
Covington	WGFS	1430	CW/TLK	Gainesville	WGGA	1240	TLK+
Crawford	WGMG	102.1	AC	Gainesville	WBCX	89.1	JAZ
Cumming	WMLB	1170	POP	Gainesville	WFOX	97.1	OLG
Cumming	WWEV	91.5	RL/CC	Gainesville	WYAY	106.7	CW
Cuthbert	WCUG	850	CW	Glennville	WKIG	1580	RL/SG
Cuthbert	WMRZ	100.7	POP	Glennville	WKIG	106.3	AC
Dahlonega	WDGR	1210	SILNT	Gordon	WBNM	1120	UR/RB
Dahlonega	WKHC	104.3	CW	Gordon	WALJ	107.1	UR/RB
Dallas	WDPC	1500	RL/G	Gray	WRNC	96.5	CW
Dalton	WBLJ	1230	TLK/S	Grayson	WPLO	610	SPN
Dalton	WDAL	1430	SPN	Greensboro	WDDK	103.9	AC
Dalton	WTTI	1530	RL/SG	Griffin	WHIE	1320	CW
Dalton	WYYU	104.5	AC	Griffin	WKEU	1450	NWS+
Darien	WYNR	107.7	CW	Griffin	WMVV	90.7	RL/CC
Decatur	WPBC	1310	SPN	Hahira	WTHV	810	RL/SG
Decatur	WATB	1420	SPN	Hampton	WHTA	107.9	UR
Dock Jct	WXMK	105.9	CHR	Harlem	WCHZ	95.1	RK
Donalsonville	WSEM	1500	CW	Hartwell	WKLY	980	CW
Donalsonville	WGMK	106.3	AC/H	Hawkinsville	WCEH	610	AC/S
Donalsonville	WWGF	107.5	RL	Hawkinsville	WQSY	103.9	AC/S
Douglas	WDMG	860	TLK+	Hazelhurst	WVOH	920	CW
Douglas	WOKA	1310	SPN	Hazelhurst	WVOH	93.5	CW
Douglas	WDMG	99.5	AC	Helen	WTFH	89.9	RL/CC
Douglas	WOKA	106.7	CW	Helen	WHEL	105.1	TLK+
Douglasville	WDCY	1520	RL/G	Hinesville	WGML	990	RL/G

CITY	STATION	FREQ	FORMAT	CITY	STATION	FREQ	FORMAT
Hinesville	WSKX	92.3	AC/S	Metter	WHCG	1360	CH
Hinesville	WHVL	104.7	AC	Metter	WBMZ	103.7	CH
Hogansville	WMAX	98.1	AC/H	Midway	WGCO	98.3	OLG
Homerville	WBTY	105.5	OLG	Milan	WMCG	104.9	CW
Irwinton	WVKX	103.7	UR/RB	Milledgeville	WMVG	1450	SPT
Jackson	WJGA	92.1	AC	Milledgeville	WGER	88.9	AP
Jasper	WYYZ	1490	CW	Milledgeville	WLRR	100.7	SILNT
Jasper	WNEE	88.3	RL/CC	Milledgeville	WKZR	102.3	CW
Jefferson	WBKZ	880	RL/G	Millen	WHKN	94.9	CW
Jeffersonville	WPEZ	93.7	AC	Monroe	WKUN	1580	CW
Jesup	WLOP	1370	CW	Montezuma	WMNZ	1050	CW
Jesup	WLPT	88.3	RL	Morrow	WSSA	1570	RL/SG
Jesup	WIFO	105.5	CW	Moultrie	WMTM	1300	CW
Kingsland	WKBX	106.3	CW	Moultrie	WMTM	93.9	CW
La Fayette	WQCH	1590	CW	Mt Vernon	WYUM	101.7	CW
La Grange	WTRP	620	TLK+	Mountain C.	WALH	1340	RL/G
La Grange	WLAG	1240	SPT	Murrayville	WGTJ	1330	RL/SG
La Grange	WOAK	90.9	RL/SG	Nashville	WNGA	1600	RL
La Grange	WJZF	104.1	JAZ	Nashville	WJYF	95.3	AC
Lagrange	WGSE	720	TLK+	Newnan	WNEA	1300	RL/SG
Lakeland	WVGA	105.9	CW	Newnan	WCOH	1400	CW
Leesburg	WJAD	103.5	RK	No. Atlanta	WCNN	680	NWS+
Louisville	WPEH	1420	CW	Ochlocknee	WJEP	1020	RL/CC
Louisville	WPEH	92.1	CW	Ocilla	WKAA	97.7	OLG
Lumpkin	WKCN	99.3	CW	Ocilla	WLPF	98.5	RL
Lyons	WBBT	1340	UR/RB	Omega	WTIF	107.5	CW
Lyons	WLYU	100.9	CW	Pchtree City	WLDA	96.7	AC/H
Mableton	WAMJ	102.5	JAZ	Pearson	WPNG	101.9	RK/CL
Macon	WBML	900	RL/SG	Perry	WPGA	980	Youth
Macon	WMAC	940	TLK+	Perry	WPGA	100.9	AC/H
Macon	WDDO	1240	RL/G	Port W'worth	WAYH	91.9	RL/CC
Macon	WLCG	1280	RL/G	Quitman	WSFB	1490	VAR
Macon	WNEX	1400	Youth	Quitman	WSTI	105.3	AC/S
Macon	WDEN	1500	CW	Reidsville	WTNL	1390	RL/SG
Macon	WMKS	92.3	CH	Reidsville	WRBX	104.1	RL/SG
Macon	WDEN	99.1	CHR	Ringgold	WMPZ	93.7	AC
Macon	WAYS	105.3	OLG	Ringold	WSGC	101.9	CH
Macon	WGEF	107.9	AC	Riverside	WMGA	580	SPN
Madison	WYTH	1250	TLK+	Rockmart	WZOT	1220	RL/SG
Manchester	WFDR	1370	POP	Rockmart	WTSH	107.1	CW
Manchester	WVFJ	93.3	RL/CC	Rome	WROM	710	RL/G
Marietta	WFTD	1080	SPN	Rome	WTSH	1360	RL/SG
Marietta	WFOM	1230	TLK/S	Rome	WLAQ	1410	TLK+
Marietta	WKHX	101.5	CW	Rome	WRGA	1470	TLK+
Marrieta	WGHR	100.7	SILNT	Rome	WKCX	97.7	AC/S
Martinez	WGOR	93.9	OLG	Rome	WQTU	102.3	CHR
Martinez	WPRW	107.7	UR/RB	Rossville	WUUS	980	RL
McDonough	WKKP	1410	POP	Rossville	WRXR	105.5	RK
McRae	WYIS	1410	OLG	Royston	WBIC	810	RL/SG
McRae	WYSC	102.7	OLG	Royston	WPUP	103.7	RK/NU

CITY	STATION	FREQ	FORMAT	CITY	STATION	FREQ	FORMAT
Sandersville	WSNT	1490	UR/RB	Thomson	WTWA	1240	AC
Sandersville	WSNT	99.9	CW	Thomson	WTHO	101.7	CW
Sasser	WEGC	107.7	AC	Tifton	WTIF	1340	CW
Savannah	WBMQ	630	NWS+	Tifton	WKZZ	92.5	CW
Savannah	WJLG	900	RL/G	Tifton	WOBB	100.3	CW
Savannah	WSOK	1230	RL/G	Tifton	WPLH	103.1	AP
Savannah	WCHY	1290	Youth	Toccoa	WNEG	630	RL/G
Savannah	WHGM	1400	RL/G	Toccoa	WLET	1420	SILNT
Savannah	WLXP	88.1	RL/CC	Toccoa	WNGC	106.1	CW
Savannah	WYFS	89.5	RL	Toccoa Falls	WTXR	89.7	RL/CC
Savannah	WHCJ	90.3	JAZ	Toccoa Falls	WRAF	90.9	RL/CC
Savannah	WEAS	93.1	UR/RB	Trenton	WKWN	1420	OLG
Savannah	WSCA	94.1	CW	Trenton	WBDX	102.7	RL/CC
Savannah	WIXV	95.5	RK/CL	Trion	WSAF	1180	SPN
Savannah	WJCL	96.5	CW	Trion	WATG	95.7	CW
Savannah	WAEV	97.3	AC/H	Unadilla	WAFI	99.9	RL
Savannah	WZAT	102.1	CHR	Valdosta	WFVR	910	TLK+
Smithville	WZIQ	106.9	RL	Valdosta	WGOV	950	UR/RB
Smyrna	WAZX	1550	RL/SP	Valdosta	WJEM	1150	RL/G
Smyrna	WSTR	94.1	CHR	Valdosta	WVLD	1450	OLG
Soperton	WKTM	106.1	RL/SP	Valdosta	WVVS	90.9	AP
Sparta	WMGZ	97.7	AC/H	Valdosta	WAAC	92.9	CW
Sparta	WPMA	102.7	RL	Valdosta	WQPW	95.7	AC/S
Springfield	WSIS	103.9	UR/RB	Valdosta	WYZK	96.7	CW
St. Marys	WECC	1190	CHR	Valdosta	WAFT	101.1	RL
St. Simons Is	WHFX	92.7	CLA	Valdosta	WWRQ	107.7	RK/CL
Statenville	WHLJ	97.5	UR/AC	Vidalia	WVOP	970	OLG
Statesboro	WPTB	850	JAZ	Vidalia	WGPH	91.5	RL
Statesboro	WWNS	1240	NWS+	Vidalia	WTCQ	97.7	AC
Statesboro	WVGS	91.9	AP	Vienna	WKTF	1550	OLG
Statesboro	WMCD	100.1	AC/H	Warner Rob.	WNNG	1350	POP
Statesboro	WPMX	102.9	AC/S	Warner Rob	WAXP	1600	SILNT
Summerville	WGTA	950	SPN	Warner Rob	WRNC	1670	CW
Swainsboro	WJAT	800	RL/G	Warner Rob	WRBV	101.7	UR/AC
Swainsboro	WXRS	1590	TLK+	Warner Rob	WLCG	102.5	RL/G
Swainsboro	WELT	98.1	AC	Warrenton	WRFN	93.1	SPT
Swainsboro	WXRS	100.5	CW	Washington	WLOV	1370	AC/S
Sylvania	WSYL	1490	TLK+	Washington	WXKT	100.1	AC/H
Sylvania	WZBX	106.5	CW	Waycross	WASW	91.9	RL/CC
Sylvester	WWSG	102.1	VAR	Waycross	WWUF	97.7	OLG
Sylvester	WRXZ	106.1	UR/RB	Waycross	WBGA	102.5	CW
Talking Rock	WNSY	100.1	OLG	Waycross	WWSN	103.3	AC/S
Tallapoosa	WKNG	1060	CW	Waynesboro	WAEJ	100.9	CHR
Tennille	WJFL	101.9	AC	Waynesboro	WYFA	107.1	RL/G
Thomaston	WTGA	1590	CW	West Point	WPLV	1310	RL/G
Thomaston	WTGA	101.1	AC/S	West Point	WDWZ	1490	SILNT
Thomasville	WSTT	730	RL/G	West Point	WCJM	100.9	CW
Thomasville	WHGH	840	UR/RB	Winder	WIMO	1300	CW
Thomasville	WPAX	1240	POP	Winder	WYFW	89.5	RL/G
Thomasville	WTLY	107.1	AC	Woodbine	WCGA	1100	TLK+
				Wrens	WAKB	96.9	UR/AC
				Wrightsville	WDBN	107.5	RK
				Yng Harris	WZCM	770	RL/SG
				Zebulon	WEKS	92.5	CW

CITY	STATION	FREQ	FORMAT	CITY	STATION	FREQ	FORMAT
Aiea	KGMZ	107.9	OLG	Kapaa	KITH	98.9	TLK+
Eleele	KUAI	720	AC	Kawaihae	KWYI	106.9	CHR
Haiku	KUAU	1570	SILNT	Kealakekua	KKON	790	POP
Hali'imaile	KUAU	105.5	ETH/P	Kealakekua	KAOY	101.5	RK/CL
Hanalei	KKCR	90.9	VAR	Kihei	KAOI	1110	TLK+
Hilo	KIPA	620	POP	Kilauea	KAQA	91.9	VAR
Hilo	KPUA	670	TLK+	Lahaina	KPOA	93.5	ET/H
Hilo	KHLO	850	OLG	Lahaina	KLHI	101.1	RK
Hilo	KAHU	1060	ETH/P	Lanai City	KONI	104.7	AC/H
Hilo	KCIF	90.3	RL/CC	Lihue	KQNG	570	RK/CL
Hilo	KHWI	92.7	RK/CL	Lihue	KQNG	93.5	CHR
Hilo	KWXX	94.7	AC	Lihue	KFMN	96.9	AC
Hilo	KPVS	95.9	AC	Lihue	KAWV	98.1	RL
Hilo	KNWB	97.1	CH	Makawao	KDLX	94.3	CW
Hilo	KKBG	97.9	AC	Paauilo	KNUQ	103.7	CHR
Hilo	KAPA	100.3	ET/O	Pearl City	KUCD	101.9	AC
Honolulu	KSSK	590	AC/G	Pukalani	KMVI	98.3	RK/CL
Honolulu	KHNR	650	NWS+	Volcano	KKOA	107.7	CH/D
Honolulu	KQMQ	690	Youth	Wailuku	KMVI	550	OLG
Honolulu	KGU	760	TLK/S	Wailuku	KAOI	95.1	RK
Honolulu	KHVH	830	TLK+	Waipahu	KJPN	940	ET/J
Honolulu	KAIM	870	RL	Waipahu	KSSK	92.3	AC
Honolulu	KIKI	990	BUS	Waipahu	KKHN	102.7	CW
Honolulu	KLHT	1040	RL	Kapaa	KITH	98.9	TLK+
Honolulu	KWAI	1080	TLK+				
Honolulu	KBNZ	1170	NWS+				
Honolulu	KNDI	1270	RL				
Honolulu	KCCN	1420	ET/H				
Honolulu	KGMZ	1460	OLG				
Honolulu	KUMU	1500	POP				
Honolulu	KISA	1540	TLK+				
Honolulu	KTUH	90.3	VA				
Honolulu	KQMQ	93.1	CH/D				
Honolulu	KIKI	93.9	CH/40				
Honolulu	KUMU	94.7	EZL				
Honolulu	KAIM	95.5	RL/CC				
Honolulu	KPOI	97.5	RK/NU				
Honolulu	KDNN	98.5	ET/O				
Honolulu	KORL	99.5	AC/S				
Honolulu	KCCN	100.3	AC				
Honolulu	KINE	105.1	ET/H				
Honolulu	KLEO	106.1	AC				
Honolulu	KZOO	1210	ET/O				
Kahului	KNUI	900	ET/O				
Kahului	KNUI	99.9	AC/H				
Kailua	KRTR	96.3	ET/K				
Kailua-Kona	KLUA	93.9	AC				
Kalaheo	KTOH	99.9	AC				
Kaneohe	KBLZ	104.3	CH/D				
Kaneohe	KXME	104.3	CHR				

Radio on the Road   Idaho

City	Station	Freq	Format	City	Station	Freq	Format
AmericanFalls	KORR	104.1	AC	Mccall	KDZY	98.3	CW
Ammon	KUPI	980	AS	Mccall	KMCL	101.1	AC
Blackfoot	KECN	690	NWS+	Meridien	KKIC	950	RL/G
Blackfoot	KLCE	97.3	AC	Montpelier	KVSI	1450	CW
Blackfoot	KCVI	101.5	RK/H	Moscow	KRPL	1400	AS
Boise	KIDO	630	TLK+	Moscow	KUOI	89.3	JAZ
Boise	KBOI	670	AC	Moscow	KZFN	106.1	CHR
Boise	KSPD	790	RL	Mtn Home	KMHI	1240	SPT
Boise	KGEM	1140	POP	Mtn Home	KTPZ	99.1	CHR
Boise	KIZN	92.3	CW	Nampa	KFXD	580	CW
Boise	KQFC	97.9	CW	Nampa	KTIK	1340	SPT
Boise	KLTB	104.3	OLG	Nampa	KFXJ	94.9	AP
Boise	KJOT	105.1	RK	Nampa	KKGL	96.9	RK/CL
BonnersFerry	KBFI	1450	TLK+	N. Plym'th	KZMG	93.1	CHR
Burley	KBAR	1230	TLK+	Orofino	KLER	1300	CW
Burley	KBSY	88.5	TLK+	Orofino	KLER	95.3	AC/S
Burley	KZDX	99.9	AC/H	Payette	KIOV	1450	SPT
Caldwell	KBGN	1060	RL	Payette	KQXR	100.3	RK/H
Caldwell	KCID	1490	CW	Pocatello	KSEI	930	TLK/S
Caldwell	KTSY	89.5	RL/CC	Pocatello	KWIK	1240	TLK/S
Caldwell	KBXL	94.1	RL	Pocatello	KOUU	1290	CW
Caldwell	KARO	103.3	RK	Pocatello	KZBQ	93.7	CW
Caldwell	KCID	107.1	AC	Pocatello	KPKY	94.9	OLG
Chubbuck	KRTK	1490	RL/CC	Pocatello	KMGI	102.5	RK/C
Chubbuck	KLLP	98.5	AC/S	Preston	KACH	1340	OLG
Coeurd'alene	KVNI	1080	OLG	Preston	KKEX	96.7	CW
Coeurd'alene	KICR	102.3	CW	Rexburg	KRXK	1230	SPT
Coeurd'alene	KCDA	103.1	AC/H	Rexburg	KWBH	91.5	RL/CC
Donnelly	KMCL	1240	AC/S	Rexburg	KADQ	94.3	AC
Eagle	KXLT	107.9	AC/S	Rexburg	KGTM	98.1	RK/H
Emmett	KJHY	101.9	SPN	Rupert	KBBK	970	SPN
Fruitland	KWEI	99.5	SPN	Rupert	KKMV	92.5	CW
Garden City	KCIX	105.9	AC/H	Salmon	KSRA	960	CW
Gooding	KRXR	1480	SPN	Salmon	KSRA	92.7	CW
Gooding	KMXM	100.7	CW	Sandpoint	KSPT	1400	TLK+
Grangeville	KORT	1230	CW	Sandpoint	KPND	95.3	AC
Grangeville	KORT	92.7	CW	Sandpoint	KIBR	102.5	CW
Hayden	KHTQ	94.5	RK	Soda Sprs	KBRV	790	CW
Idaho Falls	KID	590	TLK+	Soda Sprs	KFIS	100.1	CW
Idaho Falls	KICN	1260	TLK+	St.Anthony	KIGO	1400	SILNT
Idaho Falls	KID	96.1	CW	St. Maries	KOFE	1240	CW
Idaho Falls	KUPI	99.1	CW	Sun Valley	KWRV	91.9	CLA
Idaho Falls	KFTZ	103.3	CHR	Sun Valley	KECH	95.3	RK
Idaho Falls	KOSZ	105.5	OLG	Sun Valley	KSKI	103.7	RK/NU
Jerome	KART	1400	CW	Twin Falls	KTFI	1270	POP
Jerome	KMVX	102.9	AC/H	Twin Falls	KLIX	1310	TLK+
Ketchum	KIKX	104.7	CH	Twin Falls	KEZJ	1450	JAZ
Lewiston	KOZE	950	TLK+	Twin Falls	KAWZ	89.5	RL/CC
Lewiston	KRLC	1350	CW	Twin Falls	KCIR	90.7	RL/CC
Lewiston	KLHS	89.1	AC	Twin Falls	KEZJ	95.7	CW
Lewiston	KOZE	96.5	RK	Twin Falls	KLIX	96.5	OLG
Lewiston	KATW	101.5	AC/H	WALLACE	KWAL	620	CW
Lewiston	KVTY	105.1	CHR	Wallace	KTWD	97.5	RL
Lewiston	KMOK	106.9	CW	Weiser	KWEI	1260	SPN

Radio on the Road

City	Station	Freq	Format	City	Station	Freq	Format
Albion	WBJW	91.7	RL/CC	Centralia	WRXX	95.3	CH
Aledo	WRMJ	102.3	CW	Champaign	WDWS	1400	N/TLK
Alton	WBGZ	1570	TLK+	Champaign	WBCP	1580	URB
Alton	KATZ	100.3	URB	Champaign	WPCD	88.7	RK/C
Anna	WIBH	1440	CW	Champaign	WEFT	90.1	VAR
Anna	WKIB	96.5	CH	Champaign	WBGL	91.7	RL/CC
Arcola	WXET	107.9	AC/H	Champaign	WCFF	92.5	CH
Arlington Hts	WKIE	92.7	AS	Champaign	WLRW	94.5	AC/H
Atlanta	WLCN	96.3	CW	Champaign	WHMS	97.5	AC/S
Augusta	WAHI	98.5	SIL	Champaign	WIXY	100.3	CW
Aurora	WBIG	1280	TLK+	Champaign	WPGU	107.1	RK/C
Aurora	WKKD	1580	TLK+	Charleston	WEIC	1270	RL/G
Aurora	WERV	95.9	CH	Charleston	WEIU	88.9	CLA
Aurora	WLEY	107.9	SP/MEX	Charleston	WWGO	92.1	RK/C
Ava	WXAN	103.9	RL	Chester	KSGM	980	CW
Bartonville	WIXO	99.9	RK	Chicago	WIND	560	TLK+
Beardstown	WRMS	790	RL	Chicago	WSCR	670	SPTS
Beardstown	WRMS	94.3	CW	Chicago	WGN	720	TLK+
Belleville	WSDZ	1260	YTH	Chicago	WBBM	780	NWS+
Belvidere	WXRX	104.9	RK	Chicago	WAIT	820	RL
Benton	WQRL	106.3	OLG	Chicago	WLS	890	TLK+
Berwyn	WRLL	1690	OLG	Chicago	WNTD	950	SP/MEX
Bethalto	WFUN	95.5	URB	Chicago	WMVP	1000	SPTS
Bloomington	WJBC	1230	N/TLK	Chicago	WMBI	1110	RL/SP
Bloomington	WESN	88.1	VAR	Chicago	WYLL	1160	RL
Bloomington	WBNQ	101.5	CH	Chicago	WLXX	1200	SP/TK
Breese	WDLJ	97.5	RK/C	Chicago	WSBC	1240	N/A
Brookport	WNTX	750	N/TLK	Chicago	WGCI	1390	RL/G
Bushnell	WLMD	104.7	CW	Chicago	WCRX	88.1	URB
Cairo	WKRO	1490	URB	Chicago	WZRD	88.3	AD/A
Cairo	WBEL	88.5	RL/CC	Chicago	WXAV	88.3	RK/C
Canton	WBYS	1560	TLK+	Chicago	WHPK	88.5	JAZ
Canton	WBYS	107.9	CH	Chicago	WLUW	88.7	VAR
Carbofdale	WSIU	91.9	VAR	Chicago	WIIT	88.9	VAR
Carbondale	WCIL	1020	N/TLK	Chicago	WKKC	89.3	URB
Carbondale	WDBX	91.1	VAR	Chicago	WMBI	90.1	RL/CC
Carbondale	WCIL	101.5	CH	Chicago	WRTE	90.5	SP/OM
Carlinville	WTSG	90.1	RL/G	Chicago	WBHI	90.7	SIL
Carlinville	WIBI	91.1	RL/CC	Chicago	WBEZ	91.5	VAR
Carlinville	WCNL	95.9	RL	Chicago	WXRT	93.1	AD/A
Carlyle	WCXO	96.7	OLG	Chicago	WLIT	93.9	AC/S
Carmi	WROY	1460	OLG	Chicago	WZZN	94.7	RK/C
Carmi	WRUL	97.3	CW	Chicago	WNUA	95.5	JAZ
Carrier Mills	WBVN	104.5	RL/CC	Chicago	WBBM	96.3	CH
Carterville	WUEZ	95.1	AC/S	Chicago	WDRV	97.1	CH
Carthage	WCAZ	990	TLK+	Chicago	WLUP	97.9	RK/C
Carthage	WQKQ	92.1	CH	Chicago	WFMT	98.7	CLA
Carthage	WCEZ	93.9	AC/S	Chicago	WUSN	99.5	CW
Casey	WKZI	800	RL	Chicago	WNND	100.3	AC
Casey	WLHW	91.5	N/A	Chicago	WKQX	101.1	RK/H
Casey	WCBH	104.5	CH	Chicago	WKQX	103.5	CH
Centralia	WILY	1210	OLG	Chicago	WJMK	104.3	OLG

CITY	STATION	FREQ	FORMAT	CITY	STATION	FREQ	FORMAT
Chicago	WGCI	107.5	URB	Effingham	WXEF	97.9	AC/H
Chicago Hts	WCFJ	1470	N/A	Eldorado	WEBQ	102.3	AC
Chicago Hts	WCGO	1600	AS	Elgin	WRMN	1410	TLK+
Chillicothe	WPMJ	94.3	JAZ	Elgin	WEPS	88.9	VAR
Christopher	WXLT	103.5	RK/C	Elgin	WJKL	94.3	RL/CC
Cicero	WVON	1450	URB	Elmhurst	WJJG	1530	TLK+
Clinton	WHOW	1520	N/TLK	Elmhurst	WRSE	88.7	AD/A
Clinton	WHOW	95.9	SPTS	Elmwood	WFYR	97.3	CW
Coal City	WBVS	100.7	RK/C	Elmwood	WCKG	105.9	TLK+
Colchester	WMQZ	104.1	AC/H	Elsah	WTPC	105.3	AD/A
Colfax	WRPW	92.9	CH	Eureka	WPIA	98.5	RL/CC
Columbia	KMJM	104.9	URB	Evanston	WKTA	1330	N/A
Crest Hill	WCCQ	98.3	CW	Evanston	WONX	1590	RL
Crete	WBMF	88.1	RL	Evanston	WNUR	89.3	VAR
Crete	WYCA	102.3	RL/G	Evanston	WOJO	105.1	SP/MEX
Crystal Lake	WAIT	850	TLK+	Fairbury	WYST	107.7	AC
Danville	WITY	980	AS	Fairfield	WFIW	1390	N/TLK
Danville	WDAN	1490	TLK+	Fairfield	WFIW	104.9	AC
Danville	WRHK	94.9	RK/C	Fairfield	WOKZ	105.9	CW
Danville	WXTT	99.1	CH	Farmer City	WWHP	98.3	CW
Danville	WDNL	102.1	AC	Farmington	WWCT	96.5	RK/C
Decatur	WDZ	1050	URB	Fisher	WGNN	102.5	RL
Decatur	WSOY	1340	TLK+	Flora	WNOI	103.9	AC
Decatur	WNLD	88.1	RL/CC	Flossmoor	WHFH	88.5	AD/A
Decatur	WJMU	89.5	VAR	Freeport	WFRL	1570	AS
Decatur	WYDS	93.1	CH	Freeport	WNIE	89.1	VAR
Decatur	WDZQ	95.1	CW	Freeport	WFPS	92.1	CW
Decatur	WSOY	102.9	CH	Freeport	WXXQ	98.5	CW
Dekalb	WLBK	1360	TLK+	Galatia	WISH	98.9	AC
Dekalb	WNIJ	89.5	VAR	Galena	WDBQ	107.5	OLG
Dekalb	WDEK	92.5	CH	Galesburg	WGIL	1400	TLK+
Dekalb	WDKB	94.9	AC	Galesburg	WAIK	1590	AC/S
Des Plaines	WPPN	106.7	SP/OM	Galesburg	WVKC	90.7	VAR
Dixon	WIXN	1460	OLG	Galesburg	WLSR	92.7	RK
Dixon	WRCV	101.7	CW	Galesburg	WAAG	94.9	CW
Dorsey	WDRS	89.5	N/A	Galva	WJRE	102.5	AC/H
Downers Grv	WDGC	88.3	VAR	Genesco	WGEN	1500	N/TLK
Dundee	WWYW	103.9	OLG	Genesco	WAXR	88.1	RL/CC
Duquoin	WDQN	1580	AC/S	Geneva	WSPY	1480	AS
Duquoin	WAWJ	90.1	RL	Genoa	WOXM	106.3	URB
Duquoin	WDQN	95.9	RL	Gibson City	WGCY	106.3	EZL
Dwight	WJEZ	98.9	AC/S	Glasford	WVEL	101.1	RL/G
Earlville	WMKB	102.9	RK/C	Glen Ellyn	WDCB	90.9	JAZ
East Moline	WDLM	960	RL/G	Glenview	WGBK	88.5	AD/A
East Moline	WDLM	89.3	RL/CC	Godfrey	WLCA	89.9	AD/A
East Moline	KUUL	101.3	OLG	Golconda	WLIE	94.3	CW
E St. Louis	WESL	1490	URB	Granite City	WGNU	920	TLK+
E St. Louis	WCBW	89.7	RL/G	Granite City	WARH	106.5	CH
E St. Louis	WVRV	101.1	AC	Greenville	WGRN	89.5	RL/CC
Edwardsville	WRYT	1080	RL	Greenville	WGEL	101.7	CW
Edwardsville	WSIE	88.7	JAZ	Harrisburg	WEBQ	1240	CW
Effingham	WCRA	1090	TLK+	Harrisburg	WOOZ	99.9	CW

City	Station	Freq	Format	City	Station	Freq	Format
Harvey	WGBX	1570	RL/G	Lockport	WLRA	88.1	AD/A
Havana	WDUK	99.3	CW	Loves Park	WLUV	1520	CW
Henry	WRVY	100.5	CH	Loves Park	WGSL	91.1	RL/CC
Herrin	WJPF	1340	N/TLK	Loves Park	WKGL	96.7	RK/C
Herrin	WVZA	92.7	CH	Lynnville	WEAI	107.1	AC/H
Highland	WDID	1510	RL/G	Macomb	WLRB	1510	AS
Highland	WDID	1510	RL/G	Macomb	WIUS	88.3	URB
Highland Pk	WEEF	1430	N/A	Macomb	WIUM	91.3	CLA
Highland Pk	WVIV	103.1	SP/OM	Macomb	WNLF	95.9	RK/C
Hillsboro	WXAJ	99.7	CH	Macomb	WKAI	100.1	AC/H
Hinsdale	WHSD	88.5	AD/A	Macomb	WJEQ	102.7	RK/C
Hoopeston	WHPO	100.9	CW	Macon	WZUS	100.9	CW
Jacksonville	WLDS	1180	TLK+	Mahomet	WGKC	105.9	RK/C
Jacksonville	WJIL	1550	AS	Marion	WGGH	1150	RL/G
Jacksonville	WYMG	100.5	CH	Maroa	WDKR	107.3	OLG
Jerseyville	WJBM	1480	TLK+	Marseilles	WKOT	96.5	CH
Jerseyville	WRDA	104.1	AS	Marshall	WMMC	105.9	AC
Johnston City	WDDD	810	SPTS	Mattoon	WLBH	1170	AS
Johnston City	WDDD	107.3	CW	Mattoon	WLKL	89.9	AD/A
Joliet	WJOL	1340	TLK+	Mattoon	WLBH	96.9	AC/S
Joliet	WWHN	1510	RL/G	Mcleansboro	WMCL	1060	CW
Joliet	WCSF	88.7	AC/H	Mendota	WGLC	100.1	CW
Joliet	WJCH	91.9	RL	Metropolis	WMOK	920	CW
Joliet	WVIX	93.5	SP/OM	Metropolis	WRIK	98.3	AC
Joliet	WSSR	96.7	AC	Metropolis	WREZ	105.5	AC/H
Kankakee	WKAN	1320	TLK+	Milford	WJCZ	91.3	RL/CC
Kankakee	WAWF	88.3	RL	Moline	WFXN	1230	SPTS
Kankakee	WONU	89.7	RL/CC	Moline	WXLP	96.9	RK/C
Kankakee	WKCC	91.1	TRAV	Monee	WJCG	88.9	RL
Kankakee	WKIF	92.7	N/TLK	Monmouth	WRAM	1330	CW
Kankakee	WVLI	95.1	OLG	Monmouth	WMOI	97.7	AC
Kewanee	WKEI	1450	N/TLK	Monticello	WCZQ	105.5	URB
Kewanee	WYEC	93.9	EZL	Morris	WCSJ	1550	AS
Knoxville	WKAY	105.3	AC	Morris	WBEQ	90.7	VAR
La Grange	WRDZ	1300	YTH	Morris	WCSJ	103.1	AS
La Grange	WLTL	88.1	AD/A	Morris	WCFL	104.7	RL/CC
La Salle	WLPO	1220	OLG	Morrison	WZZT	102.7	CH
La Salle	WLPO	1220	N/TLK	Morton	WDQX	102.3	CH
La Salle	WNIW	91.3	VAR	Mt Carmel	WVMC	1360	TLK+
La Salle	WAJK	99.3	AC	Mt Carmel	WYNG	94.9	SPTS
Lake Forest	WMXM	88.9	AD/A	Mt Vernon	WMIX	940	TLK+
Lansing	WSRB	106.3	URB	Mt Vernon	WBMV	89.7	RL/CC
Lawrenceville	WAKO	910	AC	Mt Vernon	WMIX	94.1	CW
Lawrenceville	WAKO	103.1	AC	Mt Vernon	WIBV	102.1	RL/CC
Lena	WQLF	102.1	CH	Mount Zion	WXFM	99.3	AC/S
Leroy	WBWN	104.1	CW	Mt. Carmel	WVJC	89.1	RK/C
Lexington	WDQZ	99.5	CH	Mt. Sterling	WPWQ	106.7	OLG
Lincoln	WLLM	1370	RL/CC	Mt. Vernon	WVSI	88.9	VAR
Lincoln	WLNX	88.9	VAR	Mt. Vernon	WAPO	90.5	RL
Lincolnshire	WAES	88.1	VAR	Murphysboro	WINI	1420	TLK+
Litchfield	WSMI	1540	CW	Murphysboro	WTAO	105.1	RK/C
Litchfield	WSMI	106.1	CW	Naperville	WONC	89.1	RK

City	Station	Freq	Format	City	Station	Freq	Format
Nashville	WNSV	104.7	AC/H	Quincy	WCOY	99.5	CW
Neoga	WHQQ	98.9	OLG	Quincy	WQCY	103.9	SPTS
Neoga	WMCI	101.3	CW	Quincy	WGEM	105.1	TLK+
Newton	WIKK	103.5	RK/C	Ramsey	WJLY	88.3	RL/CC
Normal	WGLT	89.1	JAZ	Ramsey	WTRH	93.3	TLK+
Normal	WIHN	96.7	RK/C	Rantoul	WJCI	1460	SPTS
Oak Lawn	WNWI	1080	N/A	Rantoul	WBNB	95.3	RK/C
Oak Park	WPNA	1490	N/A	Rantoul	WQQB	96.1	CH
Oak Park	WVAZ	102.7	URB	River Grove	WRRG	88.9	RK/C
Oglesby	WALS	102.1	CW	Robinson	WTAY	1570	AC/S
Olney	WVLN	740	SPTS	Robinson	WTYE	101.7	AC/S
Olney	WPTH	88.1	RL	Rochelle	WRHL	1060	N/TLK
Olney	WSEI	92.9	CH	Rochelle	WRHL	102.3	AC
Oregon	WSEY	95.7	OLG	Rock Island	WKBF	1270	TLK+
Ottawa	WCMY	1430	AC/S	Rock Island	WVIK	90.3	CLA
Ottawa	WWGN	88.9	RL	Rock Island	WHTS	98.9	CH
Ottawa	WRKX	95.3	AC/H	Rockford	WNTA	1330	TLK+
Palatine	WHCM	88.3	VAR	Rockford	WROK	1440	N/TLK
Pana	WZRS	89.3	N/A	Rockford	WFEN	88.3	RL/CC
Pana	WMKR	94.3	CW	Rockford	WNIU	90.5	CLA
Paris	WPRS	1440	OLG	Rockford	WZOK	97.5	CH
Paris	WACF	98.5	CW	Rockford	WQFL	100.9	RL/CC
Park Forest	WRZA	99.9	CH	Rockton	WGFB	103.1	AC/S
Park Ridge	WMTH	90.5	VAR	Rushville	WKXQ	92.5	CH
Paxton	WPXN	104.9	CH	Salem	WJBD	1350	CW
Pekin	WVEL	1140	RL	Salem	WSLE	91.3	N/A
Pekin	WBNH	88.5	RL/CC	Salem	WJBD	100.1	AC/S
Pekin	WCIC	91.5	RL/CC	Sandwich	WAUR	930	RL
Pekin	WGLO	95.5	RK/C	Savanna	WCCI	100.3	CW
Pekin	WXCL	104.9	CW	Seneca	WJDK	95.7	AC
Peoria	WPEO	1020	RL	Shelbyville	WINU	870	RL/CC
Peoria	WIRL	1290	CW	Shelbyville	WEJT	105.1	AC
Peoria	WOAM	1350	AS	Sherman	WABZ	93.9	CH
Peoria	WMBD	1470	TLK+	Skokie	WTMX	101.9	AC/H
Peoria	WCBU	89.9	CLA	S Jacks'nville	WJVO	105.5	CW
Peoria	WZPW	92.3	URB	South Beloit	WTJK	1380	SPTS
Peoria	WPBG	93.3	OLG	Sparta	WHCO	1230	N/TLK
Peoria	WXMP	105.7	AC/H	Spring Vly	WSOG	88.1	RL
Peoria	WSWT	106.9	AC/S	Spring Vly	WIVQ	103.3	CH
Peru	WBZG	100.9	RK/C	Springfield	WMAY	970	N/TLK
Petersburg	WLWJ	88.1	RL/CC	Springfield	WTAX	1240	N/TLK
Petersburg	WYVR	97.7	AC	Springfield	WFMB	1450	SPTS
Pittsfield	WBBA	1580	SILNT	Springfield	WQNA	88.3	VAR
Pittsfield	WBBA	97.5	CW	Springfield	WLGM	89.7	RL/CC
Plano	WSPY	107.1	AC/S	Springfield	WSCT	90.5	RL/CC
Polo	WLLT	107.7	AC	Springfield	WUIS	91.9	VAR
Pontiac	WTRX	93.7	RK/C	Springfield	WNNS	98.7	AC
Princeton	WZOE	1490	N/TLK	Springfield	WQQL	101.9	OLG
Princeton	WZOE	98.1	OLG	Springfield	WDBR	103.7	CH
Quincy	WTAD	930	TLK+	Springfield	WFMB	104.5	CW
Quincy	WGEM	1440	SPTS	St. Joseph	WGNJ	89.3	RL/CC
Quincy	WGCA	88.5	RL/CC	Sterling	WSDR	1240	N/TLK

Radio on the Road

CITY	STATION	FREQ	FORMAT		CITY	STATION	FREQ	FORMAT
Sterling	WNIQ	91.5	VAR					
Sterling	WSSQ	94.3	AC					
Streator	WIZZ	1250	TLK+					
Streator	WSTQ	97.7	CH					
Streator	WYYS	106.1	AC/S					
Sugar Grove	WSRI	88.7	RL/CC					
Sullivan	WZNX	106.7	RK/C					
Summit	WARG	88.9	AD/A					
Sycamore	WSQR	1560	AS					
Taylorville	WIHM	1410	RL					
Taylorville	WQLZ	92.7	RK					
Taylorville	WTIM	97.3	N/TLK					
Teutopolis	WKJT	102.3	CW					
Tower Hill	WRAN	98.3	AS					
Tuscola	WEBX	93.5	RK					
Urbana	WILL	580	N/TLK					
Urbana	WILL	90.9	CLA					
Sterling	WNIQ	91.5	VAR					
Sterling	WSSQ	94.3	AC					
Streator	WIZZ	1250	TLK+					
Streator	WSTQ	97.7	CH					
Streator	WYYS	106.1	AC/S					
Sugar Grove	WSRI	88.7	RL/CC					
Sullivan	WZNX	106.7	RK/C					
Summit	WARG	88.9	AD/A					
Sycamore	WSQR	1560	AS					
Taylorville	WIHM	1410	RL					
Taylorville	WQLZ	92.7	RK					
Taylorville	WTIM	97.3	N/TLK					
Teutopolis	WKJT	102.3	CW					
Tower Hill	WRAN	98.3	AS					
Tuscola	WEBX	93.5	RK					
Urbana	WILL	580	N/TLK					
Urbana	WILL	90.9	CLA					

Radio on the Road   Indiana

City	Station	Freq	Format		City	Station	Freq	Format
Anderson	WHBU	1240	TLK+		Columbus	WINN	104.9	OLG
Anderson	WGNR	1470	RL		Connorsville	WCNB	1580	CW
Anderson	WBSB	89.5	VAR		Connorsville	WIFE	100.3	CW
Anderson	WGNR	97.9	RL		Corydon	WOCC	1550	OLG
Anderson	WQME	98.7	RL/CC		Corydon	WGZB	96.5	URB
Angola	WEAX	88.3	VAR		Corydon	WSFR	107.7	RK/C
Angola	WLKI	100.3	AC		Covington	WFOF	90.3	RL
Attica	WFWR	91.5	VAR		Covington	WKZS	103.1	CW
Attica	WLFF	95.7	CW		Crawfordsville	WCVL	1550	AS
Auburn	WGLL	1570	RL		Crawfordsville	WNDY	91.3	AD/A
Auburn	WXTW	102.3	RK/C		Crawfordsville	WIMC	103.9	CH
Aurora	WSCH	99.3	CW		Crawfordsville	WCDQ	106.3	RL/CC
Austin	WJCP	92.7	CW		Crawfordsville	WCDQ	106.3	AC
Austin	WJAA	96.3	RK/C		Crothersville	WOJC	89.9	N/A
Batesville	WRBI	103.9	CW		Crown Point	WXRD	103.9	RK/C
Battle Grnd	WASK	98.7	OLG		Danville	WEDJ	107.1	SP/MEX
Bedford	WBIW	1340	N/TLK		Decatur	WADM	1540	AC
Bedford	WQRK	105.5	OLG		Decatur	WQHK	105.1	CW
Beech Grv	WNTS	1590	RL/G		Delphi	WXXB	102.9	CH
Berne	WZBD	92.7	AC		Earl Park	WIBN	98.1	OLG
Bloomington	WGCL	1370	TLK+		Edinburgh	WYGB	102.9	CW
Bloomington	WFHB	91.3	VAR		Elkhart	WFRN	1270	RL
Bloomington	WTTS	92.3	AD/A		Elkhart	WTRC	1340	TLK+
Bloomington	WBWB	96.7	CH		Elkhart	WVPE	88.1	VAR
Bloomington	WUZR	105.7	OLG		Elkhart	WBYT	100.7	CW
Bluffton	WNUY	100.1	AC		Elkhart	WFRN	104.7	RL/CC
Boonville	WBNL	1540	AC/S		Ellettsville	WHCC	105.1	CW
Boonville	WYXY	107.1	CH		Elwood	WURK	101.7	OLG
Brazil	WSDX	1130	SPTS		Evansville	WSWI	820	AD/A
Brazil	WSDM	97.7	OLG		Evansville	WGBF	1280	N/TLK
Bremen	WHPZ	96.9	RL/CC		Evansville	WVHI	1330	RL
Brookston	WSHP	95.3	RK/C		Evansville	WEOA	1400	URB
Brownsburg	WKLU	101.9	CH		Evansville	WNIN	88.3	CLA
Cannelton	WLME	102.9	OLG		Evansville	WPSR	90.7	VAR
Carmel	WHJE	91.3	RK/C		Evansville	WUEV	91.5	JAZ
Centerville	WHON	930	TLK+		Evansville	WIKY	104.1	AC/S
Chandler	WJPS	93.5	CW		Evansville	WJLT	105.3	OLG
Charlestown	WEGK	104.3	CW		Evansville	WABX	107.5	RK/C
Chesterton	WDSO	88.3	VAR		Ferdinand	WQKZ	98.5	CW
Chesterton	WBEW	89.5	JAZ		Fishers	WISG	93.9	RL/CC
Churubusco	WNHT	96.3	CH		Fort Branch	WBGW	101.5	RL/CC
Cicero	WJCY	91.5	RL		Fort Wayne	WFCV	1090	RL
Clarkesville	WJZL	93.1	JAZ		Fort Wayne	WOWO	1190	TLK+
Clinton	WPFR	93.9	RL		Fort Wayne	WGL	1250	AS
Cloverdale	WSPM	89.1	RL		Fort Wayne	WONO	1380	SPTS
ColumbiaCty	WJHS	91.5	AD/A		Fort Wayne	WLYV	1450	RL
ColumbiaCty	WDDB	106.3	CH		Fort Wayne	WLAB	88.3	RL/CC
Columbus	WCSI	1010	N/TLK		Fort Wayne	WBCL	90.3	RL/CC
Columbus	WYGS	91.1	RL/CC		Fort Wayne	WFWI	92.3	RK
Columbus	WKKG	101.5	CW		Fort Wayne	WAJI	95.1	AC

City	Station	Freq	Format	City	Station	Freq	Format
Fort Wayne	WMEE	97.3	AC/H	Indianapolis	WBDG	90.9	AC/H
Fort Wayne	WLDE	101.7	OLG	Indianapolis	WEDM	91.1	VAR
Fort Wayne	WYLT	103.9	CH	Indianapolis	WRFT	91.5	VAR
Frankfort	WILO	1570	AS	Indianapolis	wnou	93.1	CH
Frankfort	WSHW	99.7	AC	Indianapolis	WFBQ	94.7	RK
Franklin	WFCI	89.5	VAR	Indianapolis	WFMS	95.5	CW
Franklin	WIJY	95.9	RL/CC	Indianapolis	WHHH	96.3	CH
French Lick	WFLQ	100.1	CW	Indianapolis	WRZX	103.3	RK/C
Ft. Wayne	WBOI	89.1	VAR	Indianapolis	wjjk	104.5	AS
Gary	WWCA	1270	RL	Indianapolis	WYXB	105.7	AC/S
Gary	WLTH	1370	TLK+	Indianapolis	WTPI	107.9	AC
Gary	WGVE	88.7	URB	Jasper	WITZ	990	AC
Goshen	WKAM	1460	SP/MEX	Jasper	WJKR	91.7	N/A
Goshen	WGCS	91.1	CW	Jasper	WITZ	104.7	AC
Goshen	WZOW	97.7	RK/C	Jeffersonville	WAVG	1450	CW
Greencastle	WIKL	91.5	RL/CC	Jeffersonville	WQMF	95.7	RK/C
Greencastle	WGRE	91.5	AD/A	Kendallville	WAWK	1140	OLG
Greencastle	WREB	94.3	CW	Kendallville	WBTU	93.3	CW
Greenfield	WRGF	89.7	AD/A	Kentland	WLRT	101.7	CW
Greenfield	WZPL	99.5	AC/H	Knightstown	WKPW	90.7	CW
Greensburg	WTRE	1330	CW	Knox	WKVI	1520	AC/S
Greensburg	WAUZ	89.1	RL/G	Knox	WKVI	99.3	AC/S
Greensburg	WRZQ	107.3	AC	Kokomo	WIOU	1350	TLK+
Greenwood	WYXB	105.7	AC/S	Kokomo	WMWC	91.7	RL
Greenwood	WTLC	106.7	URB	Kokomo	WZWZ	92.5	AC
Hagerstown	WBSH	91.1	VAR	Kokomo	WWKI	100.5	CW
Hammond	WJOB	1230	N/TLK	La Porte	WLOI	1540	AS
Hammond	wpwx	92.3	URB	La Porte	WCOE	96.7	CW
Hanna	WHLP	89.9	RL/CC	Ladoga	WJCJ	88.9	N/A
Hanna	WHTI	96.7	RK/C	Lafayette	wlas	1410	CW
Hardinsburg	WKLO	96.9	AC/H	Lafayette	WASK	1450	OLG
Hartford City	WHTY	93.5	RK/C	Lafayette	WQSG	90.7	N/A
Hartford City	WKLO	96.9	VAR	Lafayette	WJEF	91.9	OLG
Howe	WHWE	89.7	VAR	Lafayette	WKHY	93.5	RK/C
Howe	WQKO	91.9	RL	Lafayette	WAZY	96.5	CH
Huntingburg	WBDC	100.9	CW	Lafayette	WKOA	105.3	CW
Huntington	WBZQ	1300	AC/S	Lafay'tte Twn	WCYT	91.1	AD/A
Huntington	WVSH	91.9	VAR	Lagrange	WTHD	105.5	CW
Huntington	wxke	102.9	CH	Lebanon	WIRE	91.1	AC
Indianapolis	WSYW	810	SP/MEX	Ligonier	WLEG	102.7	AC
Indianapolis	WXLW	950	SPTS	Linton	WBTO	1600	CW
Indianapolis	WIBC	1070	TLK+	Linton	WYTJ	89.3	RL
Indianapolis	WNDE	1260	SPTS	Linton	WQTY	93.3	CW
Indianapolis	WTLC	1310	RL/G	Logansport	WSAL	1230	AC/S
Indianapolis	WXNT	1430	N/TLK	Logansport	WWTS	89.5	N/A
Indianapolis	WBRI	1500	RL	Logansport	WLHM	102.3	CH
Indianapolis	WICR	88.7	JAZ	Loogootee	WBHW	88.7	RL/CC
Indianapolis	WJEL	89.3	CH	Loogootee	WRZR	94.5	RK/C
Indianapolis	WFYI	90.1	VAR	Lowell	WTMK	88.5	N/A
Indianapolis	WFYI	90.1	VAR	Lowell	WZVN	107.1	AC

Radio on the Road   Indiana

City	Station	Freq	Format	City	Station	Freq	Format
Madison	WXGO	1270	OLG	Plymouth	WTCA	1050	OLG
Madison	WORX	96.7	AC	Plymouth	WRXH	89.3	N/A
Marengo	WBRO	89.9	VAR	Plymouth	WZOC	94.3	OLG
Marion	WGOM	860	TLK+	Portage	WNDZ	750	TLK+
Marion	WBAT	1400	OLG	Portland	WPGW	1440	AC
Marion	WBSW	90.9	VAR	Portland	WBSJ	91.7	VAR
Marion	WMRI	106.9	AC/S	Portland	WPGW	100.9	CW
Martinsville	WMCB	1540	CW	Princeton	WRAY	1250	TLK+
Martinsville	WCBK	102.3	CW	Princeton	WRAY	98.1	CW
Michig'n City	WIMS	1420	N/TLK	Princeton	WSJD	100.5	RK/C
Michig'n City	WEFM	95.9	OLG	Princeton	WSJD	100.5	RK/C
Mitchell	WMBL	88.1	RL	Rensselaer	WRIN	1560	AS
Mitchell	WWEG	102.5	OLG	Rensselaer	WPUM	90.5	RK/C
Monticello	WMRS	107.7	AC/S	Rensselaer	WLQI	97.7	CH
Montpelier	WJCO	91.3	RL/CC	Richmond	WKBV	1490	TLK/S
Morgantown	WCJL	90.9	N/A	Richmond	WVXR	89.3	VAR
Morristown	WJCF	88.1	RL/CC	Richmond	WECI	91.5	VAR
Mt Vernon	WPCO	1590	CW	Richmond	WQLK	96.1	CW
Mt Vernon	WYFX	106.7	CW	Richmond	WFMG	101.3	AC/H
Muncie	WLHN	990	RL/G	Roann	WARU	101.9	OLG
Muncie	WXFN	1340	SPTS	Roanoke	WCKZ	94.1	RK/C
Muncie	WWDS	90.5	VAR	Rochester	WHNI	88.5	N/A
Muncie	WWHI	91.3	AD/A	Rochester	WROI	92.1	OLG
Muncie	WBST	92.1	VAR	Rockville	WAXI	104.9	AS
Muncie	WLBC	104.1	AC/H	Royal Center	WHZR	103.7	CW
Muncie	WERK	104.9	OLG	Rushville	WKWH	94.3	CW
Nappanee	WYPW	95.7	CH	Salem	WSLM	1220	CW
Nashville	WVNI	95.1	RL/CC	Salem	WSLM	97.9	RL/G
New Albany	WWSZ	1570	SPTS	Salem	WZTR	98.9	CH
New Albany	WNAS	88.1	VAR	Santa Claus	WAXL	103.3	AC
New Albany	WFIA	94.7	RL/G	Scottsburg	WMPI	105.3	CW
New Carlisle	WOZW	102.3	RK/C	Seelyville	WWSY	95.9	AC/H
New Castle	WMDH	1550	AS	Seymour	WZZB	1390	AC/S
New Castle	WMDH	102.5	CW	Seymour	WJLR	91.5	RL/CC
New Haven	WJFX	107.9	CH	Seymour	WQKC	93.7	CW
NewWash'ton	WSOH	88.3	RL/CC	Shelbyville	WKWH	1520	CH
Newburgh	WGAB	1180	RL/G	Shelbyville	WLHK	97.1	CW
Newburgh	WDKS	106.1	CH	South Bend	WSBT	960	N/TLK
N Manch'ster	WBKE	89.5	VAR	South Bend	WNDV	1490	CH
N Vernon	WNVI	1460	VAR	South Bend	WDND	1580	SPTS
N Vernon	WWWY	106.1	RK/C	South Bend	WHLY	1620	AS
Notre Dame	WSND	88.9	CLA	South Bend	WUBS	89.7	RL
Orland	WBNI	91.3	CLA	South Bend	WETL	91.7	VAR
Paoli	WSEZ	1560	OLG	South Bend	WNDV	92.9	CH
Paoli	WUME	95.3	AC/H	South Bend	WNSN	101.5	AC
Pendleton	WEEM	91.7	AD/A	South Bend	WHME	103.1	RL/G
Peru	WARU	1600	CH	South Bend	WRBR	103.9	RK/C
Peru	WMYK	98.5	RK/C	South Bend	WUBU	106.3	JAZ
Petersburg	WBTO	102.3	RK/C	South Whitley	WLZQ	101.1	AC/H
Plainfield	WRDZ	98.3	YTH	Speedway	WYJZ	100.9	JAZ

73

Indiana

Radio on the Road

City	Station	Freq	Format	City	Station	Freq	Format
Spencer	WSKT	92.7	CW				
Sullivan	WNDI	1550	CW				
Sullivan	WNDI	95.3	CW				
Syracuse	WAWC	103.5	CH				
Tell City	WTCJ	1230	AC/S				
Tell City	WTCJ	105.7	CH				
Terre Haute	WBOW	1300	SPTS				
Terre Haute	WBOW	1300	SPTS				
Terre Haute	WPFR	1480	RL				
Terre Haute	WCRT	88.5	RL/CC				
Terre Haute	WISU	89.7	RK/C				
Terre Haute	WMHD	90.5	AD/A				
Terre Haute	WAPC	91.9	RL				
Terre Haute	WTHI	99.9	CW				
Terre Haute	WMGI	100.7	CH				
Terre Haute	WBOW	102.7	AC/S				
Union City	WJYW	88.9	RL/CC				
Upland	WTUR	89.7	RL/CC				
Valparaiso	WAKE	1500	AS				
Valparaiso	WVUR	95.1	VAR				
Valparaiso	WLJE	105.5	CW				
Van Buren	WCJC	99.3	CW				
Veedersburg	WSKL	92.9	OLG				
Versailles	WKRY	88.1	RL/G				
Versailles	WXCH	103.1	CW				
Vevay	WKID	95.9	CW				
Vincennes	WAOV	1450	TLK/S				
Vincennes	WATI	89.9	RL/CC				
Vincennes	WVUB	91.1	AC/H				
Vincennes	WZDM	92.1	AC				
Vincennes	WFML	96.7	CW				
Wabash	WJOT	1510	OLG				
Wabash	WKUZ	95.9	CW				
Wabash	WJOT	105.9	OLG				
Wadesville	WRFM	90.1	N/A				
Walton	WFRR	93.7	RL/CC				
Warsaw	WRSW	1480	SPTS				
Warsaw	WRSW	107.3	CH				
Washington	WAMW	1580	AS				
Washington	WWBL	106.5	CW				
Washington	WAMW	107.9	AC/S				
W Lafayette	WBAA	101.3	CLA				
W Lafayette	WBAA	920	JAZ				
W Lafayette	WHPL	89.9	RL				
W Lafayette	WGLM	106.7	AC				
W Terre H'te	WWVR	105.5	RK/C				
Winamac	WFRI	100.1	RL/CC				
Winchester	WZZY	98.3	AC				
Winchester	WZZY	98.3	AC				

74

Radio on the Road   Iowa

City	Station	Freq	Format	City	Station	Freq	Format
Adel	KIHS	88.5	RL/CC	Cedar Rap	KDAT	104.5	AC/S
Albia	KLBA	1370	CW	Centerville	KCOG	1400	AC
Albia	KLBA	1370	CW	Centerville	KMGO	98.7	CW
Albia	KLBA	96.7	CW	Chariton	KELR	105.5	AC
Algona	KLGA	1600	CW	Charles City	KCHA	1580	AS
Algona	KLGA	92.7	AC	Charles City	KCHA	95.9	AC/S
Alta	KBVU	97.5	RK	Cherokee	KCHE	1440	AS
Ames	WOI	640	N/TLK	Cherokee	KCHE	92.1	AC
Ames	KASI	1430	TLK+	Clarinda	KKBZ	99.3	AC
Ames	KURE	88.5	VAR	Clarion	KIAQ	96.9	CW
Ames	WOI	90.1	VAR	Clear Lake	KLKK	103.7	RK/C
Ames	KLTI	104.1	AC/S	Clinton	KROS	1340	AC
Ames	KCCQ	105.1	RK/C	Clinton	KCLN	1390	AS
Ankeny	KDRB	106.3	AH	Clinton	KZEG	94.7	CW
Asbury	WJOD	103.3	CW	Clinton	KMXG	96.1	AC
Atlantic	KJAN	1220	AC/S	Council Bluffs	KLNG	1560	RL
Atlantic	KJAN	95.7	AC	Council Bluffs	KIWR	89.7	AD/A
Audubon	KSOM	96.5	CW	Council Bluffs	KQKQ	98.5	AC/H
Belle Pine	KZAT	95.5	RK/C	Cresco	KCZQ	102.3	AC
Bettendorf	KQCS	93.5	AC/H	Creston	KSIB	1520	CW
Bloomfield	KOJY	106.3	RL/G	Creston	KLOX	90.9	N/A
Boone	KFFF	1260	RL/G	Creston	KSIB	101.3	CW
Boone	KWBG	1590	TLK+	Davenport	KJOC	1170	TLK+
Boone	KRKQ	98.3	TLK+	Davenport	WOC	1420	TLK+
Boone	KFFF	99.3	RL/G	Davenport	KALA	88.5	VAR
Britt	KHAM	99.5	N/A	Davenport	WLLR	103.7	CW
Brooklyn	KSKB	99.1	RL/CC	Davenport	KCQQ	106.5	CH
Burlington	KCPS	1150	TLK+	Decorah	KDEC	1240	AS
Burlington	KBUR	1490	TLK+	Decorah	KWLC	1240	JAZ
Burlington	KAYP	89.9	RL/CC	Decorah	KLNI	88.7	CLA
Burlington	KKMI	93.5	AC	Decorah	KLCD	89.5	CLA
Burlington	KDMG	103.1	CW	Decorah	KDEC	100.5	AC/H
Burlington	KGRS	107.3	AC/H	Decorah	KVIK	104.7	CH
Carroll	KCIM	1380	AC/S	Denison	KDSN	1530	CW
Carroll	KWOI	90.7	VAR	Denison	KDSN	107.1	AC
Carroll	KKRL	93.7	AC/H	Des Moines	KPSZ	940	RL/CC
Castana	KILV	107.5	RL/CC	Des Moines	WHO	1040	TLK+
Cedar Falls	KDNZ	1250	SPTS	Des Moines	KWKY	1150	RL
Cedar Falls	KCNZ	1650	SPTS	Des Moines	KRNT	1350	AS
Cedar Falls	KHKE	89.5	CLA	Des Moines	KXNO	1460	SPTS
Cedar Falls	KUNI	90.9	VAR	Des Moines	KBGG	1700	SP/N
Cedar Falls	KOEL	98.5	CW	Des Moines	KDPS	88.1	VAR
Cedar Rapids	WMT	600	N/TLK	Des Moines	KJMC	89.3	URB
Cedar Rapids	KMJM	1360	SPTS	Des Moines	KDFR	91.3	RL
Cedar Rapids	KMRY	1450	AS	Des Moines	KIOA	93.3	OLG
Cedar Rapids	KCRG	1600	SPTS	Des Moines	KGGO	94.9	RK
Cedar Rapids	KCCK	88.3	JAZ	Des Moines	KHKI	97.3	CW
Cedar Rapids	WMT	96.5	AC/H	Des Moines	KMXD	100.3	AC/H
Cedar Rapids	KHAK	98.1	CW	Des Moines	KSTZ	102.5	AC/H
Cedar Rapids	KZIA	102.9	CH	Des Moines	KKDM	107.5	CH

City	Station	Freq	Format		City	Station	Freq	Format
Dewitt	KBOB	104.9	CW		Iowa City	KRUI	89.7	VAR
Dubuque	KDTH	1370	AS		Iowa City	KSUI	91.7	CLA
Dubuque	WDBQ	1490	TLK+		Iowa City	KRNA	94.1	RK
Dubuque	KDUB	90.1	N/A		Iowa City	KKRQ	100.7	CH
Dubuque	KATF	92.9	AC		Iowa Falls	KIFG	1510	AC/S
Dubuque	KXGE	102.3	RK		Iowa Falls	KIFG	95.3	AC/S
Dubuque	KLYV	105.3	CH		Jefferson	KGRA	98.9	CW
Dyersville	KDST	99.3	CW		Keokuk	KOKX	1310	AS
Eagle Grove	KJYL	100.7	RL/CC		Keokuk	KMDY	90.9	RL
Eddyville	KKSI	101.5	RK/C		Keokuk	KOKX	95.3	OLG
Eldon	KRKN	104.3	CW		Keokuk	KRNQ	96.3	RK/C
Eldora	KDAO	99.5	AC		Knoxville	KNIA	1320	CW
Elkader	KADR	1400	AC		Knoxville	KRLS	92.1	AC
Emmetsburg	KDWD	100.1	CH		Lake City	KIKD	106.7	CW
Epworth	KGRR	97.3	CH		Lamoni	KIKD	97.9	VAR
Estherville	KILR	1070	TLK+		Le Mars	KLEM	1410	AC
Estherville	KILR	95.9	CW		Le Mars	KKMA	99.5	RK/C
Fairfield	KMCD	1570	TLK+		Madrid	KNWM	96.1	RL/CC
Fairfield	KHOE	90.5	VAR		Manchester	KMCH	94.7	AC
Fairfield	KIIK	95.9	AC/H		Maquoketa	KMAQ	1320	CW
Forest City	KZOW	91.9	RL/CC		Maquoketa	KMAQ	95.1	AC/S
Forest City	KIOW	107.3	CW		Marshalltown	KDAO	1190	AS
Fort Dodge	KWMT	540	CW		Marshalltown	KFJB	1230	N/TLK
Fort Dodge	KVFD	1400	OLG		Marshalltown	KXIA	101.1	CW
Fort Dodge	KICB	88.1	AD/A		Mason City	KGLO	1300	TLK+
Fort Dodge	KUEL	92.1	AC		Mason City	KRIB	1490	OLG
Fort Dodge	KKEZ	94.5	AC/H		Mason City	KBDC	88.5	RL/CC
Fort Madison	KBKB	1360	SPTS		Mason City	KUNY	91.5	VAR
Fort Madison	KBKB	101.7	CW		Mason City	KIAI	93.9	CW
Ft. Dodge	KEGR	89.5	N/A		Mason City	KCMR	97.9	EZL
Ft. Dodge	KTPR	91.1	VAR		Mason City	KLSS	106.1	AC
Garnavillo	KCTN	100.1	CW		Milford	KUQQ	102.1	RK/C
Glenwood	KXKT	103.7	CW		Mount Pleasant	KILJ	1130	CW
Grinnell	KGRN	1410	AC/S		Mount Pleasant	KILJ	105.5	AS
Grinnell	KDIC	88.5	VAR		Mount Vernon	KRNL	89.7	VAR
Grinnell	KRTI	106.7	AC/H		Muscatine	KWPC	860	AC
Grundy Ctr	KCRR	97.7	RK/C		Muscatine	KWCC	93.1	CW
Hampton	KLMJ	104.9	CW		Muscatine	KBEA	99.7	CH
Harlan	KNOD	105.3	OLG		New Hampton	KCZE	95.1	CW
Hiawatha	KWOF	91.1	RL/CC		New London	KKNL	97.3	AC
Hudson	KCVM	96.1	AC/H		New Sharon	KCWN	99.9	RL/CC
Humboldt	KHBT	97.7	RK/C		Newton	KCOB	1280	CW
Ida Grove	KKIA	92.9	CW		Newton	KNNU	88.3	N/A
Independence	KQMG	1220	SPTS		Newton	KCOB	95.9	CW
Independence	KQMG	95.3	AC		Northwood	KYTC	102.7	CW
Indianola	KXLQ	1490	SIL		Oelwein	KOEL	950	FRM
Indianola	KSTM	88.9	VAR		Oelwein	KKHQ	92.3	CH
Iowa City	KXIC	800	N/TLK		Okoboji	KOJI	90.7	VAR
Dewitt	KBOB	104.9	CW		Onawa	KZSR	102.3	AC/H
Dubuque	KDTH	105.3	AS		Osage	KSMA	98.7	CH

76

City	Station	Freq	Form't	City	Station	Freq	Form't
Osceola	KNWI	107.1	RL/CC	Waterloo	KWOF	850	RL/CC
Oskaloosa	KBOE	740	CW	Waterloo	KNWS	1090	RL/CC
Oskaloosa	KIGC	88.7	VAR	Waterloo	KWLO	1330	AS
Oskaloosa	KBOE	104.9	CW	Waterloo	KXEL	1540	N/TLK
Ottumwa	KBIZ	1240	TLK+	Waterloo	KBBG	88.1	VAR
Ottumwa	KLEE	1480	TLK+	Waterloo	KNWS	101.9	RL/CC
Ottumwa	KUNZ	91.1	N/A	Waterloo	KOKZ	105.7	OLG
Ottumwa	KTWA	92.7	AC	Waterloo	KFMW	107.9	RK
Ottumwa	KTWA	92.7	AC	Waukon	KNEI	1140	SPTS
Ottumwa	KOTM	97.7	CH	Waukon	KNEI	103.9	CW
Parkersburg	KQCR	98.9	AC	Waverly	KWAY	1470	CW
Pella	KCUI	89.1	AD/A	Waverly	KWAR	89.1	VAR
Pella	KAZR	103.3	RK	Waverly	KWAY	99.3	AC/H
Perry	KDLS	1310	VAR	Webster City	KQWC	1570	AS
Perry	KDLS	105.5	VAR	Webster City	KQWC	95.7	AC
Postville	KPVL	89.1	VAR	W Des Moines	KWDM	88.7	AD/A
Red Oak	KOAK	1080	CW	W Des Moines	KJJY	92.5	CW
Red Oak	KCSI	95.3	CW	Whiting	KKYY	101.3	CW
Rock Valley	KIHK	106.9	CW	Williamsburg	KEWM	88.7	N/A
Sageville	KIYX	106.1	AC	Winterset	KZZQ	99.5	RL/CC
Saint Ansgar	KJCY	95.5	RL/CC				
Sheldon	KIWA	1550	TLK+				
Sheldon	KIWA	105.3	TLK+				
Shenandoah	KYFR	920	RL				
Shenandoah	KMA	960	TLK+				
Sioux Center	KSOU	1090	RL/CC				
Sioux Center	KDCR	88.5	RL/CC				
Sioux Center	KSOU	93.9	AC				
Sioux City	KMNS	620	TLK+				
Sioux City	KSCJ	1360	N/TLK				
Sioux City	KWSL	1470	SPTS				
Sioux City	KMSC	88.3	VAR				
Sioux City	KWIT	90.3	VAR				
Sioux City	KGLI	95.5	CH				
Sioux City	KSEZ	97.9	RK				
Sioux City	KTFC	103.3	RL				
Sioux Rapids	KTFG	102.9	RL/G				
Spencer	KICD	1240	TLK+				
Spencer	KLLT	104.9	AC/S				
Spencer	KICD	107.7	CW				
Spirit Lake	KJIA	88.9	RL				
Spirit Lake	KUOO	103.9	AC				
Storm Lake	KAYL	990	AS				
Storm Lake	KAYL	101.5	AC				
Stuart	KKRF	107.9	CW				
Twin Lakes	KTLB	105.9	OLG				
Vinton	KROJ	107.1	N/A				
Wapello	KLRX	88.9	N/A				
Washington	KCII	1380	AC				
Washington	KCII	106.1	AC				

City	Station	Freq	Format	City	Station	Freq	Format
Abilene	KABI	1560	AS	Emporia	KNGM	91.9	RL/CC
Abilene	KSAJ	98.5	OLG	Emporia	KANS	96.1	AC/S
Andover	kdgs	93.9	CH	Emporia	KVOE	101.7	CW
Arkansas City	KSOK	1280	CW	Emporia	KFFX	104.9	AC/H
Arkansas City	KAXR	91.3	RL	Enterprise	KBMP	90.5	RL
Arkansas City	KACY	102.5	RK/C	Eureka	KOTE	93.5	CW
Arkansas City	KYQQ	106.5	SP/MEX	Fairway	KCNW	1380	RL
Atchison	KAIR	1470	CW	Fort Scott	KMDO	1600	OLG
Augusta	KFXJ	104.5	RK/C	Fort Scott	KOMB	103.9	OLG
Baldwin City	KNBU	89.7	VAR	Fort Scott	KVCY	104.7	RL
Baxter Springs	KMOQ	107.1	CH	Fredonia	KGGF	104.1	OLG
Belle Plaine	KANR	92.7	RK/C	Galena	KCAR	104.3	OLG
Belleville	KREP	92.1	CW	Garden City	KIUL	1240	TLK+
Beloit	KVSV	1190	AC	Garden City	KANZ	91.1	CLA
Beloit	KVSV	105.5	EZL	Garden City	KKJQ	97.3	CW
Bronson	KBJQ	88.3	RL	Girard	KSEK	99.1	RK
Burlington	KSNP	95.3	RK/C	Goodland	KLOE	730	TLK+
Caney	KEQJ	101.1	RL/CC	Goodland	KKCI	102.5	AC/S
Cawker City	KZDY	96.3	OLG	Goodland	KGCR	107.7	RL/CC
Chanute	KKOY	1460	TLK+	Great Bend	KVGB	1590	TLK+
Chanute	KKOY	105.5	AC/S	Great Bend	KBDA	89.7	RL
Clay Center	KCLY	100.9	AC	Great Bend	KHCT	90.9	VAR
Clearwater	KFH	98.7	TLK+	Great Bend	KWBI	91.9	RL/CC
Coffeyville	KGGF	690	N/TLK	Great Bend	KVGB	104.3	RK/C
Coffeyville	KKRK	98.9	RK/C	Great Bend	KZLS	107.9	AC
Colby	KXXX	790	CW	Hays	KAYS	1400	OLG
Colby	KTCC	91.9	RK/C	Hays	KPRD	88.9	RL/CC
Colby	KWGB	97.9	CW	Hays	KZAN	91.7	CLA
Colby	KQLS	100.3	AC/H	Hays	KHAZ	99.5	CW
Columbus	KJML	105.3	RK/C	Hays	KJLS	103.3	AC/H
Concordia	KNCK	1390	AC/S	Haysville	KWSJ	105.3	AC/H
Concordia	KVCO	88.3	VAR	Herington	KJRL	105.7	RL/CC
Concordia	KCKS	94.9	AC/H	Hiawatha	KNZA	103.9	CW
Copeland	KSKZ	98.1	AC/H	Hill City	KZNA	90.5	CLA
Copeland	KJIL	99.1	RL/CC	Hill City	KKQY	101.9	CW
Copeland	KHYM	103.9	RL/G	Hoisington	KHOK	100.7	CW
Dearing	KUSN	98.1	CW	Holcomb	KBUF	1030	TLK+
Derby	KZCH	96.3	CH	Horton	KAIR	93.7	CW
Dodge City	KGNO	1370	TLK+	Hugoton	KFXX	106.7	SP/MEX
Dodge City	KDCC	1550	SP/N	Humboldt	KINZ	94.3	CH
Dodge City	KONQ	91.9	VAR	Hutchinson	KWBW	1450	N/TLK
Dodge City	KZRD	93.9	RK/C	Hutchinson	KHCC	90.1	VAR
Dodge City	KOLS	95.5	AC	Hutchinson	KSKU	97.1	CH
Downs	KDNS	94.1	CW	Hutchinson	KZSN	102.1	CW
El Dorado	KAHS	1360	SP/MEX	Hutchinson	KHUT	102.9	CW
El Dorado	KBTL	88.1	VAR	Independence	KIND	1010	AS
El Dorado	KTLI	99.1	RL/CC	Independence	KBQC	88.5	RL
Emporia	KVOE	1400	TLK+	Independence	KARF	91.9	RL/CC
Emporia	KANH	89.7	VAR	Independence	KIND	101.7	AC
Emporia	KPOR	90.7	RL	Ingalls	KSSH	96.3	OLG

City	Station	Freq	Format	City	Station	Freq	Format
Ingalls	KSSA	105.9	SP/MEX	Olathe	KCCV	92.3	RL
Iola	KALN	1370	OLG	Olsburg	KANV	91.3	CLA
Iola	KIKS	99.3	AC	Osage City	KKYD	92.9	CH
Junction City	KJCK	1420	CW	Ottawa	KOFO	1220	CW
Junction City	KJCK	94.5	CH	Ottawa	KTJO	88.9	CH
Kansas City	KKHK	1250	SP/MEX	Ottawa	KRBW	90.5	RL/CC
Kansas City	KCKN	1340	TLK+	Ottawa	KCHZ	95.7	CH
Kansas City	KXTR	1660	CLA	Overland Park	KCCV	760	RL
Kansas City	KFKF	94.1	CW	Parsons	KLKC	1540	AC
Kansas City	KUDL	98.1	AC/S	Parsons	KLKC	93.5	AC
Kingman	KCVW	94.3	RL	Perry	KRRB	89.7	N/A
Kingman	KTCM	100.3	SP/MEX	Phillipsburg	KKAN	1490	AC
Larned	KNNS	1510	SPTS	Phillipsburg	KQMA	92.5	AC
Larned	KGTR	96.7	OLG	Phillipsburg	KKAN	95.3	N/A
Larned	KBGL	106.9	OLG	Pittsburg	KKOW	860	TLK+
Lawrence	KLWN	1320	TLK+	Pittsburg	KSEK	1340	SPTS
Lawrence	KJHK	90.7	JAZ	Pittsburg	KRPS	89.9	VAR
Lawrence	KANU	91.5	CLA	Pittsburg	KKOW	96.9	CW
Lawrence	KLZR	105.9	AC/H	Plainville	KFIX	96.9	RK/C
Leavenworth	KKLO	1410	RL/G	Pratt	KWLS	1290	OLG
Leavenworth	KQRC	98.9	RK	Pratt	KDGB	93.1	AC
Leoti	KWKR	98.9	RK/C	Riley	KACZ	96.3	CH
Liberal	KSCB	1270	TLK+	Rozel	KKCZ	102.5	N/A
Liberal	KYUU	1470	SP/MEX	Russell	KRSL	990	AC
Liberal	KSLS	101.5	AC/S	Russell	KCAY	95.9	AC
Liberal	KLDG	102.7	CW	Salina	KFRM	550	FRM
Liberal	KZQD	105.1	SP/RL	Salina	KINA	910	SP/MEX
Liberal	KSCB	107.5	AC/H	Salina	KSAL	1150	N/TLK
Lindsborg	KQNS	95.5	AC	Salina	KHCD	89.5	CLA
Lyons	KXKU	106.1	CW	Salina	KCVS	91.7	RL
Manhattan	KMAN	1350	N/TLK	Salina	KYEZ	93.7	CW
Manhattan	KSDB	91.9	AD/A	Salina	KSKG	99.9	CW
Manhattan	KMKF	101.5	RK	Salina	KSAL	104.9	CH
Manhattan	KXBZ	104.7	CW	Scott City	KFLA	1310	TLK+
Marysville	KNDY	1570	CW	Scott City	KSKL	94.5	OLG
Marysville	KNDY	95.5	CW	Seneca	KMZA	92.1	CW
McPherson	KNGL	1540	OLG	Silver Lake	KCVT	92.5	RL
McPherson	KBBE	96.7	AC	St. Marys	KQTP	102.9	CW
Medicine Ldge	KSNS	91.5	RL/CC	Sterling	KGGG	94.7	OLG
Medicine Ldge	KREJ	101.7	RL	Topeka	WIBW	580	TLK+
Minneapolis	KILS	92.7	RK/C	Topeka	KMAJ	1440	TLK+
Mission	KCZZ	1480	SP/MEX	Topeka	KTOP	1490	AS
Newton	KJRG	950	RL	Topeka	KJTY	88.1	RL
Newton	KMXW	92.3	AC/H	Topeka	KBUZ	90.3	RL
No Fort Riley	KBLS	102.5	AC	Topeka	WIBW	94.5	CW
North Newton	KBCU	88.1	VAR	Topeka	KWIC	99.3	CH
Norton	KQNK	1530	CH	Topeka	KDVV	100.3	RK
Norton	KQNK	106.7	CH	Topeka	KTPK	106.9	CW
Oberlin	KFNF	101.1	CW	Topeka	KMAJ	107.7	AC
Ogden	KQLA	103.5	AC/H	Ulysses	KULY	1420	CW

City	Station	Freq	Format		City	Station	Freq	Format
Warnego	KHCA	95.3	RL/CC					
Wellington	KLEY	1130	SPTS					
Wellington	KWME	93.5	OLG					
Wichita	KSGL	900	AS					
Wichita	KFDI	1070	CW					
Wichita	KNSS	1240	TLK+					
Wichita	KFH	1330	N/TLK					
Wichita	KQAM	1480	YTH					
Wichita	KQAM	1480	YTH					
Wichita	KYFW	88.3	RL					
Wichita	KMUW	89.1	VAR					
Wichita	KYWA	90.7	RL/CC					
Wichita	KCFN	91.1	RL					
Wichita	KICT	95.1	RK/C					
Wichita	KRBB	97.9	AC/S					
Wichita	KFDI	101.3	CW					
Wichita	KEYN	103.7	OLG					
Wichita	KKRD	107.3	RK/C					
Winfield	KKLE	1550	SPTS					
Winfield	KBDD	91.9	RL					
Winfield	KSOK	95.9	CW					
Winfield	KSWC	100.3	RK/C					
Winfield	KSJM	107.9	URB					

City	Station	Freq	Format	City	Station	Freq	Format
Albany	WANY	1390	AC	Columbia	WAIN	1270	OLG
Albany	WANY	106.3	AC	Columbia	WAIN	93.5	CW
Allen	WMDJ	100.1	CW	Corbin	WCTT	680	AS
Ashland	WCMI	1340	SPTS	Corbin	WCTT	680	AS
Ashland	WDGG	93.7	CW	Corbin	WKDP	1330	TLK+
Auburn	WAYD	88.1	N/A	Corbin	WEKF	88.5	CLA
Auburn	WBVR	96.7	CW	Corbin	WKDP	99.5	CW
Barbourville	WYWY	950	RL/G	Corbin	WCTT	107.3	AC
Barbourville	WKKQ	96.1	AC/H	Covington	WCVG	1320	RL/G
Bardstown	WBRT	1320	CW	Cumberland	WCPM	1280	CW
Beattyville	WLJC	102.1	RL	Cumberland	WVEK	102.7	CW
Beaver Dam	WAIA	1600	SPTS	Cynthiana	WCYN	1400	CW
Benton	WCBL	1290	CW	Cynthiana	WCYN	102.3	CW
Benton	WTRT	88.1	RL/CC	Danville	WHIR	1230	TLK+
Benton	WAAJ	89.7	RL/CC	Danville	WDFB	88.1	RL/G
Benton	WVHM	90.5	RL/G	Danville	WHIR	107.1	RL/CC
Benton	WCBL	99.1	OLG	Drakesboro	WNTC	103.9	TLK+
Berea	WKXO	1500	TLK+	Eddyville	WWLK	900	RL/SG
Berea	WLFX	106.7	CH	Edmonton	WKNK	99.1	CW
Bowling Grn	WKCT	930	TLK+	Elizabethtown	WIEL	1400	SPTS
Bowling Grn	WBGN	1340	SPTS	Elizabethtown	WKUE	90.9	CLA
Bowling Grn	WKYU	88.9	CLA	Elizabethtown	WQXE	98.3	AC/H
Bowling Grn	WCVK	90.7	RL/CC	Elkhorn City	WEKB	1460	OLG
Bowling Grn	WWHR	91.7	RK/C	Elkton	WEKT	1070	RL/G
Bowling Grn	WDNS	93.3	RK/C	Eminence	WKXF	1600	SPTS
Bowling Grn	WGGC	95.1	CW	Eminence	WXLM	105.7	SP/MEX
Brandenburg	WMMG	1140	CW	Erlanger	WIZF	100.9	URB
Brandenburg	WMMG	93.5	CW	Falmouth	WIOK	107.5	RL/G
Brownsville	WKLX	100.7	AC/H	Flemingsburg	WFLE	1060	CW
Buffalo	WXAM	1430	CW	Flemingsburg	WFLE	95.1	CW
Burkesville	WKYR	107.9	CW	Florence	WBOB	1160	TLK+
Burnside	WKEQ	910	SPTS	Ft Campbell	WKFN	1370	SPTS
Burnside	WLLK	93.9	AC	Ft Campbell	WCVQ	107.9	AC/H
Cadiz	WKDZ	1110	TLK+	Fort Knox	WLVK	105.5	CW
Cadiz	WKDZ	106.5	CW	Frankfort	WFKY	1490	OLG
Calvert City	WCCK	95.7	CW	Frankfort	WKED	103.7	AC
Campbellsville	WTCO	1450	SPTS	Frankfort	WKYW	104.9	CW
Campbellsville	WAPD	91.7	RL	Franklin	WFKN	1220	CW
Campbellsville	WCKQ	104.1	AC/H	Ft. Thomas	WAQZ	97.3	RK/C
Campton	WCBJ	103.7	RK/C	Fulton	WFUL	1270	CW
Cannonsburg	WOKT	1040	SPTS	Fulton	WWKF	99.3	CH
Carlisle	WBVX	92.1	CH	Garrison	WOKE	98.3	RL/G
Carrollton	WIKI	95.3	CW	Georgetown	WXRA	1580	SP/N
Catlettsburg	WRVC	92.7	RK/C	Georgetown	WKVO	89.9	RL/CC
Cave City	WPTQ	103.7	RK/C	Georgetown	WXZZ	103.3	RK/C
Central City	WNES	1050	SPTS	Glasgow	WCDS	1440	SPTS
Central City	WMTA	1380	RL/G	Glasgow	WCLU	1490	TLK+
Central City	WQXQ	101.9	AC/H	Glasgow	WSGP	88.3	RL/G
Clinton	WLLE	102.1	CW	Glasgow	WLYE	94.1	CW
Coal Run	WPKE	103.1	AC	Glasgow	WOVO	105.3	OLG

Kentucky

City	Station	Freq	Format	City	Station	Freq	Format
Grayson	WGOH	1370	CW	Lancaster	WKYY	1280	RL/G
Grayson	WUGO	102.3	AC/S	Lancaster	WRNZ	105.1	AC/H
Greensburg	WAKY	1540	CW	Lawrenceburg	WKYL	102.1	JAZ
Greensburg	WGRK	103.1	CW	Le Rose	WOCS	88.3	VAR
Greenup	WLGC	1520	CW	Lebanon	WLBN	1590	AS
Greenup	WLGC	105.7	CW	Lebanon	WLSK	100.9	CH
Greenville	WKYA	105.5	OLG	Lebanon Jct	WTHX	107.3	AC/H
Hardinsburg	WULF	94.3	CW	Ledbetter	WHMR	90.1	RL/CC
Hardinsburg	WXBC	104.3	CW	Leitchfield	WMTL	870	CW
Harlan	WFSR	970	RL/G	Leitchfield	WKHG	104.9	AC
Harlan	WHLN	1410	OLG	Lexington	WVLK	590	TLK+
Harlan	WTUK	105.1	CW	Lexington	WLAP	630	TLK+
Harold	WXLR	104.9	CH	Lexington	WLXG	1300	SPTS
Harrodsburg	WHBN	1420	CW	Lexington	WRFL	88.1	VAR
Harrodsburg	WHBN	99.3	RL/CC	Lexington	WUKY	91.3	N/TLK
Hartford	WXMZ	106.3	RK/C	Lexington	WLXX	92.9	CW
Hawesville	WKCM	1160	CW	Lexington	WMXL	94.5	AC/H
Hazard	WKIC	1390	AS	Lexington	WBUL	98.1	CW
Hazard	WQXY	1560	OLG	Lexington	WLKT	104.5	CH
Hazard	WEKH	90.9	CLA	Liberty	WKDO	1560	CW
Hazard	WSGS	101.1	CW	Liberty	WKDO	98.7	CW
Hazard	WJMD	104.7	RL/G	London	WGWM	980	RL/G
Henderson	WSON	860	AS	London	WFTG	1400	TLK+
Henderson	WKPB	89.5	CLA	London	WYGE	90.1	RL
Henderson	WKDQ	99.5	CW	London	WWEL	103.9	CW
Henderson	WGBF	103.1	RK	Louisa	wzaq	92.3	AC/H
Highland Hts	WNKU	89.7	AD/A	Louisville	WTMT	620	SP/MEX
Hindman	WKCB	1340	RL	Louisville	WXXA	790	SPTS
Hindman	WKCB	107.1	AC	Louisville	WHAS	840	TLK+
Hodgenville	WKMO	106.3	CW	Louisville	WFIA	900	RL
Hopkinsville	WHOP	1230	TLK+	Louisville	WGTK	970	N/TLK
Hopkinsville	WHVO	1480	OLG	Louisville	WKJK	1080	TLK+
Hopkinsville	WNKJ	89.3	RL/CC	Louisville	WLLV	1240	RL/G
Hopkinsville	WZZP	97.5	RK/C	Louisville	WLOU	1350	RL/G
Hopkinsville	WHOP	98.7	AC/S	Louisville	WUOL	90.5	CLA
Hopkinsville	WVVR	100.3	CW	Louisville	WFPK	91.9	AD/A
Horse Cave	WHHT	106.7	CW	Louisville	WAMZ	97.5	CW
Hyden	WZQQ	97.9	AC/H	Louisville	WDJX	99.7	CH
Irvine	WIRV	1550	TLK+	Louisville	WTFX	100.5	RK
Irvine	WCYO	106.1	CW	Louisville	WXMA	102.3	AC/H
Jackson	WEKG	810	RL/G	Louisville	WPTI	103.9	CW
Jackson	WJSN	106.5	CW	Louisville	WVEZ	106.9	AC
Jamestown	WJKY	1060	CW	Madisonville	WFMW	730	CW
Jamestown	WJRS	104.9	CW	Madisonville	WTTL	1310	SPTS
Jeffersontown	WMJM	101.3	URB	Madisonville	WSOF	89.9	RL/G
Jenkins	WKVG	1000	RL/G	Madisonville	WKTG	93.9	RK/C
Jenkins	WIFX	94.3	RK/C	Madisonville	WYMV	106.9	AC
Junction City	WDFB	1170	RL/G	Manchester	WKLB	1250	CW
Keavy	WVCT	91.5	RL/G	Manchester	WWXL	1450	OLG
Keene	WJMM	99.1	RL/CC	Manchester	WWLT	103.1	RL/CC

Radio on the Road   Kentucky

City	Station	Freq	Format	City	Station	Freq	Format
Manchester	WTBK	105.7	RK/C	Owensboro	WSTO	96.1	CH
Mannsville	WVLC	99.9	CW	Owingsville	WKCA	107.1	CW
Marion	WMJL	1500	OLG	Paducah	WKYX	570	TLK+
Marion	WMJL	102.7	OLG	Paducah	WDXR	1450	URB
Mayfield	WNGO	1340	TLK+	Paducah	WPAD	1560	SPTS
Mayfield	WYMC	1430	AS	Paducah	WGCF	89.3	RL/CC
Mayfield	WQQR	94.7	RK/C	Paducah	WKYQ	93.3	CW
Maysville	WFTM	1240	AS	Paducah	WDDJ	96.9	CH
Maysville	WFTM	95.9	AC/S	Paintsville	WKYH	600	TLK+
Mcdaniels	WBFI	91.5	RL/CC	Paintsville	WSIP	1490	OLG
Mckee	WWAG	107.9	CW	Paintsville	WKLW	94.7	AC
Middlesboro	WMIK	560	RL/G	Paintsville	WSIP	98.9	CW
Middlesboro	WFXY	1490	AC	Paris	WYGH	1440	SP/RL
Middlesboro	WMIK	92.7	RL	Paris	WPTJ	90.7	RL/G
Midway	WBTF	107.9	URB	Paris	WGKS	96.9	AC
Monticello	WFLW	1360	RL/G	Philpot	WBIO	94.7	CW
Monticello	WMKZ	93.1	CW	Pikeville	WLSI	900	CW
Monticello	WKYM	101.7	RK/C	Pikeville	WPKE	1240	OLG
Morehead	WMOR	1330	CW	Pikeville	WJSO	90.1	RL
Morehead	WBMK	88.5	RL	Pikeville	WDHR	93.1	CW
Morehead	WMKY	90.3	VAR	Pineville	WANO	1230	OLG
Morehead	WIVY	96.3	AC/S	Pineville	WRIL	106.3	CW
Morehead	WQXX	106.1	AC/H	Pippa Passes	WWJD	91.7	RL/CC
Morganfield	WMSK	1550	CW	Prestonburg	WPRT	960	OLG
Morganfield	WMSK	95.3	CW	Prestonburg	WDOC	1310	RL/G
Morgantown	WLBQ	1570	CW	Prestonburg	WQHY	95.5	AC/H
Mount Sterling	WMST	1150	AC/S	Prestonburg	WXKZ	105.3	OLG
Mount Sterling	WAXG	88.1	RL	Princeton	WPKY	1580	SPTS
Mount Sterling	WMKJ	105.5	OLG	Princeton	WAVJ	104.9	AC/S
Mount Vernon	WRVK	1460	CW	Providence	WWKY	97.7	OLG
Mt. Washi'gton	WLCR	1040	RL	Radcliff	WASE	103.5	OLG
Munfordville	WLOC	1150	CW	Reidland	WZZL	96.7	RK/C
Munfordville	WCLU	102.3	AC/S	Richmond	WCBR	1110	RL/G
Murray	WRKY	1130	SPTS	Richmond	WEKY	1340	SP/MEX
Murray	WNBS	1340	N/TLK	Richmond	WEKU	88.9	CLA
Murray	WKMS	91.3	VAR	Richmond	WLRO	101.5	RL/CC
Murray	WFGE	103.7	CW	Russell Sprgs	WIDS	570	RL/G
Neon	WEZC	1480	RL/G	Russell Sprgs	WHVE	92.7	AC
Newburg	WDRD	680	YTH	Russellville	WRUS	610	N/TLK
Newport	WNOP	740	RL	Russellville	WUBT	101.1	URB
Nicholasville	WCGW	770	RL/G	Salyersville	WRLV	1140	CW
Nicholasville	WWFT	1250	RL	Salyersville	WRLV	97.3	CW
Nicholasville	WLTO	102.5	CH	Scottsville	WLCK	1250	RL/G
Oak Grove	WEGI	94.3	CH	Scottsville	WVLE	99.3	CW
Okolona	WJIE	88.5	RL/CC	Shelbyville	WCND	940	OLG
Owensboro	WVJS	1420	AC/S	Shelbyville	WJZO	101.7	JAZ
Owensboro	WOMI	1490	TLK+	Shepherdsville	WLRS	105.1	RK/C
Owensboro	WKWC	90.3	CW	Smiths Grove	WUHU	107.1	AC/H
Owensboro	WJVK	91.7	RL/CC	Somerset	WKVY	88.1	N/A
Owensboro	WBKR	92.5	CW	Somerset	WDCL	89.7	CLA

85

Kentucky

City	Station	Freq	Format		City	Station	Freq	Format
Somerset	WSFC	1240	TLK+					
Somerset	WTLO	1480	AS					
Somerset	WTHL	90.5	RL/G					
Somerset	WSEK	97.1	CW					
Somerset	WHMJ	102.3	RK/C					
Springfield	WAKY	102.7	CW					
St. Mathews	WRKA	103.1	OLG					
Stamping Grd	WULV	96.1	TLK+					
Stanford	WRSL	1520	SILNT					
Stanford	WXKY	96.3	RL/CC					
Stanton	WBFC	1470	RL/G					
Stanton	WSKV	104.9	CW					
Sturgis	WEZG	101.3	N/A					
Tompkinsville	WTKY	1370	CW					
Tompkinsville	WTKY	92.1	CW					
Tompkinsville	WKWY	102.7	CW					
Upton	WJCR	90.1	RL/G					
Valley Station	WRVI	105.9	RL/CC					
Vanceburg	WKKS	1570	CW					
Vanceburg	WKKS	104.9	CW					
Vancleve	WMTC	730	RL					
Vancleve	WMTC	99.9	RL/G					
Versailles	WCDA	106.3	AC/H					
Vine Grove	WRZI	101.5	RK/C					
Virgie	WZLK	107.5	CW					
West Liberty	WLKS	1450	OLG					
West Liberty	WLKS	102.9	CW					
Whitesburg	WTCW	920	OLG					
Whitesburg	WMMT	88.7	VAR					
Whitesburg	WXKQ	103.9	CW					
Whitesville	WXCM	97.1	RK/C					
Whitley City	WHAY	105.9	CW					
Wickliffe	WBCE	1200	RL					
Wickliffe	WGKY	95.9	CW					
Williamsburg	WEKC	710	RL/G					
Williamsburg	WEZJ	1440	TLK+					
Williamsburg	WEZJ	104.3	CW					
Williamstown	WNKR	106.5	CW					
Wilmore	WVRB	95.3	RL/CC					
Winchester	WMJR	1380	RL					
Winchester	WKQQ	100.1	RK					

Radio on the Road   Louisiana

City	Station	Freq	Format	City	Station	Freq	Format
Abbeville	KROF	960	N/A	Brusly	KRVE	96.1	AC
Abbeville	KPEL	105.1	N/TLK	Bunkie	KAHJ	89.5	N/A
Alexandria	KJMJ	580	RL	Bunkie	KEZP	104.3	OLG
Alexandria	KSYL	970	TLK+	Buras	KMRL	92.7	RL/CC
Alexandria	KDBS	1410	SPTS	Clinton	WQCK	92.7	RL/CC
Alexandria	KLXA	89.9	RL/CC	Columbia	KQLQ	103.1	CH
Alexandria	KLSA	90.7	VAR	Coushatta	KRRP	950	EZL
Alexandria	KAPM	91.7	RL	Coushatta	KSBH	94.9	CW
Alexandria	KQID	93.1	CH	Covington	WASO	730	TLK+
Alexandria	KFAD	93.9	AC/S	Crowley	KSIG	1450	OLG
Alexandria	KZMZ	96.9	RK/C	Crowley	KAJN	102.9	RL/CC
Alexandria	KRRV	100.3	CW	De Quincy	KTSR	92.1	AC/H
Alexandria	KEDG	106.9	URB	De Ridder	KDLA	1010	RL/G
Amite	WABL	1570	VAR	De Ridder	KBAN	91.5	RL/G
Amite	WTGG	96.5	OLG	De Ridder	KOLK	97.9	CH
Angola	KLSP	91.7	VAR	Delhi	KGGM	93.5	TLK+
Arcadia	KHCL	92.5	RL	Delhi	KGGM	93.5	RL/G
Atlanta	KCIG	106.5	CH	Denham Spr	WSKR	1210	SPTS
Baker	WTGE	107.3	CW	Donaldsonville	KKAY	104.9	RK/C
Ball	KWDF	840	RL/CC	Dry Prong	KVDP	89.1	RL
Ball	KBKK	105.5	CW	Dubach	KPCH	97.7	OLG
Basile	KQIS	102.1	AC/H	Empire	KNOU	104.5	URB
Bastrop	KAVX	91.9	RL	Erath	KRKA	107.9	CH
Bastrop	KTRY	93.9	URB	Erwinville	KPAE	91.5	RL
Bastrop	KJMG	97.3	URB	Eunice	KEUN	1490	N/A
Bastrop	KRVV	100.1	URB	Eunice	KEUN	105.5	CW
Baton Rouge	WNDC	910	RL	Farmerville	KWJM	92.7	AC
Baton Rouge	WJBO	1150	N/TLK	Ferriday	KFNV	107.1	AC
Baton Rouge	KBRH	1260	URB	Folsom	WJSH	104.7	CW
Baton Rouge	WIBR	1300	N/TLK	Franklin	KFRA	1390	AS
Baton Rouge	WYNK	1380	TLK+	Franklin	KDDK	105.5	AS
Baton Rouge	WXOK	1460	RL/G	Franklinton	WOMN	1110	CW
Baton Rouge	WJFM	88.5	RL/G	Franklinton	WUUU	98.9	CW
Baton Rouge	WRKF	89.3	CLA	Galliano	WTIX	94.3	OLG
Baton Rouge	WRBH	90.3	JAZ	Garyville	WCKW	1010	RL/G
Baton Rouge	KLSU	91.1	AD/A	Gibsland	KBEF	104.5	RL/CC
Baton Rouge	WDGL	98.1	RK/C	Gold Meadow	KLEB	1600	N/A
Baton Rouge	WXCT	100.7	CW	Grambling	KGRM	91.5	RL/G
Baton Rouge	WYNK	101.5	CW	Gretna	KKNO	750	RL/G
Baton Rouge	WFMF	102.5	CH	Gretna	KGLA	1540	SP/OM
Bayou Vista	KQKI	95.3	URB	Hammond	WFPR	1400	CW
Belle Chase	KMEZ	102.9	URB	Hammond	KSLU	90.9	AD/A
Benton	KSYR	92.1	AC	Hammond	WBBE	103.3	AC
Berwick	KBZE	105.9	URB	Hammond	WHMD	107.1	CW
Blanchard	KDKS	102.1	URB	Haughton	KBTT	103.7	URB
Bogalusa	WBOX	920	CW	Homer	KYLA	106.7	CW
Bogalusa	WIKC	1490	RL/G	Houma	KJIN	1490	SPTS
Bossier City	KBCL	1070	RL/CC	Houma	KSTE	104.1	CH
Boyce	KBCE	102.3	URB	Houma	KCIL	107.5	CW
Breaux Bridge	KFTE	96.5	RK/C	Jackson	WNXX	104.5	RK/C

City	Station	Freq	Format	City	Station	Freq	Format
Jena	KJNA	102.7	CW	Monroe	KBMQ	88.7	RL/CC
Jennings	KJEF	1290	CW	Monroe	KYFL	89.5	RL
Jennings	KJEF	92.9	OLG	Monroe	KEDM	90.3	VAR
Jonesboro	KTOC	104.9	RL	Monroe	KXUL	91.1	RK/C
Jonesville	KTGV	105.1	URB	Monroe	KNOE	101.9	CH
Jrna	KAYT	88.1	RL/G	Monroe	KJLO	104.1	CW
Kaplan	KMDL	97.3	CW	Monroe	KLIP	105.3	CH
Kenner	WKBU	105.3	RK/C	Monroe	KXRR	106.1	RK/C
Kentwood	WEMX	94.1	URB	Moreauville	KLIL	92.1	OLG
La Place	WDVW	92.3	CH	Morgan City	KMRC	1430	N/A
Lacombe	WOPR	94.7	RL/G	Morgan City	KBZZ	96.7	AC
Lafayette	KJCB	770	URB	Natchitoches	KNOC	1450	N/TLK
Lafayette	KVOL	1330	SPTS	Natchitoches	KBIO	89.3	RL
Lafayette	KPEL	1420	SPTS	Natchitoches	KNWD	91.7	VAR
Lafayette	KFXZ	1520	RL/G	Natchitoches	KDBH	97.7	AC/S
Lafayette	KRVS	88.7	VAR	Natchitoches	KZBL	100.7	OLG
Lafayette	KIKL	90.9	RL/CC	New Iberia	KANE	1240	N/A
Lafayette	KSMB	94.5	CH	New Iberia	KNIR	1360	RL
Lafayette	KRRQ	95.5	URB	New Iberia	KRDJ	93.7	CH
Lafayette	KTDY	99.9	AC	New Iberia	KXKC	99.1	CW
Lake Arthur	KJMH	107.5	URB	New Orleans	WVOG	600	RL
Lake Charles	KAOK	1400	TLK+	New Orleans	WTIX	690	TLK+
Lake Charles	KLCL	1470	CW	New Orleans	WSHO	800	RL
Lake Charles	KXZZ	1580	RL/G	New Orleans	WWL	870	TLK+
Lake Charles	KYCL	90.3	RL	New Orleans	WYLD	940	RL/G
Lake Charles	KOJO	91.1	RL	New Orleans	WGSO	990	BIZ
Lake Charles	KYKZ	96.1	CW	New Orleans	WLNO	1060	RL/CC
Lake Charles	KBXG	99.5	CW	New Orleans	WBOK	1230	RL/G
Lake Charles	KBIU	103.3	AC/H	New Orleans	WODT	1280	SPTS
Lake Charles	KZWA	105.3	UR/RB	New Orleans	WSMB	1350	TLK+
Lk Providence	KLPL	1050	SIL	New Orleans	WBYU	1450	YTH
Lk Providence	KLPL	92.7	UR/RB	New Orleans	WRBH	88.3	READ
Larose	KLRZ	100.3	N/A	New Orleans	WBSN	89.1	RL/CC
Leesville	KLLA	1570	OLG	New Orleans	WWNO	89.9	CLA
Leesville	KJAE	92.7	CW	New Orleans	WWOZ	90.7	JAZ
Leesville	KVVP	105.7	CW	New Orleans	WTUL	91.5	VAR
Mamou	KBON	101.1	VAR	New Orleans	WQUE	93.3	URB
Mansfield	KJVC	92.7	RL/G	New Orleans	WTKL	95.7	OLG
Mansfield	KORI	104.7	CW	New Orleans	WEZB	97.1	CH
Many	KWLA	1400	TLK+	New Orleans	WRNO	99.5	URB
Many	KAVK	89.7	RL	New Orleans	WRNO	99.5	RK/C
Many	KWLV	107.1	CW	New Orleans	WNOE	101.1	CW
Marksville	KAPB	97.7	CW	New Roads	KCLF	1500	URB
Maurice	KKSJ	106.3	JAZ	New Roads	KQXL	106.5	URB
Minden	KASO	1240	AS	Norco	WFNO	830	SP/OM
Minden	KLKL	95.7	OLG	Norco	WWLA	91.1	N/A
Monroe	KNOE	540	N/TLK	North Fort Polk	KUMX	106.7	URB
Monroe	KLIC	1230	RL/G	Oak Grove	KWCL	96.7	OLG
Monroe	KMLB	1440	TLK+	Oakdale	KKST	98.7	AC
Monroe	KRJO	1680	RL/G	Oil City	KQHN	107.9	SIL

Radio on the Road Louisiana

City	Station	Freq	Format		City	Station	Freq	Format
Opelousas	KSLO	1230	CW		Vivian	KNCB	105.3	CW
Opelousas	KTSJ	105.9	JAZ		Washington	KNEK	1190	JAZ
Opelousas	KOGM	107.1	OLG		Washington	KNEK	104.7	URB
Pineville	KTLD	1110	RL		West Monroe	KMBS	1310	SPTS
Port Allen	WPFC	1550	RL/G		West Monroe	KZRZ	98.3	AC
Port Sulphur	KAGY	1510	N/A		White Castle	KKAY	1590	OLG
Port Sulphur	KKND	106.7	RK/C		Winnfield	KVCL	1270	CW
Rayne	KBEB	106.7	AC/S		Winnfield	KVCL	92.1	CW
Rayville	KMYY	92.3	CW		Winnsboro	KMAR	95.9	CW
Reserve	WPRF	94.9	RL/G		Zwolle	KTEZ	99.9	AC
Richwood	KHLL	100.9	RL/CC					
Ruston	KRUS	1490	RL/G					
Ruston	KAPI	88.3	RL					
Ruston	KLPI	89.1	AD/A					
Ruston	KNBB	99.3	AC/H					
Ruston	KXKZ	107.5	CW					
Shreveport	KEEL	710	N/TLK					
Shreveport	KOKA	980	RL/G					
Shreveport	KWKH	1130	CW					
Shreveport	KSYB	1300	RL					
Shreveport	KRMD	1340	TLK+					
Shreveport	KIOU	1480	RL/G					
Shreveport	KDAQ	89.9	VAR					
Shreveport	KSCL	91.3	AD/A					
Shreveport	KXKS	93.7	CW					
Shreveport	KRUF	94.5	CH					
Shreveport	KVKI	96.5	AC					
Shreveport	KMJJ	99.7	URB					
Shreveport	KRMD	101.1	CW					
Shreveport	KBED	102.9	URB					
Slidell	WSLA	1560	SPTS					
So. Fort Polk	KROK	95.7	AD/A					
Springhill	KBSF	1460	URB					
Springhill	KTKC	92.7	RL/G					
St. Martinville	KSJY	89.9	RL					
Sulphur	KEZM	1310	SPTS					
Sulphur	KKGB	101.3	RK					
Tallulah	KBYO	1360	SIL					
Tallulah	KBYO	1360	N/A					
Tallulah	KBYO	104.5	CH					
Thibodaux	KTIB	640	TLK+					
Thibodaux	KTLN	90.5	CLA					
Thibodaux	KNSU	91.5	AD/A					
Thibodaux	KXOR	106.3	RK/C					
Tioga	KLAA	103.5	CW					
Varnado	WBOX	92.9	CW					
Vidalia	KVLA	1400	SILNT					
Ville Platte	KVPI	1050	OLG					
Ville Platte	KVPI	92.5	OLG					
Vivian	KNCB	1320	CW					

City	Station	Freq	Format	City	Station	Freq	Format
Auburn	WTHT	99.9	CW	Hermon	WNZT	1230	N/A
Augusta	WMDR	1340	RL	Houlton	WHOU	100.1	AC
Augusta	WJZN	1400	AS	Howland	WVOM	103.9	TLK+
Augusta	WMME	92.3	CH	Islesboro	WBYA	105.5	CH
Augusta	WKCG	101.3	AC/S	Kennebunk	WBQQ	99.3	CLA
Bangor	WZON	620	SPTS	Kennebunkport	WHXQ	104.7	RK/C
Bangor	WABI	910	AC/S	Kittery	WSHK	105.3	CH
Bangor	WHCF	88.5	RL	Lewiston	WCNM	1240	NWS+
Bangor	WHSN	89.3	RK/C	Lewiston	WLAM	1470	OLG
Bangor	WMEH	90.9	VAR	Lewiston	WRBC	91.5	VAR
Bangor	WZEQ	92.9	AC/S	Lewiston	WCYI	93.9	RK/C
Bangor	WWBX	97.1	CH	Lewiston	WFNK	107.5	CH
Bar Harbor	WLKE	99.1	CW	Lincoln	WHMX	105.7	RL/CC
Bar Harbor	WBQI	107.7	CLA	Machias	WALZ	95.3	CH
Bath	WJTO	730	AS	Madawawaska	WCXX	102.3	AC
Bath	WBCI	105.9	RL	Madison	WIGY	97.5	SPTS
Belfast	WBFB	104.7	CW	Mexico	WTBM	100.7	CW
Biddleford	WVAE	1400	AS	Millbridge	WRMO	93.7	RL
Biddleford	WCYY	94.3	RK/C	Millinocket	WSYY	1240	SPTS
Blue Hill	WERU	89.9	VAR	Millinocket	WSYY	94.9	CW
Boothbay Harb	WCME	96.7	N/TLK	Monticello	WREM	710	TLK+
Brewer	WKIT	100.3	RK/C	No. Windham	WHXR	106.7	RK/C
Brewer	WQCB	106.5	CW	Norway	WOXO	92.7	CW
Brunswick	WJJB	900	SPTS	Oakland	WMDR	88.9	N/A
Brunswick	WBOR	91.1	VAR	Old Town	WBZN	107.3	CH
Brunswick	WCLZ	98.9	AD/A	Orono	WMEB	91.9	AD/A
Calais	WMED	89.7	VAR	Pittsfield	WJCX	99.5	RL/CC
Calais	WQDY	92.7	CH	Portland	WGAN	560	N/TLK
Camden	WMEP	90.5	VAR	Portland	WZAN	970	TLK+
Camden	WQSS	102.5	CH	Portland	WLOB	1310	TLK+
Caribou	WFST	600	RL	Portland	WBAE	1490	AS
Caribou	WCXU	97.7	AC	Portland	WMEA	90.1	VAR
Dennysville	WCRQ	102.9	AC/H	Portland	WMGX	93.1	AC/H
Dexter	WGUY	102.1	OLG	Portland	WJBQ	97.9	CH
Dover-Foxcroft	WDME	103.1	AD/A	Portland	WPOR	101.9	CW
Eastport	WSHD	91.7	VAR	Portland	WBLM	102.9	RK/C
Ellsworth	WDEA	1370	AS	Presque Isle	WEGP	1390	TLK+
Ellsworth	WKSQ	94.5	AC	Presque Isle	WUPI	92.1	AD/A
Ellsworth	WWMJ	95.7	CH	Presque Isle	WQHR	96.1	AC/H
Fairfield	WCTB	93.5	CH	Presque Isle	WBPW	96.9	CW
Farmington	WKTJ	99.3	AC	Presque Isle	WOZI	101.7	OLG
Farmington	WUMF	100.5	AD/A	Presque Isle	WMEM	106.1	VAR
Fort Kent	WUFK	92.1	AC	Rockland	WRKD	1450	SPTS
Freeport	WMSJ	89.3	RL/CC	Rockland	WMCM	103.3	CW
Ft. Kent	WMEF	106.5	VAR	Rumford	WLLB	780	RL
Gardiner	WFAU	1280	SPTS	Rumford	WLOB	96.3	TLK+
Gardiner	WABK	104.3	OLG	Saco	WRED	95.9	CH
Gorham	WLAM	870	TLK+	Sanford	WPHX	1220	SPTS
Gorham	WMPG	90.9	VAR	Sanford	WSEW	88.5	RL
Harpswell	WYFP	91.9	RL	Sanford	WPHX	92.1	RK/C

City	Station	Freq	Format	City	Station	Freq	Format
Scarborough	WBQW	106.3	CLA				
Searsport	WFZX	101.7	RK/C				
Showhegan	WSKW	1160	SPTS				
Showhegan	WTOS	105.1	RK/C				
Showhegan	WHQO	107.9	AC				
South Paris	WKTQ	1450	RL				
Standish	WSJB	91.5	VAR				
Thomaston	WBQX	106.9	CLA				
Topsham	WJJB	95.5	SPTS				
Veazie	WNZS	1340	N/TLK				
Veazie	WWNZ	1400	TLK+				
Waterville	WTVL	1490	AS				
Waterville	WMHB	89.7	VAR				
Waterville	WEBB	98.5	CW				
Westbrook	WJAE	1440	SPTS				
Westbrook	WYNZ	100.9	OLG				
Winslow	WWWA	95.3	RL/CC				
Winter Harbor	WNSX	97.7	RK/C				
Yarmouth	WYAR	88.3	AS				
York Center	WUBB	95.3	CW				

Radio on the Road   Maryland

City	Station	Freq	Format	City	Station	Freq	Format
Aberdeen	WAMD	970	OLG	Federalsburg	WTDK	107.1	OLG
Annapolis	WYRE	810	SP/TK	Frederick	WXTR	820	N/TLK
Annapolis	WBIS	1190	BIZ	Frederick	WFMD	930	N/TLK
Annapolis	WNAV	1430	TLK+	Frederick	WYPF	88.1	JAZ
Annapolis	WHFS	99.1	SP/OM	Frederick	WFRE	99.9	CW
Annapolis	WFSI	107.9	RL	Frostburg	WFRB	560	AS
Baltimore	WCAO	600	RL/G	Frostburg	WFWM	91.9	VAR
Baltimore	WCBM	680	TLK+	Frostburg	WLIC	97.1	RL/G
Baltimore	WBMD	750	RL	Frostburg	WFRB	105.3	CW
Baltimore	WBGR	860	RL	Fruitland	WKHI	107.5	AC/S
Baltimore	WOLB	1010	TLK+	Gaithersburg	WMET	1160	TLK+
Baltimore	WBAL	1090	N/TLK	Glen Burnie	WFBR	1590	N/A
Baltimore	WITH	1230	TLK+	Glen Burnie	WWIN	95.9	URB
Baltimore	WJFK	1300	SPTS	Grantsville	WAIJ	90.3	RL/CC
Baltimore	WWIN	1400	RL/G	Grasonville	WRNR	103.1	AD/A
Baltimore	WBJC	91.5	CLA	Hagerstown	WJEJ	1240	EZL
Baltimore	WERQ	92.3	URB	Hagerstown	WARK	1490	TLK+
Baltimore	WPOC	93.1	CW	Hagerstown	WETH	89.1	N/TLK
Baltimore	WRBS	95.1	RL/CC	Hagerstown	WAYZ	104.7	CW
Baltimore	WIYY	97.9	RK	Hagerstown	WWEG	106.9	CH
Baltimore	WLIF	101.9	AC/S	Halfway	WHAG	1410	AS
Baltimore	WQSR	102.7	AH	Halfway	WDLD	96.7	CH
Baltimore	WSMJ	104.3	JAZ	Havre/Grace	WASA	1330	RL
Baltimore	WWMX	106.5	AC/H	Havre/Grace	WXCY	103.7	CW
Bel Air	WHFC	91.1	AD/A	Hurlock	WAAI	100.9	CW
Berlin	WOCQ	103.9	CH	Indian Head	WWGB	1030	SP/RL
Bethesda	WTNT	570	TLK+	La Plata	WKIK	1560	CW
Bethesda	WARW	94.7	RK/C	Laurel	WILC	900	SP/OM
Bethesda	WMMJ	102.3	URB	Lexington Park	WPTX	1690	TLK+
Braddock Hts	WWVZ	103.9	AC/H	Lexington Park	WMDM	1690	TLK/S
Brunswick	WTRI	1520	AS	Lexington Park	WMDM	97.7	OLG
California	WKIK	102.9	CW	Mechanicsville	WSMD	98.3	CH
Cambridge	WCEM	1240	SPTS	Middletown	WAFY	103.1	AC
Cambridge	WINX	94.3	CW	Morningside	WPGC	1580	RL/G
Cambridge	WCEM	106.3	AC/H	Morningside	WPGC	95.5	CH
Catonsville	WHFS	105.7	TLK+	Mtn Lake Park	WKHJ	104.5	AC
Chestertown	WCTR	1530	AC/S	Oakland	WMSG	1050	AS
College Park	WMUC	88.1	AD/A	Oakland	WWHC	92.3	CW
Crisfield	WBEY	97.9	CW	Ocean City	WKHZ	1590	CH
Cumberland	WCMD	1230	AC/S	Ocean City	WWFG	99.9	CW
Cumberland	WCBC	1270	N/TLK	Ocean City	WQHQ	104.7	AC
Cumberland	WTBO	1450	TLK+	Ocean City	WRXS	106.9	RK/C
Cumberland	WROG	102.9	CW	Ocean Pines	WQJC	97.1	JAZ
Cumberland	WKGO	106.1	AC	Pikesville	WWLG	1370	OLG
Denton	WKDI	840	RL	Pocomoke City	WGOP	540	AS
Easton	WEMD	1460	AS	Pocomoke City	WXMD	92.5	AC
Easton	WCEI	96.7	AC	Pocomoke City	WKHW	106.5	CW
Elkton	WSER	1550	SPTS	Potomoc	WCTN	950	SP/OM
Elkton	WOEL	89.9	RL	Prin. Frederick	WBZS	92.7	SP/OM
Emmitsburg	WMTB	89.9	VAR	Princess Anne	WESM	91.3	JAZ

93

City	Station	Freq	Format		City	Station	Freq	Format
Rockville	WLXE	1600	N/A					
Salisbury	WTGM	960	SPTS					
Salisbury	WICO	1320	TLK+					
Salisbury	WJDY	1470	RL/G					
Salisbury	WSCL	89.5	CLA					
Salisbury	WDIH	90.3	RL/G					
Salisbury	WICO	97.5	CW					
Salisbury	WSBY	98.9	URB					
Salisbury	WDKZ	105.5	CH					
Silver Spring	WFED	1050	TLK+					
Snow Hill	WQMR	101.1	TLK+					
Takoma Park	WGTS	91.9	RL/CC					
Thurmont	WTHU	1450	AS					
Towson	WNST	1570	SPTS					
Towson	WTMD	89.7	AD/A					
Waldorf	WWZZ	104.1	AC/H					
Walkersville	WDMV	700	TLK+					
Westernport	WWPN	101.1	RL/CC					
Westminster	WTTR	1470	OLG					
Westminster	WZBA	100.7	RK/C					
Wheaton	WACA	1540	SP/TK					
Williamsport	WYPR	88.1	JAZ					
Williamsport	WEAA	88.9	JAZ					
Williamsport	WCRH	90.5	RL/CC					
Williamsport	WICL	95.9	OLG					
Worton	WKHS	90.5	AD/A					

City	Station	Freq	Format	City	Station	Freq	Format
Acton	WHAB	89.1	VAR	Dudley	WXRB	95.1	OLG
Amherst	WPNI	1430	NWS+	E Longmeadow	WHNP	1600	TLK+
Amherst	WFCR	89.1	VAR	Easton	WSHL	91.3	AD/A
Amherst	WAMH	89.3	AD/A	Everett	WXKS	1430	TLK+
Amherst	WMUA	91.1	VAR	Fairhaven	WFHN	107.1	CH
Amherst	WRNX	100.9	AD/A	Fall River	WHTB	1400	N/A
Ashland	WSRO	650	TLK+	Fall River	WSAR	1480	TLK+
Athol	WNYN	99.9	RK/C	Falmouth	WFPB	91.9	VAR
Attleboro	WARL	1320	TLK+	Falmouth	WCIB	101.9	CH
Barnstable	WQRC	99.9	AC	Fitchburg	WFGL	960	RL
Beverly	WNSH	1570	TLK+	Fitchburg	WEIM	1280	TLK+
Boston	WEZE	590	RL	Fitchburg	WXPL	91.3	AD/A
Boston	WRKO	680	TLK+	Fitchburg	WXLO	104.5	AC/H
Boston	WEEI	850	SPTS	Framingham	WKOX	1200	TLK+
Boston	WROL	950	RL	Framingham	WDJM	91.3	AD/A
Boston	WBZ	1030	N/TLK	Framingham	WROR	105.7	CH
Boston	WILD	1090	URB	Franklin	WGAO	88.3	RK/C
Boston	WTTT	1150	TLK+	Gardner	WGAW	1340	N/TLK
Boston	WMKI	1260	YTH	Gardner	WJWT	91.7	N/A
Boston	WWZN	1510	SPTS	Gloucester	WBOQ	104.9	OLG
Boston	WERS	88.9	VAR	Grt Barrington	WSBS	860	AC
Boston	WGBH	89.7	VAR	Grt Barrington	WAMQ	105.1	N/TLK
Boston	WBUR	90.9	N/TLK	Greenfield	WHMQ	1240	TLK+
Boston	WUMB	91.9	VAR	Greenfield	WIZZ	1520	AS
Boston	WJMN	94.5	CH	Greenfield	WPVQ	95.3	CW
Boston	WTKK	96.9	TLK+	Greenfield	WHAI	98.3	AC
Boston	WBMX	98.5	AC/H	Harwich	WCCT	90.3	N/TLK
Boston	WZLX	100.7	RK/C	Harwich Port	WDVT	93.5	OLG
Boston	WODS	103.3	OLG	Haverhill	WCCM	1490	TLK+
Boston	WBCN	104.1	TLK+	Haverhill	WXRV	92.5	AD/A
Boston	WRBB	104.9	VAR	Holliston	WHHB	99.9	VAR
Boston	WMJX	106.7	AC	Holyoke	WCCH	103.5	VAR
Boxford	WBMT	88.3	RK/C	Hyannis	WPXC	102.9	RK
Brewster	WZAI	90.1	N/A	Hyannis	WCOD	106.1	AC
Bridgewater	WBIM	91.5	VAR	Lawrence	WNNW	800	SP/OM
Brockton	WMSX	1410	SP/OM	Lawrence	WMKK	93.7	AH
Brockton	WBET	1460	N/TLK	Leicester	WVNE	760	RL
Brockton	WBOT	97.7	URB	Leominster	WCMX	1000	RL/CC
Brookline	WUNR	1600	SP/OM	Lowell	WCAP	980	TLK+
Brookline	WBOS	92.9	AD/A	Lowell	WLLH	1400	SP/OM
Cambridge	WJIB	740	AS	Lowell	WJUL	91.5	VAR
Cambridge	WMBR	88.1	VAR	Lowell	WKLB	99.5	CW
Cambridge	WHRB	95.3	VAR	Lynn	WLYN	1360	N/A
Charlton	WBPV	90.1	RL/CC	Lynn	WFNX	101.7	RK/C
Chatham	WFCC	107.5	CLA	Marion	WWTA	88.5	VAR
Chicopee	WACE	730	RL	Marshfield	WATD	95.9	AC
Concord	WBNW	1120	BIZ	Mashpee	WTWV	101.1	OLG
Concord	WIQH	88.3	VAR	Maynard	WAVM	91.7	VAR
Dedham	WAMG	890	SP/OM	Medford	WMFO	91.5	VAR
Deerfield	WGAJ	91.7	VAR	Medford	WXKS	107.9	CH

City	Station	Freq	Format	City	Station	Freq	Format
Middleborough	WVBF	1530	N/A	Southbridge	WESO	970	CW
Milford	WMRC	1490	AC	Southbridge	WWFX	100.1	CH
Milton	WMLN	91.5	VAR	Springfield	WHYN	560	TLK+
Nantucket	WNCK	89.5	CLA	Springfield	WSPR	1270	SP/OM
Nantucket	WNAN	91.1	N/TLK	Springfield	WMAS	1450	TLK+
Nantuckett	WRZE	96.3	CH	Springfield	WSCB	89.9	VAR
Natick	WBIX	1060	TLK+	Springfield	WTCC	90.7	VAR
New Bedford	WNBH	1340	AS	Springfield	WAIC	91.9	CH
New Bedford	WBSM	1420	N/TLK	Springfield	WHYN	93.1	AC/H
New Bedford	WFHL	88.1	N/A	Springfield	WMAS	94.7	AC
New Bedford	WJFD	97.3	N/A	Springfield	WAQY	102.1	RK/C
New Bedford	WCTK	98.1	CW	Springfield	WNEK	105.1	VAR
Newburyport	WNBP	1450	AS	Sudbury	WYAJ	97.7	VAR
Newburyport	WNEF	91.7	VAR	Taunton	WPEP	1570	TLK+
Newton	WNTN	1550	VAR	Taunton	WSNE	93.3	AC/H
Newton	WZBC	90.3	RK/C	Tisbury	WMVY	92.7	AD/A
Norfolk	WDIS	1170	TLK+	Truro	WCDJ	102.3	N/TLK
North Adams	WNAW	1230	AC	Turners Falls	WRSI	93.9	AD/A
North Adams	WJJW	91.1	AD/A	Waltham	WRCA	1330	N/A
North Adams	WMNB	100.1	OLG	Waltham	WBRS	100.1	VAR
No Dartmouth	WUMD	89.3	VAR	Waltham	WCRB	102.5	CLA
No Dartmouth	WSMU	91.1	VAR	Ware	WARE	1250	OLG
Northampton	WHMP	1400	TLK+	Watertown	WAZN	1470	N/A
Northampton	WOZQ	91.9	VAR	Webster	WGFP	940	CW
Northampton	WHMP	99.3	RK/C	Webster	WORC	98.9	OLG
Northampton	WEIB	106.3	JAZ	Wellesley	WZLY	91.5	VAR
Northfield	WNMH	91.5	VAR	Wellfleet	WWTE	90.7	N/A
Orange	WJOE	700	TLK+	W Barnstable	WKKL	90.7	VAR
Orange	WJDF	97.3	AC	W Springfield	WACM	1490	SP/OM
Orleans	WFPB	1170	VAR	W Yarmouth	WBUR	1240	N/TLK
Orleans	WKPE	104.7	RK/C	W Yarmouth	WXTK	95.1	N/TLK
Pittsfield	WUHN	1110	AS	Westborough	WAAF	107.3	RK/C
Pittsfield	WBRK	1340	AC	Westfield	WNNZ	640	SPTS
Pittsfield	WBEC	1420	TLK+	Westfield	WSKB	89.5	AD/A
Pittsfield	WTBR	89.7	VAR	Williamstown	WCFM	91.9	RK/C
Pittsfield	WUPE	95.9	OLG	Winchendon	WKMY	91.1	RL/CC
Pittsfield	WBRK	101.7	AC/H	Winchendon	WOQL	97.7	OLG
Pittsfield	WBEC	105.5	AC/H	Woods Hole	WCAI	90.1	N/TLK
Plymouth	WPLM	1390	BIZ	Worcester	WTAG	580	TLK+
Plymouth	WPLM	99.1	AC/S	Worcester	WCRN	830	OLG
Provincetown	WOMR	92.1	VAR	Worcester	WNEB	1230	RL
Quincy	WJDA	1300	TLK+	Worcester	WORC	1310	SP/OM
Rockland	WRPS	88.3	AC/H	Worcester	WWTM	1440	SPTS
Salem	WESX	1230	N/TLK	Worcester	WCHC	88.1	AD/A
Salem	WMWM	91.7	AD/A	Worcester	WICN	90.5	JAZ
Sandwich	WSDH	91.5	AD/A	Worcester	WCUW	91.3	VAR
Scituate	WSMA	90.5	N/A	Worcester	WBPR	91.9	VAR
Sheffield	WBSL	91.7	VAR	Worcester	WSRS	96.1	AC/S
Scuth Hadley	WMHC	91.5	VAR	Southbridge	WESO	970	CW
So Yarmouth	WOCN	103.9	AS	Southbridge	WWFX	100.1	CH

City	Station	Freq	Format	City	Station	Freq	Format
Ada	WDSS	1680	YTH	Berrien Spr	WAUS	90.7	CLA
Adrian	WABJ	1490	TLK+	Beulah	WOUF	92.1	AC
Adrian	WQTE	95.3	CW	Beulah	WOUF	92.1	CW
Adrian	WLEN	103.9	AC	Big Rapids	WBRN	1460	TLK+
Adrian	WVAC	107.9	VAR	Big Rapids	WBRN	100.9	RK/C
Albion	WUFN	96.7	RL/CC	Big Rapids	WYBR	102.3	AC/H
Allegan	WZUU	92.3	RK/C	Birmingham	WCSX	94.7	RK/C
Allendale	WGVU	88.5	JAZ	Bloomfield Hls	WBFH	88.1	VAR
Alma	WFYC	1280	SPTS	Boyne City	WBCM	93.5	CW
Alma	WQAC	90.9	AD/A	Bridgman	WYTZ	97.5	CW
Alma	WQBX	104.9	AC/H	Bronson	WCVM	94.7	RL/CC
Alpena	WATZ	1450	TLK+	Brooklyn	WKHM	105.3	CH
Alpena	WCML	91.7	VAR	Buchanan	WSMK	99.1	URB
Alpena	WATZ	99.3	CW	Burton	WTAC	89.7	RL/CC
Alpena	WHSB	107.7	AC/H	Cadillac	WATT	1240	TLK+
Ann Arbor	WTKA	1050	SPTS	Cadillac	WLJW	1370	RL
Ann Arbor	WAAM	1600	TLK+	Cadillac	WOLW	91.1	RL/CC
Ann Arbor	WCBN	88.3	VAR	Cadillac	WJZQ	92.9	JAZ
Ann Arbor	WUOM	91.7	N/TLK	Cadillac	WLXV	96.7	AC/H
Ann Arbor	WWWW	102.9	CW	Cadillac	WCKC	107.1	RK
Ann Arbor	WQKL	107.1	AD/A	Caro	WKYO	1360	CW
Ashley	WJSZ	92.5	CH	Caro	WIDL	92.1	AC/H
Atlanta	WFDX	92.5	CH	Carrollton	WXQL	100.5	OLG
Auburn Hts	WAHS	89.5	VAR	Cassopolis	WGTO	910	OLG
Auburn Hills	WXOU	88.3	VAR	Charlevoix	WMKT	1270	TLK+
Bad Axe	WLEW	1340	CW	Charlevoix	WTCK	90.9	N/A
Bad Axe	WLEW	102.1	AC/H	Charlevoix	WKHQ	105.9	CH
Baraga	WVCN	104.3	RL	Charlevoix	WCZW	107.9	OLG
Baraga	WVCN	104.3	RL	Charlotte	WLCM	1390	RL/CC
Battle Creek	WBCK	930	TLK+	Charlotte	WQTX	92.7	SPTS
Battle Creek	WRCC	1400	AS	Cheboygan	WCBY	1240	AS
Battle Creek	WOLY	1500	RL	Cheboygan	WGFM	105.1	RK/C
Battle Creek	WBXX	95.3	AC/S	Clare	WCFX	95.3	CH
Battle Creek	WKFR	103.3	CH	Coldwater	WTVB	1590	OLG
Bay City	WKNX	1250	POP	Coldwater	WNWN	98.5	CW
Bay City	WNEM	1250	TLK+	Coleman	WPRJ	101.7	RL/CC
Bay City	WMAX	1440	RL	Crystal Falls	WOBE	100.7	OLG
Bay City	WTRK	89.1	RL/CC	Crystal Falls	WOBE	100.7	CH
Bay City	WUCX	90.1	JAZ	De Witt	WQHH	96.5	URB
Bay City	WCHW	91.3	RK	Dearborn	WXDX	1310	TLK+
Bay City	WHNN	96.1	OLG	Dearborn	WHFR	89.3	VAR
Bay City	WIOG	102.5	CH	Dearborn	WNIC	100.3	AC
Bear Crk Twn	WTLI	89.3	RL/CC	Dearborn Hts	WNZK	690	N/A
Bear Crk Twn	WTLI	89.3	RL/CC	Detroit	WJR	760	N/TLK
Bear Lake	WCUZ	100.1	AC/H	Detroit	WWJ	950	N/TLK
Beaverton	WMRX	97.7	AS	Detroit	WDFN	1130	SPTS
Benton Harbor	WHFB	1060	TLK+	Detroit	WXYT	1270	SPTS
Benton Harbor	WCNF	94.9	AC/H	Detroit	WDTK	1400	TLK+
Benton Harbor	WCNF	94.9	AC/H	Detroit	WLQV	1500	RL
Benton Harbor	WHFB	99.9	CW	Detroit	WRCJ	90.9	VAR

Michigan

Radio on the Road

City	Station	Freq	Format	City	Station	Freq	Format
Detroit	WMXD	92.3	URB	Freeland	WWZP	90.9	N/A
Detroit	WDRQ	93.1	AH	Fremont	WSHN	1550	SPTS
Detroit	CIDR	93.9	AP	Gaylord	WSNQ	900	SILNT
Detroit	WKQI	95.5	CH	Gaylord	WBLW	88.1	RL/G
Detroit	WDVD	96.3	AC/H	Gaylord	WPHN	90.5	RL/G
Detroit	WKRK	97.1	TLK+	Gaylord	WMJZ	101.5	OLG
Detroit	WJLB	97.9	URB	Gaylord	WMJZ	101.5	OLG
Detroit	WVMV	98.7	JAZ	Gaylord	WKPK	106.7	AC/S
Detroit	WYCD	99.5	CW	Gladstone	WGKL	105.5	OLG
Detroit	WRIF	101.1	RK	Gladstone	WGKL	105.5	OLG
Detroit	WDET	101.9	VAR	Gladwin	WGDN	1350	TLK+
Detroit	WMUZ	103.5	RL/CC	Gladwin	WGDN	103.1	CW
Detroit	WOMC	104.3	OLG	Glen Arbor	WJZJ	95.5	RK/C
Detroit	WMGC	105.1	AC	Glen Arbor	WGFN	98.1	RK/C
Detroit	WDTJ	105.9	URB	Good Hart	WJOG	91.3	N/A
Detroit	WDTW	106.7	CH	Grand Haven	WGHN	1370	AC/S
Detroit	WGPR	107.5	JAZ	Grand Haven	WGHN	92.1	AC/S
Dimondale	WXLA	1180	URB	Grand Rapids	WTKG	1230	TLK+
Dowagiac	WDOW	1440	SPTS	Grand Rapids	WTKG	1230	TLK+
Dowagiac	WDOW	92.1	RL/CC	Grand Rapids	WTKG	1230	TLK+
Eagle	WJOM	88.5	N/A	Grand Rapids	WOOD	1300	TLK+
East Jordan	WICV	100.9	CLA	Grand Rapids	WBBL	1340	SPTS
East Lansing	WVFN	730	SPTS	Grand Rapids	WNWZ	1410	SP/MEX
East Lansing	WKAR	870	N/TLK	Grand Rapids	WFUR	1570	RL
East Lansing	WDBM	88.9	AD/A	Grand Rapids	WBLU	88.9	VAR
East Lansing	WKAR	90.5	CLA	Grand Rapids	WAYG	89.9	RL/CC
East Lansing	WMMQ	94.9	RK/C	Grand Rapids	WCSG	91.3	RL/CC
East Lansing	WFMK	99.1	AC	Grand Rapids	WBCT	93.7	CW
Elmwood Twn	WLJN	1400	RL	Grand Rapids	WLHT	95.7	AC/S
Escanaba	WCHT	600	TLK+	Grand Rapids	WLAV	96.9	RK/C
Escanaba	WDBC	680	TLK+	Grand Rapids	WGRD	97.9	RK/C
Escanaba	WGLQ	97.1	CH	Grand Rapids	WFGR	98.7	OLG
Escanaba	WYKX	104.7	CW	Grand Rapids	WCUZ	101.3	RK/C
Essexville	WMJO	97.3	AC/H	Grand Rapids	WFUR	102.9	RL/G
Fenton	WCXI	1160	RL/G	Grand Rapids	WOOD	105.7	AC/S
Flint	WSNL	600	RL/CC	Grayling	WGRY	1230	AS
Flint	WFDF	910	YTH	Grayling	WGRY	100.3	CW
Flint	WTRX	1330	SPTS	Greenville	WSCG	1380	TLK+
Flint	WFLT	1420	RL/G	Greenville	WKLQ	107.3	RK/C
Flint	WFNT	1470	AS	Gulliver	WCMM	102.5	CW
Flint	WWCK	1570	TLK+	Hancock	WMPL	920	TLK+
Flint	WGRI	88.9	RL/CC	Hancock	WKMJ	93.5	AC/H
Flint	WAKL	88.9	RL/CC	Hancock	WGLI	98.7	RK/C
Flint	WFUM	91.1	N/TLK	Harbor Beach	WCZE	103.7	N/A
Flint	WDZZ	92.7	URB	Harbor Spr	WCMW	103.9	VAR
Flint	WFBE	95.1	CW	Harrietta	WKAD	93.7	OLG
Flint	WWCK	105.5	CH	Harrison	WVXH	92.1	VAR
Flint	WCRZ	107.9	AC	Harrisville	WJOJ	89.7	RL/CC
Frankenmuth	WRCL	93.7	CH	Hart	WWKR	94.1	RK/C
Frankfort	WBNZ	99.3	AC/H	Hart	WCXT	105.3	AC/S

98

City	Station	Freq	Format	City	Station	Freq	Format
Hartford	WZTY	103.7	CW	Jackson	WJXQ	106.1	RK
Hartford	WSPZ	103.7	OLG	Kalamazoo	WKZO	590	N/TLK
Hastings	WBCH	1220	TLK+	Kalamazoo	WKMI	1360	TLK+
Hastings	WBCH	100.1	CW	Kalamazoo	WKPR	1420	RL
Hemlock	WCEN	94.5	CW	Kalamazoo	WKLZ	1470	TLK+
Highland Park	WHPR	88.1	UR/RB	Kalamazoo	WQSN	1660	SPTS
Highland Park	WHPR	88.1	URB	Kalamazoo	WAYK	88.3	RL/CC
Hillman	WKJZ	94.9	RK/C	Kalamazoo	WAYK	88.3	RL/CC
Hillsdale	WCSR	1340	AC/S	Kalamazoo	WIDR	89.1	VAR
Hillsdale	WCSR	92.1	AC/S	Kalamazoo	WKDS	89.9	CH
Holland	WHTC	1450	AC	Kalamazoo	WMUK	102.1	VAR
Holland	WTHS	89.9	AD/A	Kalamazoo	WQLR	106.5	AC/S
Holland	WTNR	94.5	CW	Kalkaska	WKLT	97.5	RK
Holland	WVTI	96.1	AC/H	Kentwood	WJNZ	1140	URB
Holton	WVIB	100.1	URB	Kentwood	WGVU	1480	N/TLK
Honor	WKVK	100.7	AC/S	Kingsford	WEUL	98.1	RL/G
Houghton	WCCY	1400	AS	Kingsley	WJZZ	1210	NWS+
Houghton	WGGL	91.9	CLA	Kingsley	WLDR	1210	CW
Houghton	WMTU	91.9	VAR	Lake City	WAIR	104.9	RL/CC
Houghton	WOLV	97.7	CH	Lakeview	WPLB	106.3	CW
Houghton	WHKB	102.3	CW	Lakeview	wscg	106.3	CW
Houghton Lake	WUPS	98.5	CH	L'anse	WCUP	105.7	CW
Howell	WHMI	93.5	CH	L'anse	WCUP	105.7	CW
Hubbard Lake	WKHN	88.1	N/A	Lansing	WJIM	1240	TLK+
Hudson	WMXE	102.5	AC	Lansing	WILS	1320	AS
Hudson	WBZV	102.5	RK/C	Lansing	WLNZ	89.7	JAZ
Imlay City	WWKM	89.1	RL	Lansing	WJIM	97.5	OLG
Inkster	WMKM	1440	RL/G	Lansing	WITL	100.7	CW
Intrerlochen	WIAA	88.7	CLA	Lansing	WHZZ	101.7	CH
Ionia	WION	1430	AC	Lapeer	WMPC	1230	RL/CC
Iron Mountain	WMIQ	1450	TLK+	Lapeer	WLSP	1530	AS
Iron Mountain	WVCM	91.5	RL	Lapeer	WQUS	103.1	RK
Iron Mountain	WIMK	93.1	RK/C	Leland	WFCX	94.3	CH
Iron Mountain	WJNR	101.5	CW	Leroy Twnsp	WLGH	88.1	RL/CC
Iron Mtn	WHTO	106.7	AH	Leroy Twp	WLGH	88.1	RL/CC
Iron River	WIKB	1230	OLG	Lexington	WBTI	96.9	AC
Iron River	WIKB	99.1	OLG	Livonia	WCAR	1090	RL
Ironwood	WJMS	590	CW	Ludington	WKLA	1450	TLK+
Ironwood	WIMI	99.7	AC/H	Ludington	Wkla	106.3	Ac
Ironwood	WUPM	106.9	AC/H	Luna Pier	WTWR	98.3	CH
Ishpeming	WZAM	970	NWS+	Mackinaw City	WDQV	88.3	RL/CC
Ishpeming	WIAN	1240	TLK+	Mackinaw City	WDQV	88.5	RL/CC
Ishpeming	WJPD	92.3	CW	Mackinaw City	WLJZ	94.5	RK/C
Ishpeming	WMQT	107.7	CH	Manistee	WMTE	1340	TLK+
Jackson	WKHM	970	TLK+	Manistee	WVXM	97.7	VAR
Jackson	WIBM	1450	SPTS	Manistee	WXYQ	101.5	OLG
Jackson	WJKN	1510	RL/CC	Manistique	WPIQ	99.9	TLK+
Jackson	WJKQ	88.5	OLG	Manistique	WMII	650	N/A
Jackson	WJCQ	89.7	N/A	Manistique	WTIQ	1490	OLG
Jackson	WVIC	94.1	CH	Manistique	WTIQ	1490	OLG

City	Station	Freq	Format	City	Station	Freq	Format
Marine City	WHLX	1590	AS	Norway	WZNL	94.3	AC
Marine City	WHLX	1590	AS	Novi	WOVI	89.5	AD/A
Marlette	WMSQ	89.3	OLG	Olivet	WOCR	89.7	VAR
Marlette	WBGV	92.5	CW	Onsted	WAAQ	88.3	OLG
Marlette	WBGV	92.5	CW	Ontonagan	WOAS	88.5	VAR
Marquette	WDMJ	1320	TLK+	Ontonogon	WUPY	101.1	CW
Marquette	WNMU	90.1	CLA	Orchard Lake	WBLD	89.3	RK/C
Marquette	WUPX	91.5	VAR	Oscoda	WCMB	95.7	VAR
Marquette	WUPK	94.1	RK	Oscoda	WWTH	100.7	CW
Marquette	WHWL	95.7	RL/G	Otsego	WAKV	980	AS
Marquette	WFXD	103.3	CW	Otsego	WQXC	100.9	OLG
Marshall	WWKN	104.9	CH	Ovid	WOES	91.3	VAR
Mason	WUNN	1110	RL/G	Owosso	WOAP	1080	RL
Menominee	WAGN	1340	AS	Owosso	WAHV	103.9	RK/C
Menominee	WHYB	103.7	CW	Pentwater	WWKR	94.1	RK/CL
Midland	WMPX	1490	AS	Pentwater	WMOM	102.7	AC/H
Midland	WKQZ	93.3	RK	Pentwater	WMOM	102.7	AC/H
Midland	WUGN	99.7	RL/CC	Petoskey	WWKK	750	N/A
Mio	WAVC	93.9	CW	Petoskey	WWKK	750	TLK+
Mio	WAVC	93.9	CW	Petoskey	WJML	1110	TLK+
Monroe	WRDT	560	RL	Petoskey	WMBN	1340	AS
Monroe	WDTR	88.1	RL/CC	Petoskey	WLXT	96.3	AC/S
Monroe	WYDM	97.5	OLG	Petoskey	WKLZ	98.9	RK
Mt Clemens	WDMK	102.7	URB	Pickford	WADW	105.5	OLG
Mt Pleasant	WCMU	89.5	VAR	Pickford	WADW	105.5	RL
Mt Pleasant	WMHW	91.5	RK/C	Pinconning	WYLZ	100.9	RK/C
Mt Pleasant	WCZY	104.3	AC/S	Pinconning	WSAG	104.1	AC/S
Munising	WQXO	1400	AS	Pittsford	WPCJ	91.1	RL/CC
Munising	WRUP	98.3	RK/C	Plymouth	WSDP	88.1	RK/C
Muskegon	WGVS	850	N/TLK	Port Huron	WPHM	1380	N/TLK
Muskegon	WKBZ	1090	TLK+	Port Huron	WHLS	1450	AS
Muskegon	WKBZ	1520	SILNT	Port Huron	WNFA	88.3	RL/CC
Muskegon	WMHG	1600	AS	Port Huron	WSGR	91.3	JAZ
Muskegon	WPQZ	88.1	N/A	Port Huron	WORW	91.9	VAR
Muskegon	WMCQ	91.7	RL	Port Huron	WGRT	102.3	AC
Muskegon	WSNX	104.5	CH	Port Huron	WSAQ	107.1	CW
Muskegon	WMUS	106.9	CW	Portage	WNWN	1560	URB
Muskegon	WSHZ	107.9	AC	Portage	WFAT	96.5	CH
Muskegon Hts	WMRR	101.7	CH	Portage	WRKR	107.7	RK
Negaunee	WNGE	99.5	OLG	Raco	WJOH	91.5	VAR
Negaunee	WKQS	101.9	AC/S	Reed City	WDEE	1500	N/A
Negaunee	WKQS	101.9	AC/S	Reed City	WDEE	97.3	OLG
Newaygo	WODJ	92.5	N/A	Reed City	WDEE	97.3	OLG
Newberry	WNBY	1450	CW	Rockford	WMJH	810	SP/MEX
Newberry	WNBY	93.7	AC	Rogers City	WHAK	960	CW
Newberry	WIHC	97.9	RK/C	Rogers City	WXVA	96.7	VAR
Niles	WNIL	1290	AC/S	Rogers City	WHAK	99.9	OLG
Niles	WAOR	95.3	RK	Rogers Hts	WAQQ	88.1	OLG
N Muskegon	WLCS	98.3	OLG	Roscommon	WQON	101.1	AC/S

Radio on the Road Michigan

City	Station	Freq	Format	City	Station	Freq	Format
Royal Oak	WEXL	1340	RL/G	Taylor	WCHB	1200	RL/G
Rust Twnsp	WKKM	88.5	VAR	Three Rivers	WLKM	1520	CW
Saginaw	WSGW	790	N/TLK	Three Rivers	WLKM	95.9	AC/S
Saginaw	WSAM	1400	AC/S	Traverse City	WTCM	580	N/TLK
Saginaw	WKCQ	98.1	CW	Traverse City	WCCW	1310	SPTS
Saginaw	WILZ	104.5	RK/C	Traverse City	WLJN	89.9	RL/CC
Saginaw	WGER	106.3	AC/S	Traverse City	WNMC	90.7	JAZ
Saginaw	WTLZ	107.1	URB	Traverse City	WICA	91.5	N/TLK
Salem	WSDS	1480	CW	Traverse City	WLDR	101.9	AC
Salem Twn	WSDS	1480	CW	Traverse City	WLDR	101.9	CW
Saline	WCAS	1290	TLK+	Traverse City	WTCM	103.5	CW
Sandusky	WMIC	660	TLK+	Traverse City	WCCW	107.5	OLG
Sandusky	WNFR	90.7	RL/CC	Trout Lake	WHWG	89.9	RL/CC
Sandusky	WTGV	97.7	AC/S	Trout Lake	WHWG	89.9	RL
Saugatuck	WVYN	92.7	CH	Tuscola	WWBN	101.5	RK
S. Ste. Marie	WSOO	1230	AC/S	Tuscola	WWBN	101.5	RK/C
S. Ste. Marie	WKNW	1400	TLK+	Twin Lake	WBLV	90.3	VAR
S. Ste. Marie	WLSO	90.1	AD/A	Vassar	WOWE	98.9	OLG
S. Ste. Marie	WCMZ	98.3	VAR	Walker	WTRV	100.5	AC/S
S. Ste. Marie	WYSS	99.5	CH	Walled Lake	WPON	1460	TLK+
S. Ste. Marie	WSUE	101.3	RK/C	Warren	WPHS	89.1	VAR
S. Ste. Marie	WTHN	102.3	RL/CC	Warren	WPHS	89.1	VAR
Schoolcraft	WOFR	89.5	RL	West Branch	WBMI	105.5	AC/H
Scottville	WKZC	94.9	CW	White Star	WEJC	88.3	RL/CC
Shepherd	WMMI	830	TLK+	Whitehall	WUBR	1490	SPTS
South Haven	WSPZ	940	OLG	Whitehall	WGVS	95.3	JAZ
South Haven	WCSY	98.3	AC/S	Whitehall	WEFG	97.5	CW
South Haven	WCSY	98.3	AC	Wyoming	WYGR	1530	SP/MEX
Southfield	WSHJ	88.3	URB	Wyoming	WYCE	88.1	AD/A
Spring Arbor	KTGG	1540	RL/CC	Ypsilanti	WDEO	990	RL
Spring Arbor	WJKN	89.3	RL/CC	Ypsilanti	WEMU	89.1	JAZ
Spring Arbor	WSAE	106.9	RL/CC	Zeeland	WMFN	640	TLK+
St. Ignace	WIDG	940	SPTS	Zeeland	WPNW	1260	TLK+
St. Ignace	WMKC	102.9	CW	Zeeland	WGNB	89.3	RL
St. Johns	WWSJ	1580	RL/G	Zeeland	WJQK	99.3	RL/CC
St. Johns	WTXQ	92.1	SPTS				
St. Joseph	WSJM	1400	TLK+				
St. Joseph	WIRX	107.1	RK				
St. Louis	WMLM	1520	CW				
Standish	WWCM	96.9	VAR				
Stephenson	WMXG	106.3	CHR				
Stephenson	WMXG	106.3	AC/H				
Sterling Hts	WUFL	1030	RL/CC				
Sturgis	WMSH	1230	SPTS				
Sturgis	WMSH	99.3	OLG				
Tawas City	WIOS	1480	RL/G				
Tawas City	WQLB	103.3	RK/CL				
Tawas City	WQLB	103.3	RK/C				
Tawas City	WKJC	104.7	CW				

City	Station	Freq	Format	City	Station	Freq	Format
Ada	KRJB	106.3	CW	Browerville	KXDL	99.7	OLG
Aitkin	KKIN	930	AS	Buffalo	KRWC	1360	VAR
Aitkin	KKIN	94.3	CW	Buhl	WIRN	92.5	N/TLK
Albany	KASM	1150	CW	Caledonia	KCLH	94.7	CH
Albany	KDDG	105.5	OLG	Cambridge	WGVY	105.3	AD/A
Albert Lea	KATE	1450	CW	Cloquet	WKLK	1230	AS
Albert Lea	KCPI	94.9	CH	Cloquet	WKLK	96.5	RK/C
Albert Lea	KQPR	96.1	RK/C	Cloquet	WSCN	100.5	N/TLK
Alexandria	KXRA	1490	VAR	Cold Spring	KMXK	94.9	AC/H
Alexandria	KBHG	89.5	RL	Coleraine	KGPZ	96.1	CW
Alexandria	KXRA	92.3	RK/C	Collegeville	KNSR	88.9	N/TLK
Alexandria	KULO	94.3	OLG	Collegeville	KSJR	90.1	CLA
Alexandria	KXRZ	99.3	AC	Coon Rapids	WFMP	107.1	TLK+
Anoka	KQQL	107.9	OLG	Coon Rapids	WFMP	107.1	TLK+
Appleton	KNCM	88.5	N/TLK	Crookston	KROX	1260	VAR
Appleton	KRSU	91.3	CLA	Crookston	KQHT	96.1	CH
Atwater	KKLN	94.1	RK/C	Crookston	KYCK	97.1	CW
Austin	KNFX	970	SPTS	Crosby	KTCF	101.5	CW
Austin	KAUS	1480	N/TLK	Dassel	KARP	106.9	CW
Austin	KNSE	89.5	N/TLK	Deer River	KBAJ	105.5	RK/C
Austin	KMSK	91.3	VAR	Detroit Lakes	KDLM	1340	AC
Austin	KAUS	99.9	CW	Detroit Lakes	KRVI	95.1	AC/S
Babbitt	KAOD	106.7	RK/C	Detroit Lakes	KRCQ	102.3	CW
Bagley	KKCQ	96.7	CW	Duluth	WEBC	560	SPTS
Baxter	WWWI	1270	TLK+	Duluth	KDAL	610	TLK+
Bemidji	KKBJ	1360	TLK+	Duluth	WWJC	850	RL/G
Bemidji	KBUN	1450	SPTS	Duluth	KQDS	1490	TLK+
Bemidji	KCRB	88.5	CLA	Duluth	WJRF	89.5	RL/CC
Bemidji	KBSB	89.7	CH	Duluth	KDNI	90.5	RL/CC
Bemidji	KNBJ	91.3	N/TLK	Duluth	WSCD	92.9	CLA
Bemidji	KKZY	95.5	AC	Duluth	KQDS	94.9	RK/C
Bemidji	KBHP	101.1	CW	Duluth	KDAL	95.7	AC/S
Bemidji	KKBJ	103.7	AC/H	Duluth	KDNW	97.3	RL/CC
Benson	KBMO	1290	AS	Duluth	KTCO	98.9	CW
Benson	KSCR	93.5	AC	Duluth	KLDJ	101.7	OLG
Blackduck	WBJI	98.3	CW	Duluth	KUMD	103.3	VAR
Blooming Pr.	KOWZ	100.9	AC	Duluth	KKCB	105.1	CW
Blue Earth	KBEW	1560	FRM	E Grand Forks	KCNN	1590	TLK+
Blue Earth	KBEW	98.1	CW	E Grand Forks	KZLT	104.3	AC/S
Blue Earth	KJLY	104.5	RL	Eden Prairie	WGVZ	105.7	AD/A
Brainerd	KVBR	1340	TLK+	Elk River	KLCI	106.1	SP/MEX
Brainerd	KLIZ	1380	SPTS	Ely	WELY	1450	VAR
Brainerd	KBPR	90.7	CLA	Ely	WELY	92.1	VAR
Brainerd	KUAL	103.5	OLG	Eveleth	KRBT	1340	TLK+
Brainerd	WJJY	106.7	AC	Eveleth	WEVE	97.9	AC
Brainerd	KLIZ	107.5	RK/C	Fairmont	KSUM	1370	CW
Breckenridge	KBMW	1450	CW	Fairmont	KFMC	106.5	RK/C
Breckenridge	KLTA	105.1	AC	Faribault	KDHL	920	CW
Breezy Point	KLKS	104.3	AS	Faribault	KQCL	95.9	RK/C
Brooklyn Park	KLBP	1470	AS	Faribault	KBGY	107.5	SP/MEX

City	Station	Freq	Format	City	Station	Freq	Format
Fergus Falls	KJJK	1020	OLG	Mankato	KYSM	1230	SPTS
Fergus Falls	KBRF	1250	CW	Mankato	KTOE	1420	TLK+
Fergus Falls	KCMF	89.7	CLA	Mankato	KMSU	89.7	VAR
Fergus Falls	KNWF	91.5	N/TLK	Mankato	KEEZ	99.1	AC/H
Fergus Falls	KJJK	96.5	CW	Mankato	KYSM	103.5	CW
Fergus Falls	KZCR	103.3	RK/C	Maplewood	WCTS	1030	RL
Forest Lake	WLKX	95.9	RL/CC	Marshall	KMHL	1400	TLK+
Fosston	KKCQ	1480	TLK+	Marshall	KKCK	99.7	CH
Fosston	KKEQ	107.1	RL/CC	Marshall	KARZ	107.5	CH
Glencoe	KTTB	96.3	CH	Minneapolis	KFXN	690	SPTS
Glenwood	KMGK	107.1	AC/S	Minneapolis	KUOM	770	AD/A
Golden Valley	KDIZ	1440	YTH	Minneapolis	WCCO	830	N/TLK
Golden Valley	KYCR	1570	TLK+	Minneapolis	KTIS	900	RL/CC
Golden Valley	KQRS	92.5	RK/C	Minneapolis	KFAN	1130	TLK+
Grand Marais	WMLS	88.7	CLA	Minneapolis	WWTC	1280	TLK+
Grand Marais	WLSN	89.7	N/TLK	Minneapolis	WMNN	1330	RL
Grand Marais	WTIP	90.7	VAR	Minneapolis	KLBB	1400	AS
Grand Marais	WXXZ	95.3	RK/C	Minneapolis	KSTP	1500	N/TLK
Grand Rapids	KOZY	1320	OLG	Minneapolis	KBEM	88.5	JAZ
Grand Rapids	KAXE	91.7	N/TLK	Minneapolis	KMOJ	89.9	URB
Grand Rapids	KMFY	96.9	AC	Minneapolis	KFAI	90.3	VAR
Granite Falls	KKRC	93.9	OLG	Minneapolis	WMCN	91.7	JAZ
Hastings	KDWA	1460	N/TLK	Minneapolis	KXXR	93.7	RK/C
Hermantown	WWAX	92.1	CH	Minneapolis	KSTP	94.5	AC
Hibbing	WMFG	1240	AS	Minneapolis	KTCZ	97.1	AD/A
Hibbing	KADU	90.1	RL/CC	Minneapolis	KTIS	98.5	RL/CC
Hibbing	WTBX	93.9	AC/H	Minneapolis	KSJN	99.5	CLA
Hibbing	WMFG	106.3	OLG	Minneapolis	kjzi	100.3	JAZ
Hutchinson	KDUZ	1260	TLK+	Minneapolis	WLTE	102.9	AC/S
Internat'l Falls	KGHS	1230	AC	Minneapolis	KNOF	95.3	RL/G
Internat'l Falls	KXBR	91.9	RL/CC	Minneapolis	KEEY	102.1	CW
Internat'l Falls	KBHW	99.5	RL/CC	Montevideo	KDMA	1460	CW
Internat'l Falls	KSDM	104.1	CW	Montevideo	KBPG	89.5	RL
Jackson	KKOJ	1190	CW	Montevideo	KMGM	105.5	AC/S
Jackson	KRAQ	105.7	OLG	Moorhead	KVOX	1280	SPTS
La Crescent	KXLC	91.1	N/TLK	Moorhead	KCCD	90.3	N/TLK
La Crescent	KQEG	102.7	OLG	Moorhead	KCCM	91.1	CLA
Lake City	KLCH	94.9	AC	Moorhead	KQWB	98.7	RK
Lake City	KMFX	102.5	CW	Moorhead	KVOX	99.9	CW
Lake Crystal	KLCH	95.7	N/A	Moose Lake	WMOZ	106.9	OLG
Lakeville	WGVX	105.1	AD/A	Mora	KBEK	95.5	AS
Litchfield	KLFD	1410	CW	Morris	KMRS	1230	TLK+
Little Falls	KLTF	960	TLK+	Morris	KUMM	89.7	AD/A
Little Falls	WYRQ	92.1	CW	Morris	KKOK	95.7	CW
Little Falls	KFML	94.1	AC	Nashwauk	WNMT	650	N/TLK
Long Prairie	KEYL	1400	CW	Nashwauk	KMFG	102.9	RK/C
Luverne	KQAD	800	AC/S	New Prague	KCHK	1350	CW
Luverne	KLQL	101.1	CW	New Prague	KRDS	95.5	OLG
Madison	KLQP	92.1	CW	New Ulm	KNUJ	860	CW
Mahnomen	KRJM	101.5	OLG	New Ulm	KXLP	93.1	RK/C

Radio on the Road   Minnesota

City	Station	Freq	Format	City	Station	Freq	Format
Nissa	KBPQ	93.3	CW	Roseau	KCAJ	102.1	AC
North Branch	KMKL	90.3	RL/CC	Rushford	KWNO	99.3	CW
North Mankato	KDOG	96.7	AC/H	Sartell	KKSR	96.7	CH
Northfield	KYMN	1080	AS	Sauk Centre	KIKV	100.7	CW
Northfield	KRLX	88.1	VAR	Sauk Rapids	WBHR	660	CW
Northfield	KCMP	89.3	AD/A	Sauk Rapids	WVAL	800	CW
Olivia	KOLV	101.7	CW	Sauk Rapids	WHMH	101.7	RK
Ortonville	KDIO	1350	CW	Sebeka	KOPJ	89.3	N/A
Ortonville	KCGN	101.5	RL/CC	Shakopee	KSMM	1530	RL/CC
Ortonville	KPHR	106.3	RK/C	Slayton	KJOE	106.1	CW
Osakis	KBHL	103.9	RL/CC	Sleepy Eye	KNUJ	107.3	AC
Owatonna	KRFO	1390	OLG	Spring Grove	KQYB	98.3	CW
Owatonna	KRFO	104.9	CW	Spring Valley	KVGO	104.3	OLG
Park Rapids	KPRM	870	CW/TLK	Springfield	KNSG	94.7	AC
Park Rapids	KXKK	92.5	CW	St. Charles	KLCX	107.7	CH
Park Rapids	KDKK	97.5	AS	St. Cloud	WJON	1240	N/TLK
Paynesville	KZPK	98.9	CW	St. Cloud	KNSI	1450	N/TLK
Pelican Rapids	KBOT	104.1	CW	St. Cloud	KVSC	88.1	AD/A
Pequot Lakes	WZFJ	100.1	RL/CC	St. Cloud	KCFB	91.5	RL/CC
Pequot Lakes	KTIG	102.7	RL/CC	St. Cloud	WWJO	98.1	CW
Perham	KPRW	99.5	AC	St. Cloud	KCLD	104.7	CH
Pillager	WWWI	95.9	TLK+	St. James	KXAC	100.5	OLG
Pine City	WCMP	1350	AS	St. James	KRRW	101.5	CW
Pine City	WCMP	100.9	CW	St. Joseph	KKJM	92.9	RL/CC
Pipestone	KLOH	1050	CW	St. Joseph	KCML	99.9	AC/S
Pipestone	KISD	98.7	OLG	St. Louis Park	KTNF	950	TLK+
Preston	KFIL	1060	CW	St. Louis Park	KNOW	91.1	N/TLK
Preston	KFIL	103.1	CW	St. Louis Park	KZJK	104.1	AH
Princeton	WQPM	1300	CW	St. Louis Park	KUOM	106.5	AD/A
Proctor	KUSZ	107.7	AC/H	St. Peter	KRBI	1310	AC
Red Wing	KCUE	1250	N/TLK	St. Peter	KGAC	90.5	CLA
Red Wing	KWNG	105.9	CH	St. Peter	KRBI	105.5	AH
Redwood Falls	KLGR	1490	CW	St. Peter	KRBI	105.5	AH
Redwood Falls	KLGR	97.7	OLG	Staples	KNSP	1430	CW
Richfield	KKMS	980	RL	Staples	KSKK	94.7	AC
Richfield	KDWB	101.3	CH	Starbuck	KRVY	97.3	AC
Rochester	KWEB	1270	SPTS	Stewartville	KYBA	105.3	AC/S
Rochester	KROC	1340	TLK+	Stillwater	WMGT	1220	TLK+
Rochester	KOLM	1520	SPTS	Sunburg	KLFN	106.5	CH
Rochester	KMSE	88.7	CLA	Thief River Fls	KTRF	1230	AC
Rochester	KRPR	89.9	AP	Thief River Fls	KKAQ	1460	CW
Rochester	KRPR	89.9	RK/C	Thief River Fls	KSRQ	90.1	AD/A
Rochester	KZSE	90.7	N/TLK	Thief River Fls	KQMN	91.5	CLA
Rochester	KLSE	91.7	CLA	Thief River Fls	KKDQ	99.3	CW
Rochester	KFSI	92.9	RL/CC	Thief River Fls	KSNR	100.3	OLG
Rochester	KWWK	96.5	CW	Thief River Fls	KNTN	102.7	N/TLK
Rochester	KNXR	97.5	EZL	Tracy	KARL	105.1	CW
Rochester	KRCH	101.7	RK/C	Two Harbors	KZIO	104.3	RK/C
Rochester	KROC	106.9	CH	Verndale	KVKK	1070	N/A
Roseau	KRWB	1410	RK/C	Virginia	WEEP	1400	RL

105

City	Station	Freq	Format	City	Station	Freq	Format
Virginia	WUSZ	99.9	CW				
Virginia/Hibbing	WIRR	90.9	CLA				
Wabasha	KMFX	1190	CW				
Wadena	KWAD	920	CW				
Wadena	KKWS	105.9	CW				
Waite Park	KXSS	1390	SPTS				
Waite Park	KLZZ	103.7	RK/C				
Walker	KAKK	1570	OLG				
Walker	KLLZ	99.1	RK/C				
Walker	KQKK	101.9	AC/S				
Warroad	KKWQ	92.5	CW				
Waseca	KOWO	1170	TLK+				
Waseca	KRUE	92.1	OLG				
Watertown	KZGX	1600	SP/MEX				
Willmar	KWLM	1340	N/TLK				
Willmar	KDJS	1590	OLG				
Willmar	KKLW	90.9	RL/CC				
Willmar	KBHZ	91.9	RL/CC				
Willmar	KDJS	95.3	CW				
Willmar	KQIC	102.5	AC/H				
Windom	KDOM	1580	CW				
Windom	KQRB	89.9	RL				
Windom	KDOM	94.3	CW				
Winona	KWNO	1230	TLK+				
Winona	KAGE	1380	CW				
Winona	KQAL	89.5	VAR				
Winona	KSMR	92.5	AD/A				
Winona	KAGE	95.3	AC/H				
Winona	KHME	101.1	AC/S				
Worthington	KWOA	730	N/TLK				
Worthington	KBOJ	88.1	RL				
Worthington	KRSW	89.3	CLA				
Worthington	KNSW	91.7	N/TLK				
Worthington	KITN	93.5	CH				
Worthington	KWOA	95.1	AC/H				
Virginia	WUSZ	99.9	CW				
Virginia/Hibbing	WIRR	90.9	CLA				
Wabasha	KMFX	1190	CW				
Wadena	KWAD	920	CW				
Wadena	KKWS	105.9	CW				
Waite Park	KXSS	1390	SPTS				
Waite Park	KLZZ	103.7	RK/C				
Walker	KAKK	1570	OLG				
Walker	KLLZ	99.1	RK/C				
Walker	KQKK	101.9	AC/S				

106

Radio on the Road Mississippi

City	Station	Freq	Format	City	Station	Freq	Format
Aberdeen	WWZQ	1240	TLK+	Coldwater	WVIM	95.3	OLG
Aberdeen	WWKZ	105.3	CHR	Collins	WKNZ	107.1	RL/CC
Ackerman	WFCA	107.9	RL/G	Columbia	WFFF	1360	CW
Amory	WAMY	1580	TLK+	Columbia	WCJU	1450	RL
Amory	WAFM	95.3	OLG	Columbia	WPRG	89.5	RL
Artesia	WSMS	99.9	RK/C	Columbia	WFFF	96.7	AC
Baldwyn	WESE	92.5	URB	Columbus	WACR	1050	RL/G
Batesville	WJBI	1290	CW	Columbus	WJWF	1400	TLK+
Batesville	WBLE	100.5	CW	Columbus	WKOR	94.9	CW
Bay St Louis	WBSL	1190	JAZ	Columbus	WMBC	103.1	CH
Bay St Louis	WZKX	107.9	CW	Columbus	WACR	103.9	URB
Bay Springs	WIZK	1570	CW	Como	WRBO	103.5	JAZ
Bay Springs	WKZW	94.3	AC	Corinth	WTKN	1230	RL
Belzoni	WELZ	1460	RL/G	Corinth	WKCU	1350	RL/CC
Belzoni	WBYP	107.1	CW	Corinth	WXRZ	94.3	TLK+
Biloxi	WXBD	1490	SPT	Corinth	WADI	95.3	CW
Biloxi	WTNI	1640	TLK+	Crenshaw	WHKL	106.9	OLG
Biloxi	WMAH	90.3	VAR	De Kalb	WJXM	105.7	URB
Biloxi	WMJY	93.7	AC	Drew	WRKG	95.3	RK/C
Booneville	WBIP	1400	CW	Duck Hill	WAUM	91.9	RL
Booneville	WMAE	89.5	VAR	Durant	WLIN	101.1	AC
Booneville	WBVV	99.3	RL/G	Ellisville	WJKX	102.5	URB
Brandon	WRKN	970	SIL	Eupora	WLZA	96.1	AC/H
Brandon	WRJH	97.7	URB	Fayette	WTYJ	97.7	RL/G
Brookhaven	WCHJ	1470	RL/G	Flora	WFMN	97.3	TLK+
Brookhaven	WBKN	92.1	CW	Flowood	WPBQ	1240	SPTS
Brooksville	WAJV	98.9	RL/G	Forest	WQST	850	CW
Bruce	WCMR	94.5	CW	Forest	WMBU	89.1	RL
Bude	WMAU	88.9	VAR	Forest	WSQH	91.7	RL
Bude	WMJU	104.3	AC/S	Forest	WQST	92.5	RL/G
Burnsville	WOWL	91.9	RK/C	Fulton	WFTA	101.9	AC
Byhalia	WKVF	94.9	RL/CC	Gluckstadt	WYOY	101.7	CHR
Canton	WONG	1150	URB	Greenville	WGVM	1260	SPTS
Canton	WMGO	1370	RL/G	Greenville	WNIX	1330	URB
Carthage	WCKK	98.3	CW	Greenville	WFBI	91.5	N/A
Centreville	WPAE	89.7	RL	Greenville	WBAQ	97.9	EZL
Centreville	WZFL	104.9	CW	Greenville	WDMS	100.7	CW
Charleston	WTGY	95.7	RL	Greenville	WJIW	104.7	RL
Clarksdale	WROX	1450	JAZ	Greenwood	WABG	960	TLK+
Clarksdale	WKXY	92.1	JAZ	Greenwood	WGRM	1240	RL/G
Clarksdale	WKDJ	96.5	CW	Greenwood	WKXG	1540	RL/G
Clarksdale	WWUN	101.5	RL/CC	Greenwood	WGRM	93.9	RL/G
Clarksdale	WAID	106.5	URB	Greenwood	WYMX	99.1	RK/C
Cleveland	WDSK	1410	TLK+	Greenwood	WGNL	104.3	URB
Cleveland	WCLD	1490	RL/G	Grenada	WYKC	1400	CW
Cleveland	WDTL	92.9	CW	Grenada	WOHT	92.3	OLG
Cleveland	WDFX	98.3	RL/G	Grenada	WQXB	100.1	CW
Cleveland	WCLD	103.9	URB	Grenada	WMUT	101.3	RK/C
Clinton	WTWZ	1120	RL	Gulfport	WQFX	1130	RL/G
Clinton	WHJT	93.5	RL/CC	Gulfport	WGCM	1240	CW

City	Station	Freq	Format	City	Station	Freq	Format
Gulfport	WROA	1390	AS	Leland	WESY	1580	RL/G
Gulfport	WAOY	91.7	RL	Leland	WBAD	94.3	URB
Gulfport	WUJM	96.7	CW	Leland	WIQQ	102.3	AC/H
Gulfport	WGCM	102.3	OLG	Lexington	WXTN	1000	RL/G
Gulfport	WXYK	107.1	CHR	Lexington	WAGR	102.5	CW
Hattiesburg	WHSY	950	N/TLK	Liberty	WAZA	107.7	OLG
Hattiesburg	WFOR	1400	SPTS	Long Beach	WJZD	94.5	URB
Hattiesburg	WORV	1580	RL/G	Lorman	WPRL	91.7	RL/G
Hattiesburg	WUSM	88.5	VAR	Louisville	WLSM	1270	CW
Hattiesburg	WAII	89.3	RL/CC	Louisville	WLSM	107.1	RL
Hattiesburg	WJMG	92.1	URB	Lucedale	WRBE	1440	RL/G
Hattiesburg	WUSW	103.7	RK/C	Lucedale	WRBE	106.9	RL
Hattiesburg	WXRR	104.5	RK/C	Lumberton	WZNF	95.3	CH
Hazlehurst	WOEG	1220	RL/G	Magee	WSJC	810	RL
Hazlehurst	WDXO	92.9	OLG	Magee	WKXI	107.5	URB
Heidleberg	WHER	99.3	CW	Marion	WYYW	95.1	CW
Hickory	WGTC	91.3	N/A	Marks	WQMA	1520	OLG
Holly Springs	WKRA	1110	RL/G	Mccomb	WAPF	980	SPTS
Holly Springs	WURC	88.1	VAR	Mccomb	WAKK	1140	RL
Holly Springs	WKRA	92.7	URB	Mccomb	WHNY	1250	TLK+
Horn Lake	WHAL	95.7	RL/G	Mccomb	WAQL	90.5	RL
Houston	WCPC	940	RL/G	Mccomb	WAKH	105.7	CW
Houston	WJZB	88.9	RL/G	Mclain	WXAB	96.9	OLG
Houston	WSYE	93.3	AC/S	Meridian	WALT	910	TLK+
Indianola	WNLA	1380	RL/G	Meridian	WMOX	1010	TLK/S
Indianola	WTCD	96.9	TLK+	Meridian	WNBN	1290	RL/G
Indianola	WNLA	105.5	AC	Meridian	WMER	1390	RL/G
Itta Bena	WVSD	91.7	VAR	Meridian	WFFX	1450	SPTS
Iuka	WFXO	104.9	CW	Meridian	WMAW	88.1	VAR
Jackson	WJDX	620	SPTS	Meridian	WOKK	97.1	CW
Jackson	WSFZ	930	SPTS	Meridian	WJDQ	101.3	CH
Jackson	WOAD	1300	RL/G	Meridian	WMMZ	102.1	JAZ
Jackson	WKXI	1400	JAZ	Miss. State	WMAB	89.9	VAR
Jackson	WZRX	1590	URB	Monticello	WMLC	1270	SPTS
Jackson	WJSU	88.5	JAZ	Monticello	WRQO	102.1	CW
Jackson	WMPR	90.1	JAZ	Moss Point	WBUV	104.9	URB
Jackson	WMPN	91.3	VAR	Mound Bayou	WZYQ	101.9	URB
Jackson	WTYX	94.7	AH	Natchez	WMIS	1240	RL/G
Jackson	WHLH	95.5	RL/G	Natchez	WNAT	1450	TLK/S
Jackson	WUSJ	96.3	CW	Natchez	WASM	91.1	RL
Jackson	WJMI	99.7	URB	Natchez	WQNZ	95.1	CW
Jackson	WMSI	102.9	CW	Natchez	WKSO	97.3	AC/H
Kosciusko	WKOZ	1340	N/TLK	New Albany	WNAU	1470	OLG
Kosciusko	WZKR	103.3	RK/C	New Albany	WWZD	106.7	CW
Kosciusko	WQJQ	105.1	OLG	Newton	WMYQ	1100	RL/G
Laurel	WEEZ	890	JAZ	Newton	WMSO	97.9	CH
Laurel	WAML	1340	RL/G	Ocean Springs	WQYZ	92.5	RL/G
Laurel	WATP	90.7	RL	Ocean Springs	WOSM	103.1	RL/G
Laurel	WMXI	98.1	SPTS	Olive Branch	KJMS	101.1	URB
Laurel	WNSL	100.3	RL/G	Oxford	WMAV	90.3	VAR

Radio on the Road

Mississippi

City	Station	Freq	Format		City	Station	Freq	Format
Oxford	WAVI	91.5	RL		Tupelo	WZLQ	98.5	RK/C
Oxford	WQLJ	93.7	AC/H		Tylertown	WTYL	1290	CW
Oxford	WOXD	95.5	CH		Tylertown	WTYL	97.7	CW
Oxford	WWMS	97.5	CW		Union	WZKS	104.1	URB
Pascagoula	WZZJ	1580	N/TLK		University	WUMS	92.1	AD/A
Pascagoula	WPAS	91.5	RL		Utica	WJXN	100.9	RL/CC
Pascagoula	WKNN	99.1	CW		Vicksburg	WQBC	1420	TLK+
Pascagoula	WXRG	105.9	RK/C		Vicksburg	WRTM	1490	RL/G
Pearl	WJNT	1180	TLK+		Vicksburg	WJKK	98.7	AC/S
Pearl	WRXW	93.9	RK/C		Vicksburg	WBBV	101.3	CW
Petal	WZLD	106.3	URB		Vicksburg	WSTZ	106.7	RK/C
Philadelphia	WHOC	1490	AS		Walnut	WLRC	850	RL/G
Philadelphia	WWSL	102.3	AC		Water Valley	WTNM	105.5	TLK+
Picayune	WRJW	1320	CW		Waynesboro	WABO	990	CW
Picayune	WKSY	106.1	AC/S		Waynesboro	WZKM	89.7	RL/G
Pickins	WYJS	105.9	RL/G		Waynesboro	WABO	105.5	CW
Pontotoc	WSEL	1440	RL/G		Wesson	WCLL	90.7	SIL
Pontotoc	WSEL	96.7	RL/G		West Point	WROB	1450	JAZ
Poplarville	WRPM	1530	RL/G		West Point	WKBB	100.9	TLK+
Port Gibson	WATU	89.3	RL		Wiggins	WIGG	1420	CW
Port Gibson	WRTM	100.5	URB		Wiggins	WCPR	97.9	RK/C
Potts Camp	WCNA	95.9	RK/C		Winona	WONA	1570	SPTS
Prentiss	WJDR	98.3	CW		Winona	WONA	95.1	CW
Prentiss	WCJU	104.9	OLG		Yazoo City	WYAZ	89.5	RL
Quitman	WQMS	1500	SPTS		Yazoo City	WJNS	92.1	RL
Quitman	WYKK	98.9	RL/CC		Yazoo City	WYAB	93.1	OLG
Redwood	WVBG	105.5	N/A		Yazoo City	WJNS	92.1	RL
Richton	WXHB	96.5	RL/G					
Ridgeland	WIIN	720	AS					
Ripley	WKZU	102.3	CW					
Rosedale	WMJW	107.5	CW					
Sardis	KBUD	102.1	RL/G					
Senatobia	WSAO	1140	RL/G					
Senatobia	WKNA	88.9	N/TLK					
Southhaven	WAVN	1240	RL/G					
Starkville	WKOR	980	RL/G					
Starkville	WSSO	1230	TLK+					
Starkville	WMSU	91.1	URB					
Starkville	WMXU	106.1	URB					
State College	WQJB	104.5	CLA					
Stonewall	WMLV	106.9	AC/S					
Sumrall	WFMM	97.3	TLK+					
Taylorsville	WBBN	95.9	CW					
Tchula	WGNG	106.3	URB					
Tunica	WYYL	96.1	SP/MEX					
Tupelo	WELO	580	AS					
Tupelo	WKMQ	1060	TLK+					
Tupelo	WTUP	1490	SPTS					
Tupelo	WAFR	88.3	RL					
Tupelo	WAQB	90.9	RL/G					

109

City	Station	Freq	Format	City	Station	Freq	Format
Arcadia	KTNX	103.9	N/A	Carthage	KMXL	95.1	AC
Arnold	KGNA	89.9	RL	Caruthersville	KCRV	1370	CW
Asbury	KWXD	103.5	CW	Caruthersville	KLOW	105.1	OLG
Ash Grove	KZRQ	104.1	N/TLK	Cassville	KRMO	990	CW
Aurora	KSWM	940	TLK+	Cedar Hill	KNLH	89.5	RL/G
Aurora	KGMY	100.5	CW	Centralia	KMFC	92.1	RL/CC
Ava	KKOZ	1430	CW	Chaffee	KREZ	104.7	AC/S
Ava	KKOZ	92.1	CW	Charleston	KCHR	1350	TLK+
Ballwin	KYMC	89.7	VAR	Charleston	KWKZ	106.1	CW
Bethany	KAAN	870	CW	Chillicothe	KCHI	1010	OLG
Bethany	KAAN	95.5	CW	Chillicothe	KRNW	89.9	VAR
Birch Tree	KBMV	1310	OLG	Chillicothe	KCHI	103.9	OLG
Birch Tree	KBMV	107.1	AC/H	Clayton	KFUO	850	RL
Bismarck	KHCR	99.5	RL/G	Clayton	KSIV	1320	RL
Blue Springs	KCWJ	1030	RL/CC	Clayton	KWUR	90.3	VAR
Bolivar	KYOO	1200	CW	Clayton	KFUO	99.1	CLA
Bonne Terre	KDBB	104.3	AD/A	Cleveland	KCTO	1160	N/A
Boonville	KWRT	1370	CW	Clinton	KDKD	1280	OLG
Boonville	KWRT	93.1	AS	Clinton	KDKD	95.3	CW
Boonville	KCLR	99.3	CW	Clinton	KLRQ	96.1	RL/CC
Bowling Green	KPCR	1530	N/TLK	Columbia	KFRU	1400	TLK+
Bowling Green	KPVR	94.1	RL/CC	Columbia	KTGR	1580	SPTS
Branson	KOMC	1220	RL/G	Columbia	KCOU	88.1	AD/A
Branson	KLFC	88.1	RL/CC	Columbia	KOPN	89.5	VAR
Branson	KOZO	89.7	RL/G	Columbia	KWWC	90.5	JAZ
Branson	KRZK	106.3	CW	Columbia	KBIA	91.3	CLA
Brookfield	KZBK	1470	AC/H	Columbia	KCMQ	96.7	RK
Brookfield	KZBK	96.9	AC/H	Columbia	KPLA	101.5	AC
Brookline	KQRA	102.1	RK/C	Columbia	KBXR	102.3	AD/A
Buffalo	KBFL	99.9	AS	Concordia	KYRV	88.1	RL/G
Butler	KMAM	1530	CW	Country Club	KJCV	91.3	RL
Butler	KMOE	92.1	CW	Crestwood	KSHE	94.7	RK
Cabool	KFFW	89.9	RL/CC	Cuba	KGNN	90.3	RL
Cabool	KOZX	98.1	CH	Cuba	KNLQ	91.9	RL
California	KRLL	1420	CW	Cuba	KESY	107.3	CW
California	KATI	94.3	CW	De Soto	KRFT	1190	SPTS
Camdenton	KCVO	91.7	RL/CC	De Soto	KDJR	100.1	RL
Cameron	KMRN	1360	BIZ	Dexter	KDEX	1590	CW
Cameron	KKWK	100.1	AS	Dexter	KDEX	102.3	CW
Campbell	KFEB	107.5	RK/C	Dixon	KCZV	107.3	RL/CC
Canton	KRRY	100.9	AC/H	Doniphan	KDFN	1500	OLG
Cape Gir'deau	KZIM	960	TLK+	Doniphan	KOEA	97.5	CW
Cape Gir'deau	KGIR	1220	SPTS	East Prairie	KYMO	1080	AS
Cape Gir'deau	KAPE	1550	AS	East Prairie	KYMO	105.3	OLG
Cape Gir'deau	KRCU	89.9	VAR	El Dorado Spr	KESM	1580	CW
Cape Gir'deau	KGMO	100.7	RK/C	El Dorado Spr	KESM	105.5	CW
Cape Gir'deau	KEZS	102.9	CW	Eldon	KLOZ	92.7	AC/H
Carrollton	KAOL	1430	CW	Ellington	KAUL	106.7	RL/G
Carrollton	KMZU	100.7	CW	Excelsior Spr	KEXS	1090	RL
Carthage	KDMO	1490	AS	Farmington	KREI	800	TLK+

City	Station	Freq	Format	City	Station	Freq	Format
Farmington	KSEF	88.9	N/A	Kansas City	KKFI	90.1	VAR
Farmington	KTJJ	98.5	CW	Kansas City	KMXV	93.3	CHR
Fayette	KSSZ	93.9	TLK+	Kansas City	KCMO	94.9	OLG
Ferguson	KCFV	89.5	CH	Kansas City	KYYS	99.7	RK/C
Festus	KXEN	1010	RL/CC	Kansas City	KSRC	102.1	AC/S
Festus	KJFF	1400	N/TLK	Kansas City	KPRS	103.3	URB
Festus	KJFF	1400	N/TLK	Kansas City	KBEQ	104.3	CW
Florissant	KFTK	97.1	TLK+	Kennett	KOTC	830	CW
Fredericktown	KYLS	1450	CH	Kennett	KBOA	1540	TLK+
Fulton	KFAL	900	CW	Kennett	KAUF	89.9	RL
Fulton	KKCA	100.5	OLG	Kennett	KXOQ	104.3	CH
Gainesville	KMAC	99.7	AD/A	Kimberling City	KOMC	100.1	AS
Gallatin	KGOZ	101.7	CW	Kirksville	KIRX	1450	TLK+
Garden City	KCJK	105.1	AH	Kirksville	KTRM	88.7	AD/A
Gladstone	KGGN	890	RL/G	Kirksville	KKTR	89.7	CLA
Gordonville	KCGQ	99.3	RK/C	Kirksville	KHGN	90.7	RL
Halfway	KYOO	99.1	AC	Kirksville	KTUF	93.7	CW
Hannibal	KHMO	1070	N/TLK	Kirksville	KRXL	94.5	RK/C
Hannibal	KJIR	91.7	RL/G	Kirksville	KLTE	107.9	RL
Hannibal	KGRC	92.9	AC/H	Knob Noster	KCVQ	89.7	RL/CC
Harrisonville	KCFX	101.1	RK/C	Knob Noster	KXKX	105.7	CW
High Point	KMCV	89.9	RL	La Monte	KPOW	97.1	RK/C
Hollister	KBCV	1570	JAZ	Lake Ozark	KQUL	102.7	OLG
Houston	KBTC	1250	CW	Lamar	KHST	101.7	RK/C
Houston	KUNQ	99.3	CW	Lebanon	KBNN	750	BIZ
Independence	KCTE	1510	TLK+	Lebanon	KLWT	1230	TLK+
Ironton	KYLS	95.9	CW	Lebanon	KTTK	90.7	RL/G
Jackson	KUGT	1170	RL/CC	Lebanon	KJEL	103.7	CW
Jefferson City	KWOS	950	N/TLK	Lebanon	KCLQ	107.9	CW
Jefferson City	KLIK	1240	TLK+	Lee's Summit	KZPL	97.3	AC
Jefferson City	KJLU	88.9	JAZ	Lexington	KLEX	1570	RL
Jefferson City	KBBM	100.1	RK/C	Lexington	KNRX	107.3	URB
Jefferson City	KJMO	104.1	OLG	Liberty	KCXL	1140	TLK+
Jefferson City	KTXY	106.9	CHR	Liberty	KWJC	91.9	RK/C
Joplin	KWAS	1230	RL/G	Liberty	WDAF	106.5	CW
Joplin	KOCR	1310	RL/CC	Louisiana	KJFM	102.1	CW
Joplin	KQYX	1450	TLK+	Lutesville	KMHM	104.1	RL/G
Joplin	WMBH	1560	AS	Macon	KLTI	1560	CW
Joplin	KXMS	88.7	CLA	Macon	KIRK	99.9	EZL
Joplin	KOBC	90.7	RL/CC	Malden	KMAL	1470	N/TLK
Joplin	KSYN	92.5	CHR	Malden	KLSC	92.9	AC
Joplin	KIXQ	102.5	CW	Malta Bend	KRLI	103.9	AS
Kansas City	WDAF	610	SPTS	Mansfield	KTRI	95.9	CH
Kansas City	KCMO	710	TLK+	Marble Hill	KYRX	97.3	OLG
Kansas City	WHB	810	SPTS	Marshall	KMMO	1300	CW
Kansas City	KMBZ	980	N/TLK	Marshall	KMVC	91.7	VAR
Kansas City	KPHN	1190	YTH	Marshall	KMMO	102.9	CW
Kansas City	KPRT	1590	RL/G	Marshfield	KMRF	1510	RL/G
Kansas City	KLJC	88.5	RL/G	Marshfield	KNLM	91.9	RL/G
Kansas City	KCUR	89.3	CLA	Marshfield	KKLH	104.7	RK/C

Radio on the Road

Missouri

City	Station	Freq	Format	City	Station	Freq	Format
Maryville	KNIM	1580	N/TLK	Poplar Bluff	KPPL	92.5	CW
Maryville	KXCV	90.5	CLA	Poplar Bluff	KKLR	94.5	CW
Maryville	KNIM	97.1	RK/C	Poplar Bluff	KJEZ	95.5	RK/C
Memphis	KMEM	100.5	CW	Poplar Bluff	KAHR	96.7	AC/H
Mexico	KXEO	1340	AC	Poplar Bluff	KZMA	103.5	AC
Mexico	KJAB	88.3	RL/G	Portageville	KMIS	1050	SPTS
Mexico	KWWR	95.7	CW	Potosi	KYRO	1280	CW
Miner	KBHI	107.1	AC	Potosi	KNLP	89.7	RL/G
Moberly	KWIX	1230	TLK+	Potosi	KHZR	97.7	RL/CC
Moberly	KBKC	90.1	RL	Republic	KADI	99.5	RL/CC
Moberly	KRES	104.7	CW	Rolla	KTTR	1490	TLK+
Moberly	KZZT	105.5	OLG	Rolla	KMOZ	1590	AS
Monett	KKBL	95.9	CH	Rolla	KUMR	88.5	VAR
Monroe City	KWBZ	107.5	CH	Rolla	KMNR	89.7	VAR
Montgo'ry City	KMCR	103.9	AC/H	Rolla	KDAA	97.5	AC/S
Mount Vernon	KZRQ	106.7	RK/C	Rolla	KZNN	105.3	CW
Mtn Grove	KELE	1360	TLK+	Salem	KSMO	1340	N/TLK
Mtn Grove	KELE	92.5	CW	Salem	KSMO	1349	RL/CC
Mountain View	KUPH	96.9	AC/H	Salem	KKID	92.9	CH
Naylor	KZMA	99.9	N/A	Savannah	KSJQ	92.7	CW
Neosho	KBTN	1420	CW	Scott City	KGKS	93.9	AC/H
Neosho	KNEO	91.7	RL	Sedalia	KSIS	1050	TLK+
Neosho	KBTN	99.7	CW	Sedalia	KDRO	1490	CW
Nevada	KNEM	1240	CW	Sedalia	KSDL	92.1	AC/H
Nevada	KNMO	97.5	CW	Seligman	KIGL	93.3	RK/C
New Bloomf'ld	KNLG	90.3	RL/G	Shell Knob	KQMO	97.7	SP/MEX
New London	KZZK	105.9	RK/C	Sikeston	KSIM	1400	N/TLK
New Madrid	KTMO	106.5	CW	Sikeston	KRHW	1520	CW
Nixa	KGBX	105.9	AC	Sikeston	KBXB	97.9	CW
Osage Beach	KRMS	1150	TLK+	South West City	KLTK	1140	N/TLK
Osage Beach	KRMS	93.5	RK/C	South West City	KURM	100.3	N/TLK
Osceola	KCVJ	100.3	RL/CC	Sparta	KSPW	96.5	CH
Otterville	KCVK	107.7	RL/CC	Springfield	KWTO	560	TLK+
Overland	KRHS	90.1	VAR	Springfield	KTOZ	1060	AS
Owensville	KXMO	95.3	OLG	Springfield	KSGF	1260	N/TLK
Ozark	KOMG	92.9	CW	Springfield	KIDS	1340	TLK+
Palmyra	KICK	97.9	CW	Springfield	KGMY	1400	SPTS
Park Hills	KMFO	1240	N/TLK	Springfield	KLFJ	1550	TRAV
Park Hills	KBGM	91.1	RL/CC	Springfield	KWND	88.3	RL/CC
Parkville	KGSP	90.3	VAR	Springfield	KWFC	89.1	RL/G
Perryville	KBDZ	93.1	CW	Springfield	KSCV	90.1	RL
Piedmont	KPWB	1140	OLG	Springfield	KSMU	91.1	CLA
Piedmont	KPWB	104.9	CW	Springfield	KTTS	94.7	CW
Pleasant Hope	KTOZ	95.5	AC	Springfield	KXUS	97.3	RK/C
Point Lookout	KSMS	90.5	CLA	Springfield	KWTO	98.7	SPTS
Point Lookout	KCOZ	91.7	JAZ	Springfield	KTXR	101.3	AC/S
Poplar Bluff	KWOC	930	TLK+	St. Charles	KHOJ	1460	RL
Poplar Bluff	KLID	1340	OLG	St. Charles	KCLC	89.1	AD/A
Poplar Bluff	KOKS	89.5	RL/G	St. James	KTTR	99.7	TLK+
Poplar Bluff	KLUH	90.3	RL/G	St. Joseph	KFEQ	680	FRM

City	Station	Freq	Format	City	Station	Freq	Format
Waynesville	KJPW	102.3	AC				
Webb City	KKLL	1100	RL/G				
Webb City	KJMK	93.9	AC				
Webb City	KXDG	97.9	RK/C				
West Plains	KWPM	1450	TLK+				
West Plains	KSMW	90.9	CLA				
West Plains	KSPQ	93.9	RK				
West Plains	KKDY	102.5	CW				
Wheeling	KULH	105.9	RL/CC				
Willard	KOSP	105.1	OLG				
Willow Springs	KUKU	1330	TLK+				
Willow Springs	KUKU	100.3	OLG				
Windsor	KWKJ	98.5	CH				

Radio on the Road   Montana

City	Station	Freq	Format	City	Station	Freq	Format
Anaconda	KANA	580	OLG	Conrad	KTZZ	93.7	CH
Anaconda	KGLM	97.7	AC/H	Deer Lodge	Kbck	1400	CW
Arlee	KJFT	90.3	N/A	Deer Lodge	KQRV	96.9	CW
Baker	KJJM	100.5	RK/CL	Dillon	KDBM	1490	CW
Baker	KFLN	960	CW	Dillon	KDWG	90.9	VAR
Belgrade	KGVW	640	RK/C	Dillon	KBEV	98.3	AC
Belgrade	KGVW	640	RK/C	Dutton	KVVR	97.9	AC
Belgrade	KISN	96.7	CH	East Helena	KKGR	680	OLG
Belgrade	KCMM	99.1	RL/CC	East Helena	KHKR	104.1	CW
Big Fork	KIBG	100.7	VAR	East Missoula	KLCY	930	AS
Big Sky	KBZM	104.7	CH	Florence	KDTR	103.3	N/A
Billings	KURL	730	RL	Forsyth	KIKC	1250	OLG
Billings	KGHL	790	CW	Forsyth	KIKC	101.3	CW
Billings	KBLG	910	TLK+	Ft. Belknap	KGVA	88.1	VAR
Billings	KBUL	970	N/TLK	Glasgow	KLTZ	1240	CW
Billings	KMZK	1240	RL/CC	Glasgow	KLAN	93.5	AC
Billings	KLMT	89.3	RL/CC	Glendive	KGLE	590	RL
Billings	KBLW	90.1	RL	Glendive	KXGN	1400	OLG
Billings	KLRV	90.9	N/A	Glendive	KDZN	96.5	CW
Billings	KEMC	91.7	CLA	Great Falls	KMON	560	CW
Billings	KYYA	93.3	AC	Great Falls	KEIN	1310	AS
Billings	KRKX	94.1	RK/C	Great Falls	KXGF	1400	AS
Billings	KRZN	96.3	RK/C	Great Falls	KQDI	1450	N/TLK
Billings	KKBR	97.1	OLG	Great Falls	KGFC	88.9	RL
Billings	KGHL	98.5	CW	Great Falls	KGPR	89.9	VAR
Billings	KCTR	102.9	CW	Great Falls	KLFM	92.9	OLG
Billings	KBBB	103.7	AC/S	Great Falls	KMON	94.5	CW
Billings	KBEX	105.1	CH	Great Falls	KAAK	98.9	AC/H
Billings	KNDZ	105.1	OLG	Great Falls	KLSK	100.3	RL/CC
Billings	KZRV	107.5	AC/H	Great Falls	KQDI	106.1	RK/C
Bozeman	KBOZ	1090	TLK+	Great Falls	KINX	107.3	RK/C
Bozeman	KOBB	1230	AS	Hamilton	KLYQ	1240	N/TLK
Bozeman	KMMS	1450	TLK+	Hamilton	KMZO	90.3	RL/CC
Bozeman	KRFR	89.3	N/A	Hamilton	KUFN	91.9	VAR
Bozeman	KGLT	91.9	VAR	Hamilton	Kbaz	96.3	RK/C
Bozeman	KOBB	93.7	OLG	Hamilton	KXDR	98.7	CH
Bozeman	KMMS	95.1	AD/A	Hardin	KHDN	1230	TLK+
Bozeman	Kboz	99.9	CW	Hardin	kmhk	95.5	RK/C
Bozeman	KBMZ	102.1	CLA	Havre	KOJM	610	AC
Bozeman	KZMY	103.5	AC/S	Havre	KNMC	90.1	CLA
Butte	KBOW	550	CW	Havre	KPQX	92.5	CW
Butte	KXTL	1370	TLK+	Havre	KXEI	95.1	RL
Butte	KFRD	88.3	RL	Helena	KMTX	950	AS
Butte	KNFR	88.9	RL	Helena	KBLL	1240	TLK+
Butte	KAPC	91.3	N/TLK	Helena	KCAP	1340	N/TLK
Butte	KAAR	92.5	CW	Helena	KHLV	90.1	N/A
Butte	KOPR	94.1	RK/C	Helena	KUHM	91.7	VAR
Butte	KMBR	95.5	RK/C	Helena	KBLL	99.5	CW
Butte	KMSM	107.1	VAR	Helena	KMZT	101.1	RK/C
Cascade	KIKF	104.9	CW	Helena	KVCM	103.1	RL
Chinook	KRYK	101.3	AC/H	Helena	KMTX	105.3	AC
Colstrip	KCMJ	99.5	RL	Kalispell	KGEZ	600	TLK+
Columbia Falls	KKMT	95.9	AC/S	Kalispell	KOFI	1180	TLK+

115

City	Station	Freq	Format		City	Station	Freq	Format
Kalispell	KUKL	89.9	VAR					
Kalispell	KSPL	90.9	RL					
Kalispell	KALS	97.1	RL/CC					
Kalispell	KBBZ	98.5	RK/C					
Kalispell	KZMN	103.9	RK/C					
Kalispell	KDBR	106.3	CW					
Laurel	KBSR	1490	TLK+					
Laurel	KRSQ	101.7	CH					
Lewiston	KLCM	95.9	CH					
Lewistown	KXLO	1230	CW					
Lewistown	KLEU	91.1	RL					
Libby	KLCB	1230	CW					
Libby	KTNY	101.7	AS					
Livingston	KPRK	1340	CH					
Livingston	KOZB	97.5	RK/C					
Lockwood	KYLW	1450	AC/S					
Malta	KMMR	100.1	CW					
Miles City	KATL	770	AC					
Miles City	KMTA	1050	RK/C					
Miles City	KECC	90.7	CLA					
Miles City	KKRY	92.3	CW					
Missoula	KGVO	1290	N/TLK					
Missoula	KYLT	1340	TLK+					
Missoula	KGRZ	1450	SPTS					
Missoula	KUFM	89.1	N/TLK					
Missoula	KMZL	91.1	RL/CC					
Missoula	KGGL	93.3	CW					
Missoula	KYSS	94.9	CW					
Missoula	KZOQ	100.1	RK/C					
Missoula	KMSO	102.5	AC/H					
Missoula	KKNS	105.9	N/A					
Pinedale	KBQQ	106.7	OLG					
Plains	KPLG	91.5	RL					
Plentywood	KATQ	1070	CW					
Plentywood	KATQ	100.1	CW					
Polson	KERR	750	CW					
Pryor	KPGB	88.3	RL					
Red Lodge	KMXE	99.3	RK/C					
Ronan	KQRK	92.3	CH					
Scobey	KCGM	95.7	CW					
Shelby	KSEN	1150	OLG					
Shelby	KZIN	96.3	CW					
Superior	KLTC	107.5	CH					
Sydney	KGCX	93.1	RK/C					
Sydney	KGCX	93.1	RK/C					
Sydney	KTHC	95.1	AC					
W Yellowstone	KWYS	920	OLG					
W Yellowstone	KEZQ	92.9	AC/S					
Whitefish	KJJR	880	TLK+					
Wolf Point	KVCK	1450	AC/S					
Wolf Point	KVCK	92.7	CW					

Radio on the Road — Nebraska

City	Station	Freq	Format	City	Station	Freq	Format
Ainsworth	KBRB	1400	TLK+	Grand Island	KSYZ	107.7	AC
Ainsworth	KBRB	92.7	CW	Hastings	KHAS	1230	AC/S
Albion	KUSO	92.7	CW	Hastings	KICS	1550	SPTS
Alliance	KCOW	1400	TLK+	Hastings	KCNT	88.1	VAR
Alliance	KTNE	91.1	CLA	Hastings	KHNE	89.1	CLA
Alliance	KPNY	102.1	AC/H	Hastings	KFKX	90.1	AD/A
Alliance	KAAQ	105.9	CW	Hastings	KNHA	90.9	N/A
Auburn	KNCY	94.7	CW	Hastings	KLIQ	94.5	AC
Aurora	KRGY	97.3	AC/H	Hastings	KEZH	101.5	RK/C
Bassett	KMNE	90.3	CLA	Holdrege	KUVR	1380	AC/S
Beatrice	KWBE	1450	AC	Holdrege	KMTY	97.7	OLG
Beatrice	KQIQ	88.3	VAR	Hubbard	KAYA	91.3	RL
Beatrice	KTGL	92.9	RK/CL	Imperial	KADL	102.9	OLG
Bellevue	KYDZ	1180	Youth	Kearney	KGFW	1340	TLK+
Bellevue	KOZN	1620	SPTS	Kearney	KXPN	1460	SPTS
Bennington	KHUS	93.3	CW	Kearney	KLPR	91.3	JAZ
Blair	KDVC	91.1	VAR	Kearney	KKPR	98.9	OLG
Blair	KBLR	97.3	URB	Kearney	KKPR	98.9	OLG
Bridgeport	KOLT	101.3	CW	Kearney	KRNY	102.3	CW
Broken Bow	KCNI	1280	CW	Kearney	KQKY	105.9	CH
Broken Bow	KBBN	98.3	RK/C	Kimball	KIMB	1260	CW
Central City	KZEN	100.3	CW	Kimball	KBFZ	100.1	OLG
Chadron	KCSR	610	CW	Lexington	KRVN	880	FRM
Chadron	KCNE	91.9	CLA	Lexington	KLNE	88.7	CLA
Chadron	KQSK	97.5	CW	Lexington	KRVN	93.1	CW
Columbus	KJSK	900	N/TLK	Lincoln	KFOR	1240	TLK+
Columbus	KTTT	1510	TLK+	Lincoln	KLIN	1400	N/TLK
Columbus	KTLX	91.9	RL	Lincoln	KLMS	1480	SPTS
Columbus	KKOT	93.5	CH	Lincoln	KLCV	88.5	RL
Columbus	KLIR	101.1	AC/S	Lincoln	KZUM	89.3	VAR
Cozad	KAMI	1580	RL	Lincoln	KRNU	90.3	RK/C
Cozad	KAMI	104.5	RL	Lincoln	KUCV	91.1	CLA
Crete	KDNE	91.9	VAR	Lincoln	KRKR	95.1	RK/C
Crete	KKNB	104.1	RK/C	Lincoln	KLTQ	101.9	AC/S
Crookston	KINI	96.1	VAR	Lincoln	KFRX	102.7	CH
Dakota City	KTFJ	1250	RL/G	Lincoln	KKUL	105.3	OLG
Fairbury	KGMT	1310	AC	Lincoln	KLMY	106.3	AC
Fairbury	KUTT	99.5	CW	Lincoln	KBBK	107.3	AC/H
Falls City	KTNC	1230	OLG	Mccook	KNAX	700	N/A
Falls City	KLZA	101.3	AC/H	Mccook	KBRL	1300	OLG
Fremont	KHUB	1340	TLK+	Mccook	KNGN	1360	RL
Fremont	KFMT	105.5	CH	Mccook	KSWN	93.9	TLK+
Gering	KOZY	103.9	AC	Mccook	KICX	96.1	AC
Gordon	KSDZ	95.5	CW	Mccook	KRKU	98.5	RK/C
Grand Island	KMMJ	750	TLK+	Mccook	KIOD	105.3	CW
Grand Island	KRGI	1430	TLK+	Merriman	KRNE	91.5	CLA
Grand Island	KLNB	88.3	RL/CC	Milford	KFGE	98.1	CW
Grand Island	KNFA	90.7	N/A	Mitchell	KOSJ	89.5	N/A
Grand Island	KROA	95.7	RL/CC	Nebraska City	KNCY	1600	CW
Grand Island	KRGI	96.5	CW	Nebraska City	KBBX	97.7	SP/MEX

City	Station	Freq	Format	City	Station	Freq	Format
Norfolk	WJAG	780	TLK+	So Sioux City	KSFT	107.1	AC/S
Norfolk	KXNE	89.3	CLA	Superior	KRFS	1600	AS
Norfolk	KPNO	90.9	RL/G	Superior	KRFS	103.9	CW
Norfolk	KNEN	94.7	AC	Terrytown	KOAQ	690	OLG
Norfolk	KEXL	106.7	AC	Terrytown	KCMI	96.9	RL/CC
North Platte	KJLT	970	RL	Valentine	KVSH	940	AC
North Platte	KODY	1240	TLK+	Wayne	KTCH	1590	CW
North Platte	KOOQ	1410	OLG	Wayne	KWSC	91.9	AD/A
North Platte	KPNE	91.7	CLA	Wayne	KTCH	104.9	OLG
North Platte	KJLT	94.9	RL/CC	West Point	KTIC	840	CW
North Platte	KELN	97.1	AC/H	West Point	KWPN	107.9	AC
North Platte	KXNP	103.5	CW	Wilber	KFLV	89.9	RL/CC
Ogallala	KOGA	930	AS	Winnebago	KSUX	105.7	CW
Ogallala	KOGA	99.7	RK/C	York	KAWL	1370	OLG
Ogallala	KMCX	106.5	CW	York	KTMX	104.9	AC
Omaha	KXSP	590	SPTS				
Omaha	KCRO	660	RL/CC				
Omaha	KFAB	1110	TLK+				
Omaha	KKAR	1290	TLK+				
Omaha	KHLP	1420	TLK+				
Omaha	KOSR	1490	AS				
Omaha	KVSS	88.9	RL				
Omaha	KVNO	90.7	CLA				
Omaha	KIOS	91.5	CLA				
Omaha	KEZO	92.3	RK				
Omaha	KQCH	94.1	CH				
Omaha	KEFM	96.1	AC				
Omaha	KGOR	99.9	OLG				
Omaha	KGBI	100.7	RL/CC				
Omaha	KSRZ	104.5	AC/H				
Omaha	KKCD	105.9	RK/C				
O'neill	KBRX	1350	CH				
O'neill	KBRX	102.9	CW				
Orchard	KGRD	105.3	RL/CC				
Ord	KNLV	1060	OLG				
Ord	KNLV	103.9	CW				
Plattsmouth	KOIL	1000	CW				
Plattsmouth	KCTY	106.9	AH				
Ralston	KMLV	88.1	RL/CC				
Ravenna	KKJK	103.1	N/A				
Scottsbluff	KNEB	960	CW				
Scottsbluff	KOLT	1320	N/TLK				
Scottsbluff	KLJV	88.3	RL/CC				
Scottsbluff	KLJV	88.3	RL/CC				
Scottsbluff	KMOR	92.9	RK/C				
Scottsbluff	KNEB	94.1	CW				
Seward	KZKX	96.9	CW				
Sidney	KSID	1340	CW				
Sidney	KSID	98.7	AC				

City	Station	Freq	Format	City	Station	Freq	Format
Armagosa Vly	KPKK	101.1	JAZ	Las Vegas	KSNE	106.5	AC/S
Boulder City	KSTJ	102.7	OLG	Laughlin	KNYE	95.1	OLG
Carlin	KHIX	96.7	AC/H	Laughlin	KVGS	107.9	URB
Carson City	KPTL	1300	TLK+	Logandale	KADD	93.5	AC/H
Carson City	KNIS	91.3	RL/CC	Lund	KWPR	88.7	N/TLK
Carson City	KZTQ	97.3	AD/A	Mesquite	KEKL	88.5	N/A
Carson City	KBUL	98.1	CW	Mesquite	KAIZ	91.1	N/A
Elko	KELK	1240	AC	Mesquite	KVEG	97.5	CH
Elko	KTSN	1340	TLK+	Moapa Valley	KBHQ	104.7	SP/TO
Elko	KNCC	91.5	CLA	No Las Vegas	KXNT	840	TLK+
Elko	KLKO	93.5	RK	No Las Vegas	KSFN	840	TLK+
Elko	KRJC	95.3	CW	No Las Vegas	KSHP	1400	TLK+
Ely	KELY	1230	CH	No Las Vegas	KJUL	104.3	AS
Ely	KDSS	92.7	CW	Pahrump	KNYE	95.1	TLK+
Ely	KELY	101.7	AC	Pahrump	KXTE	107.5	RK/C
Fallon	KHWG	750	N/A	Panaca	KLNR	91.7	N/TLK
Fallon	KVLV	980	CW	Paradise	KNUU	970	N/TLK
Fallon	KVLV	99.3	AC	Pioche	KBZB	98.9	CW
Fallon	KRNG	101.3	RL/CC	Reno	KPLY	630	SPTS
Gardnerville	KCMY	99.1	CW	Reno	KKOH	780	TLK+
Hawthorne	KQMC	90.1	N/A	Reno	KIHM	920	RL
Henderson	KDOX	1280	SP/TO	Reno	KJFK	1230	TLK+
Henderson	KMXB	94.1	AC/H	Reno	KXEQ	1340	SP/MEX
Henderson	KWNR	95.5	CW	Reno	KHIT	1450	AS
Henderson	KMZQ	100.5	AC	Reno	KXTO	1550	SP/RL
Incline Village	KTHX	100.1	AD/A	Reno	KUNR	88.7	CLA
Incline Village	KRNO	106.9	AC/S	Reno	KURK	92.9	CH
Indian Springs	KQMR	99.3	SP/TO	Reno	KNEV	95.5	AC
Jackpot	KBSJ	91.3	N/TLK	Reno	KRNV	102.1	SP/MEX
Las Vegas	KBTB	670	N/A	Reno	KDOT	104.5	RK/C
Las Vegas	KDWN	720	TLK+	Reno	KOZZ	105.7	RK/C
Las Vegas	KBAD	920	SPTS	Sparks	KBZZ	1270	TLK+
Las Vegas	KKVV	1060	RL	Sparks	KLRH	88.3	RL/CC
Las Vegas	KSFN	1140	TLK+	Sparks	KJZS	92.1	JAZ
Las Vegas	KLAV	1230	TLK+	Sparks	KRZQ	100.9	RK/C
Las Vegas	KRLV	1340	SP/TK	Sparks	KBDB	1400	AS
Las Vegas	KENO	1460	SPTS	Sun Valley	KQLO	1590	SP/OM
Las Vegas	KCEP	88.1	URB	Sun Valley	KWNZ	93.7	CH
Las Vegas	KNPR	88.9	N/TLK	Sun Valley	KUUB	94.5	CW
Las Vegas	KCNV	89.7	CLA	TonOpah	KTPH	91.7	N/TLK
Las Vegas	KSOS	90.5	RL/CC	Tonopah	KHWK	92.7	SIL
Las Vegas	KUNV	91.5	JAZ	Wendover	KVUW	102.3	N/A
Las Vegas	KOMP	92.3	RK	Whitney	KLSQ	870	SP/OM
Las Vegas	KQOL	93.1	OLG	Winchester	KBET	790	N/A
Las Vegas	KKLZ	96.3	RK/C	Winnemucca	KWNA	1400	OLG
Las Vegas	KXPT	97.1	CH	Winnemucca	KWNA	92.7	CW
Las Vegas	KLUC	98.5	CH				
Las Vegas	KWID	101.9	SP/OM				
Las Vegas	KISF	103.5	SP/MEX				
Las Vegas	KVBC	105.1	SP/MEX				

City	Station	Freq	Format		City	Station	Freq	Format
Bedford	WMLL	96.5	RK/C		Lebanon	WVFA	90.5	RL
Belmont	WNHW	93.3	CW		Lebanon	WXXK	100.5	CW
Berlin	WMOU	1230	AC/S		Lisbon	WLTN	96.7	AC
Campton	WVFM	105.7	AD/A		Littleton	WLTN	1400	OLG
Claremont	WTSV	1230	SPTS		Littleton	WMTK	106.3	CH
Claremont	WHDQ	106.1	RK/C		Madbury	WWNH	1340	RL
Concord	WKXL	1450	N/TLK		Manchester	WGIR	610	N/TLK
Concord	WEVO	89.1	N/TLK		Manchester	WKBR	1250	OLG
Concord	WSPS	90.5	VAR		Manchester	WFEA	1370	AS
Concord	WVNH	91.1	RL/CC		Manchester	WLMW	90.7	RL/CC
Concord	WQTX	102.3	RK/C		Manchester	WZID	95.7	AC
Concord	WJYY	105.5	CH		Manchester	WGIR	101.1	RK/C
Conway	WBNC	1050	TLK+		Meredith	WWHQ	101.5	RK/C
Conway	WMWV	93.5	AD/A		Moultonborough	WSCY	106.9	CW
Conway	WBNC	104.5	AC/H		Mt. Washington	WHOM	94.9	AC/S
Derry	WDER	1320	RL		Nashua	WSNH	900	TLK+
Dover	WTSN	1270	N/TLK		Nashua	WSMN	1590	SIL
Dover	WOKQ	97.5	CW		Nashua	WEVS	88.3	N/TLK
Durham	WUNH	91.3	AD/A		Nashua	WFNQ	106.3	CH
Exeter	WGIP	1540	N/TLK		New Durham	WWPC	91.7	RL
Exeter	WPEA	90.5	VAR		New London	WSCS	90.9	VAR
Exeter	WERZ	107.1	CH		New London	WNTK	99.7	TLK+
Farmington	WZEN	106.5	OLG		Newport	WNTK	1020	CW
Franklin	WFTN	1240	AS		Newport	WVRR	101.7	RK/C
Franklin	WFTN	94.1	AC/H		North Conway	WPKQ	103.7	CW
Gorham	WEVC	107.1	N/TLK		Peterborough	WFEX	92.1	RK/H
Hampton	WSAK	102.1	RK/C		Plymouth	WPNH	1300	AS
Hanover	WQTH	720	N/A		Plymouth	WPCR	91.7	AD/A
Hanover	WDCR	1340	AD/A		Plymouth	WPNH	100.1	RK/C
Hanover	WTSL	1400	N/TLK		Portsmouth	WMYF	1380	AS
Hanover	WGXL	92.3	AC/H		Rochester	WGIN	930	N/TLK
Hanover	WFRD	99.3	RK/C		Rochester	WQSO	96.7	OLG
Hanover	WEVH	107.1	N/TLK		Salem	WCEC	1110	SP/TK
Haverhill	WYKR	101.3	CW		Somersworth	WBYY	98.7	AC/S
Henniker	WNEC	91.7	AD/A		Walpole	WLPL	96.3	OLG
Henniker	WNNH	99.1	OLG		Winchester	WINQ	98.7	CW
Hillsboro	WTPL	107.7	N/TLK		Wolfeboro	WASR	1420	AC
Hinsdale	WYRY	104.9	CW		Wolfeboro	WLZK	104.9	OLG
Jackson	WEVJ	99.5	N/TLK		Lebanon	WVFA	90.5	RL
Jaffrey	WXNH	540	N/A					
Keene	WZBK	1220	AS					
Keene	WKBK	1290	TLK+					
Keene	WEVN	90.7	N/TLK					
Keene	WKNH	91.3	AD/A					
Keene	WKNE	103.7	AC/H					
Laconia	WEZS	1350	EZL					
Laconia	WEMJ	1490	N/TLK					
Laconia	WLNH	98.3	AC/H					
Lancaster	WXXS	102.3	AC/H					
Lebanon	WUVR	1490	TLK+					

City	Station	Freq	Format	City	Station	Freq	Format
Asbury Park	WADB	1310	AS	Hackettstown	WRNJ	1510	OLG
Asbury Park	WADB	1310	AS	Hackettstown	WNTI	91.9	AD/A
Asbury Park	WYGG	88.1	N/A	Hammonton	WGYM	1580	N/TLK
Asbury Park	WJLK	94.3	AC/H	Hazlet	WDDM	89.3	N/A
Atlantic City	WMID	1340	OLG	Jersey City	WSNR	620	N/A
Atlantic City	WKXW	1450	SPTS	Lakewood Twn	WOBM	1160	AS
Atlantic City	WAJM	88.9	VAR	Lawrenceville	WRRC	107.7	VAR
Atlantic City	WNJN	89.7	N/TLK	Lincroft	WBJB	90.5	AD/A
Atlantic City	WAYV	95.1	CH	Long Branch	WWZY	107.1	AC/S
Atlantic City	WFPG	96.9	AC	Madison	WMNJ	88.9	VAR
Atlantic City	WMGM	103.7	RK/C	Mahwah	WRPR	90.3	RK/H
Atlantic City	WPUR	107.3	CW	Manahawkin	WNJM	89.9	JAZ
Avalon	WILW	94.3	OLG	Manahawkin	WYRS	90.7	RL
Beach Haven	WVBH	88.3	RL/CC	Manahawkin	WJRZ	100.1	OLG
Belvidere	WWYY	107.1	AC/S	Manahawkin	WCHR	105.7	RK/C
Berlin	WNJS	88.1	JAZ	Margate City	WTTH	96.1	URB
Berlin	WNJB	89.3	JAZ	Medford Lakes	WVBV	90.5	RL
Blackwood	WDBK	91.5	AD/A	Millville	WMVB	1440	VAR
Blairstown	WHCY	106.3	CH	Millville	WIXM	97.3	TLK+
Brick	WBGD	91.9	RK/C	Morristown	WMTR	1250	OLG
Bridgeton	WSNJ	1240	VAR	Morristown	WJSV	90.5	AD/A
Bridgewater	WWTR	1170	OLG	Mt. Holly	WWJZ	640	YTH
Brigantine	WWFP	90.5	N/A	New Brunsw'k	WCTC	1450	N/TLK
Camden	WTMR	800	RL	New Brunsw'k	WRSU	88.7	AD/A
Camden	WEMG	1310	SP/OM	New Brunsw'k	WMGQ	98.3	AC
Camden	WKDN	106.9	RL	Newark	WNSW	1430	N/A
Canton	WJKS	101.7	URB	Newark	WBGO	88.3	JAZ
Cape May	WWCJ	89.1	CLA	Newark	WBGO	88.3	JAZ
Cape May	WAIV	102.3	CH	Newark	WFME	94.7	RL
Cape May	WGBZ	105.5	CH	Newark	WHTZ	100.3	CH
Cape May CH	WJPG	88.1	RL/CC	Newark	WCAA	105.9	SP/OM
Cape May CH	WNJZ	90.3	JAZ	Newton	WNNJ	1360	OLG
Cherry Hill	WSJI	89.5	RL/CC	Newton	WNNJ	103.7	CH
Delaware Twn	WDVR	89.7	VAR	N Cape May	WSJQ	106.7	CH
Dover	WWNJ	91.1	CLA	Oakland	WVNJ	1160	AS
Dover	WDHA	105.5	RK/C	Ocean Acres	WBBO	98.5	RK/H
East Orange	WFMU	91.1	VAR	Ocean City	WIBG	1020	RL/CC
Eatontown	WHTG	1410	OLG	Ocean City	WRTQ	91.3	CLA
Eatontown	WHTG	106.3	RK/C	Ocean City	WTKU	98.3	OLG
Egg Harbor	WXGN	90.5	RL/CC	Ocean City	WKOE	106.3	AC/S
Egg Harbor	WSJO	104.9	AC/H	Parsippany	WXMC	1310	SP/RL
Elizabeth	WJDM	1530	SP/RL	Paterson	WPAT	930	SP/OM
Ewing	WIMG	1300	RL/G	Paterson	WPAT	93.1	SP/OM
Flemington	WCHR	1040	RL	Pemberton	WBZC	88.9	VAR
Flemington	WCVH	90.5	VAR	Pennsauken	WRNB	107.9	URB
Florence	WIFI	1460	RL/CC	Petersburg	WJSE	102.7	RK/C
Franklin	WSUS	102.3	AC	Piscataway	WVPH	90.3	VAR
Freehold Twp	WRDR	89.7	RL	Pleasantville	WOND	1400	N/TLK
Glassboro	WGLS	89.7	VAR	Pleasantville	WUSS	1490	OLG
Hackensack	WWDJ	970	RL	Pleasantville	WZBZ	99.3	CH

City	Station	Freq	Format	City	Station	Freq	Format
Point Pleasant	WRAT	95.9	RK/C				
Pomona	WLFR	91.7	VAR				
Pompton Lks	WGHT	1500	OLG				
Port Republic	WXXY	88.7	RL/G				
Princeton	WHWH	1350	BIZ				
Princeton	WTTM	1680	N/A				
Princeton	WPRB	103.3	VAR				
Princeton Jct	WWPH	107.9	RK/C				
Salem	WFAI	1510	RL/G				
South Orange	WSOU	89.5	RK/C				
Stirling	WKMB	1070	RL/G				
Sussex	WNJP	88.5	AD/A				
Teaneck	WFDU	89.1	URB				
Teaneck	WFDU	89.1	AD/A				
Toms River	WOBM	92.7	AC				
Trenton	WPHY	920	SPTS				
Trenton	WBUD	1260	OLG				
Trenton	WNJT	88.1	AD/A				
Trenton	WWFM	89.1	CLA				
Trenton	WTSR	91.3	AD/A				
Trenton	WPST	94.5	CH				
Trenton	WTHK	97.5	CH				
Trenton	WKXW	101.5	TLK+				
Tuckerton	WBHX	99.7	AC/S				
Union Twn	WKNJ	90.3	VAR				
Upper Montclair	WMSC	101.5	VAR				
Villas	WCZT	98.7	AC				
Vineland	WMIZ	1270	SP/OM				
Vineland	WVLT	92.1	OLG				
Wash'gton Twn	WNJC	1360	VAR				
Wayne	WPSC	88.7	RK/C				
W Long Branch	WMCX	88.9	RK/C				
Wildwood	WCMC	1230	AS				
Wildwood	WZXL	100.7	RK/C				
Wildwood Crst	WDTH	93.1	URB				
Woodbine	WJPH	89.9	RL/CC				
Zarephath	WAWZ	99.1	RL/CC				
Point Pleasant	WRAT	95.9	RK/C				
Pomona	WLFR	91.7	VAR				
Pompton Lks	WGHT	1500	OLG				
Port Republic	WXXY	88.7	RL/G				
Princeton	WHWH	1350	BIZ				
Princeton	WTTM	1680	N/A				
Princeton	WPRB	103.3	VAR				

City	Station	Freq	Format		City	Station	Freq	Format
Alamo Com'ity	KABR	1500	SPN		Carlsbad	KATK	92.1	CW
Alamogordo	KRSY	1230	TLK+		Carlsbad	KCDY	104.1	AC
Alamogordo	KINN	1270	TLK+		Chama	KZRM	95.9	CH
Alamogordo	KUPR	91.7	VAR		Clayton	KLMX	1450	CW
Alamogordo	KYEE	94.3	CH		Cloudcraft	KHII	88.9	SIL
Alamogordo	KNMZ	103.7	RK/C		Cloudcraft	KNMB	96.7	CW
Alamogordo	KZZX	105.5	CW		Clovis	KWKA	680	OLG
Alamogordo	KKBO	107.9	N/A		Clovis	KICA	980	TLK+
Albuquerque	KNML	610	SPTS		Clovis	KCLV	1240	SPTS
Albuquerque	KDAZ	730	RL		Clovis	KKCC	90.3	N/A
Albuquerque	KKOB	770	N/TLK		Clovis	KAQF	91.1	RL
Albuquerque	KSVA	920	RL		Clovis	KCLV	99.1	CW
Albuquerque	KKIM	1000	RL		Clovis	KTQM	99.9	AC
Albuquerque	KDEF	1150	SPTS		Clovis	KRMQ	101.5	OLG
Albuquerque	KXKS	1190	RL		Clovis	KKYC	102.3	CW
Albuquerque	KABQ	1350	TLK+		Clovis	KSMX	107.5	AC/H
Albuquerque	KRZY	1450	SP/TK		Corrales	KKNS	1310	SPTS
Albuquerque	KKJY	1550	AS		Corrales	KSYU	95.1	URB
Albuquerque	KRKE	1600	OLG		Deming	KOTS	1230	CW
Albuquerque	KLYT	88.3	RL/CC		Deming	KZPI	91.7	SP/RL
Albuquerque	KANW	89.1	N/TLK		Deming	KDEM	94.3	AC
Albuquerque	KUNM	89.9	N/TLK		Dulce	KCIE	90.5	VAR
Albuquerque	KFLQ	91.5	RL/CC		Espanola	KDCE	950	SP/MEX
Albuquerque	KRST	92.3	CW		Espanola	KYBR	92.9	SP/MEX
Albuquerque	KKOB	93.3	CH		Eunice	KEJL	100.9	RK/C
Albuquerque	KZRR	94.1	RK		Farmington	KNDN	960	N/A
Albuquerque	KBZU	96.3	RK/C		Farmington	KRZE	1280	SP/MEX
Albuquerque	KMGA	99.5	AC		Farmington	KENN	1390	TLK+
Albuquerque	KPEK	100.3	AC		Farmington	KNMI	88.9	RL/CC
Albuquerque	KJFA	101.3	SP/MEX		Farmington	KSJE	90.9	CLA
Albuquerque	KDRF	103.3	AH		Farmington	KRWN	92.9	RK/C
Albuquerque	KBQI	107.9	CW		Farmington	KPCL	95.7	RL/CC
Angel Fire	KKIT	99.1	RK/C		Farmington	KDAG	96.9	RK/C
Armijo	KNKT	107.1	RL/CC		Farmington	KTRA	102.1	CW
Artesia	KSVP	990	TLK+		Fruitland	KTGW	101.7	RL
Artesia	KTZA	92.9	CW		Gallup	KYVA	1230	CW
Aztec	KCQL	1340	SPTS		Gallup	KGAK	1330	N/A
Aztec	KWYK	94.9	AC		Gallup	KGLP	91.7	N/TLK
Bayard	KNFT	950	TLK+		Gallup	KXXI	93.7	RK/C
Bayard	KNFT	102.9	CW		Gallup	KKOR	94.5	AC/H
Belen	KARS	860	CW		Gallup	KGLX	99.1	CW
Belen	KVLK	90.9	RL/CC		Gallup	KFMQ	106.1	RK/C
Belen	KLVO	97.7	SP/MEX		Grants	KMIN	980	RK/H
Bloomfield	KKFG	104.5	OLG		Grants	KLGQ	90.3	N/A
Bosque Farms	KTEG	104.7	RK/C		Grants	KDSK	92.7	OLG
Bosque Farms	KOLV	105.5	RL/CC		Grants	KXXQ	100.7	CW
Cannon Afb	KKJC	90.7	N/A		Grants	KYVA	103.7	OLG
Carlsbad	KATK	740	AS		Hatch	KVLC	101.1	OLG
Carlsbad	KCCC	930	OLG		Hobbs	KYKK	1110	TLK+
Carlsbad	KAMQ	1240	RL/CC		Hobbs	KHOB	1390	OLG

City	Station	Freq	Format	City	Station	Freq	Format
Hobbs	KZOR	94.1	AC/H	Reserve	KLBZ	104.5	TLK+
Hobbs	KPER	95.7	CW	Rio Rancho	KAJZ	101.7	JAZ
Hobbs	KLMA	96.5	SP/MEX	Roswell	KBIM	910	N/TLK
Hobbs	KIXN	102.9	CW	Roswell	KINF	1020	TLK+
Jal	KPZA	103.7	SP/MEX	Roswell	KPSA	1230	SP/OM
Kirtland	KAZX	102.9	CH	Roswell	KRDD	1320	SP/MEX
La Luz	KRSY	92.7	CW	Roswell	KCRX	1430	OLG
Las Cruces	KSNM	570	AS	Roswell	KRSR	89.1	N/A
Las Cruces	KOBE	1450	TLK+	Roswell	KRLU	90.1	RL/CC
Las Cruces	KRUC	88.9	SP/RL	Roswell	KBIM	94.9	AC
Las Cruces	KMBN	89.7	RL	Roswell	KBCQ	97.1	CH
Las Cruces	KRWG	90.7	CLA	Roswell	KWFL	99.5	RL/CC
Las Cruces	KRUX	91.5	VAR	Roswell	KSFX	100.5	RK/C
Las Cruces	KROL	99.5	RL/CC	Roswell	KMOU	104.7	CW
Las Cruces	KHQT	103.1	CH	Roswell	KEND	106.5	RK/C
Las Cruces	KGRT	103.9	CW	Ruidoso	KBUY	1360	OLG
Las Vegas	KMNX	540	SP/MEX	Ruidoso	KWES	93.5	CW
Las Vegas	KFUN	1230	SP/TO	Ruidoso	KIDX	101.5	RK/C
Las Vegas	KDEP	91.1	RK/C	Ruidoso Dwns	KRUI	1490	TLK+
Las Vegas	KMDZ	96.7	RK/C	Santa Clara	KNUW	95.1	SP/MEX
Las Vegas	KBAC	98.1	AD/A	Santa Fe	KSWV	810	SP/MEX
Las Vegas	KLVF	100.7	AC	Santa Fe	KTRC	1260	TLK+
Lordsburg	KPSA	97.9	RK/C	Santa Fe	KTRC	1400	SPTS
Lordsburg	KWNM	105.5	CW	Santa Fe	KSFR	90.7	VAR
Los Alamos	KRSN	1490	TLK+	Santa Fe	KBOM	94.7	JAZ
Los Alamos	KTMN	98.5	OLG	Santa Fe	KHFM	95.5	CLA
Los Alamos	KKPL	106.7	SP/OM	Santa Fe	KKSS	97.3	CH
Los Alamos	KQBA	107.5	SP/MEX	Santa Fe	KABQ	104.1	URB
Los Lunas	KIOT	102.5	CH	Santa Fe	KKRG	105.1	CW
Los Lunas	KAGM	106.3	TLK+	Santa Fe	KRZY	105.9	SP/OM
Los Ranchos	KTBL	1050	TLK+	Santa Rosa	KSSR	1340	CH
Ro Albuq'rque	KTBL	1050	TLK+	Santa Rosa	KNLK	91.9	N/TLK
Ro Albuq'rque	KALY	1240	YTH	Santa Rosa	KIVA	95.9	CH
Lovington	KLEA	630	AC/S	Silver City	KSCQ	94.5	AC
Lovington	KLEA	101.7	OLG	Socorro	KQRI	89.5	N/A
Maljamar	KMTH	98.7	VAR	Socorro	KMXQ	92.9	CW
Maljamar	KWMW	105.1	CW	Taos	KXMT	99.1	SP/MEX
Mesilla Park	KMVR	104.9	AC	Taos	KTAO	101.5	AD/A
Mesquite	KELP	89.3	RL	Tatum	KTUM	107.1	RK/C
Milan	KQNM	1100	SIL	Thoreau	KXTC	99.9	CH
Pecos	KWRP	101.5	VAR	Truth/ Conseq	KCHS	1400	CW
Pecos	KLBU	102.9	AD/A	Truth/ Conseq	KKVS	98.7	SP/MEX
Portales	KSEL	1450	N/TLK	Tse Bonito	KHAC	880	RL
Portales	KSEL	95.3	CW	Tucumcari	KTNM	1400	TLK+
Portales	KENW	102.9	VAR	Tucumcari	KQAY	92.7	AC
Ramah	KTDB	89.7	N/TLK	White Rock	KSFQ	101.1	JAZ
Raton	KRTN	1490	AC	Zuni	KSHI	90.9	VAR
Raton	KRTN	93.9	OLG	Reserve	KLBZ	104.5	TLK+
Raton	KBKZ	96.5	CW	Rio Rancho	KAJZ	101.7	JAZ
Red River	KRDR	90.1	AD/A	Roswell	KBIM	910	N/TLK

City	Station	Freq	Format		City	Station	Freq	Format
Albany	WROW	590	TLK+		Binghamton	WHRW	90.5	VAR
Albany	WAMC	1400	N/TLK		Binghamton	WHWK	98.1	CW
Albany	WDDY	1460	YTH		Binghamton	WAAL	99.1	RK/C
Albany	WDCD	1540	RL/CC		Blue Mtn Lake	WXLH	91.3	VAR
Albany	WAMC	90.3	N/TLK		Boonville	WBRV	900	CW
Albany	WCDB	90.9	AD/A		Boonville	WBRV	101.3	CW
Albany	WYJB	95.5	AC/S		Brentwood	WXBA	88.1	VAR
Albany	WCPT	100.9	POP		Brewster	WPUT	1510	AS
Albany	WKLI	100.9	EZL		Briarcliff Manor	WXPK	107.1	AD/A
Albany	WHRL	103.1	RK/H		Bridgehampton	WBAZ	102.5	AC/S
Albany	WPYX	106.5	RK/C		Bridgeport	WTKW	99.5	RK/C
Albany	WGNA	107.7	CW		Brighton	WZNE	94.1	RK/H
Albany Area	WROW	590	TLK/S		Brockport	WASB	1590	RL/G
Albion	WJCA	102.1	RL		Brockport	WBSU	89.1	VAR
Alfred	WETD	90.7	VAR		Brockport	WMJQ	105.5	RL
Alfred	WETD	90.7	VAR		Brooklyn	WKRB	90.9	CH
Alfred	WZKZ	101.9	CW		Brookville	WCWP	88.1	JAZ
Altamont	WZMR	104.9	CW		Buffalo	WGR	550	SPTS
Amherst	WUFO	1080	RL/G		Buffalo	WBEN	930	N/TLK
Amsterdam	WCSS	1490	AC/S		Buffalo	WNED	970	N/TLK
Amsterdam	Wvtl	1570	TLK+		Buffalo	WMNY	1120	RL/G
Amsterdam	WBKK	97.7	CLA		Buffalo	WWWS	1400	URB
Arcade	WCOF	89.5	RL/CC		Buffalo	WWKB	1520	OLG
Argyle	WNGN	91.9	RL		Buffalo	WBFO	88.7	VAR
Arlington	WRRB	96.9	RK/H		Buffalo	WFBF	89.9	RL
Attica	WLOF	101.7	RL		Buffalo	WBNY	91.3	AD/A
Auburn	WKGJ	1340	YTH		Buffalo	WBUF	92.9	AH
Auburn	WAUB	1590	TLK+		Buffalo	WNED	94.5	CLA
Auburn	WDWN	88.9	AD/A		Buffalo	WJYE	96.1	AC/S
Auburn	Wphr	106.9	URB		Buffalo	WGRF	96.9	RK/CL
Avon	WYSL	1040	N/TLK		Buffalo	WDCX	99.5	RL/CC
Babylon	WNYG	1440	RL/CC		Buffalo	WMJQ	102.5	AC/H
Babylon	WBAB	102.3	RK/C		Buffalo	WEDG	103.3	RK/S
Baldwinsville	WSEN	1050	OLG		Buffalo	WHTT	104.1	OLG
Baldwinsville	WBXL	90.5	VAR		Buffalo	WYRK	106.5	CW
Baldwinsville	WSEN	92.1	OLG		Calverton	WDRE	105.3	CH
Ballston Spa	WXCR	102.3	CH		Canajoharie	WCAN	93.3	N/TLK
Batavia	WBTA	1490	AC/S		Canandaigua	WRSB	1310	RL/G
Batavia	WGCC	90.7	RK/C		Canandaigua	WCGR	1550	TLK+
Bath	WABH	1380	OLG		Canandaigua	WCIY	88.9	RL/CC
Bath	WVIN	98.3	AC/S		Canandaigua	WISY	102.3	AC/S
Bath	WCIK	103.1	RL/CC		Canton	WSLU	89.5	N/TLK
Bay Shore	WBZO	103.1	OLG		Canton	WRCD	101.5	CH
Beacon	WBNR	1260	AS		Canton	WNCQ	102.9	CW
Big Flats	WCBA	97.7	OLG		Cape Vincent	WMHI	94.7	RL
Binghamton	WINR	680	AS		Cape Vincent	WBDR	102.7	CW
Binghamton	WNBF	1290	TLK+		Carthage	WTOJ	103.1	AC/S
Binghamton	WYOS	1360	TLK+		Catskill	WCKL	560	OLG
Binghamton	WSKG	89.3	CLA		Catskill	WCTW	98.5	AC/S
Binghamton	WJIK	90.1	RL		Cazenovia	WITC	88.9	AD/A

City	Station	Freq	Format	City	Station	Freq	Format
Center Mor'es	WLVG	96.1	AC/S	Fenner	WXXE	90.5	VAR
Champlain	WCHP	760	RL	Fort Plain	WBUG	101.1	CW
Chateaugay	WYUL	94.7	CH	Frankfort	WKLL	94.9	RK/H
Cheektowaga	WECK	1230	AS	Fredonia	WCVF	88.9	VAR
Chenango Brg	WWYL	104.4	CH	Fredonia	WBKX	96.5	AC
Cherry Valley	WJIV	101.9	RL	Freeport	WGBB	1240	VAR
Clifton Park	WPTR	96.7	RL/CC	Friendship	WCID	89.1	RL/CC
Clinton	WHCL	88.7	VAR	Fulton	WAMF	1300	CW
Clyde	WCOV	93.7	RL/CC	Fulton	WBBS	104.7	CW
Cobleskill	WSDE	1190	TLK+	Garden City	WHPC	90.3	VAR
Cobleskill	WQBJ	103.5	RK/H	Garden City	WLIR	92.7	SP/OM
Conklin	WKGB	92.5	RK/C	Genesco	WGSU	89.3	RK/H
Copenhagen	WBDI	106.7	CH	Geneva	WGVA	1240	TLK+
Corinth	WFFG	107.1	CW	Geneva	WEOS	89.7	N/TLK
Corning	WCBA	1350	AS	Geneva	WFLK	101.7	CW
Corning	WCBA	1450	TLK+	Glens Falls	WMML	1230	SPTS
Corning	WCEB	91.9	RK	Glens Falls	WMML	1230	SPTS
Corning	WGMM	98.7	AS	Glens Falls	WWSC	1450	TLK+
Corning	WNKI	106.1	AC/H	Glens Falls	WLJH	90.9	RL/CC
Cornwall	WWLE	1170	N/TLK	Glens Falls	WGFR	92.7	AD/A
Cortland	WKRT	920	TLK+	Glens Falls	WCQL	95.9	RK/C
Cortland	WSUC	90.5	AD/A	Gloversville	WFNY	1440	CH
Cortland	WIII	99.9	RK/C	Gouverneur	WGIX	95.3	OLG
Dannemora	WKVJ	89.7	N/A	Grand Gorge	WGKR	105.3	RL/CC
Dansville	WDNY	1400	AS	Greece	WGMC	90.1	JAZ
Dansville	WDNY	93.9	AC	Hamilton	WRCU	90.1	AD/A
De Pew	WBLK	93.7	URB	Hampton Bays	WLIR	107.1	RK/C
De Ruyter	WWDG	105.1	RK/C	Hempstead	WHLI	1100	AS
Delhi	WDHI	100.3	OLG	Hempstead	WRHU	88.7	VAR
Deposit	WIYN	94.7	OLG	Hempstead	WKJY	98.3	AC
Dewitt	WVOA	720	N/A	Henderson	WOTT	100.7	RK/C
Dundee	WFLR	1570	TLK+	Henrietta	WITR	89.7	VAR
Dundee	WFLR	95.9	CW	Herkimer	WNRS	1420	SPTS
Dunkirk	WDOE	1410	AS	Herkimer	WVHC	91.5	VAR
East Hampton	WHBE	96.7	BIZ	Herkimer	WXUR	92.7	OLG
East Syracuse	WSIV	1540	RL/G	Highland	WRWD	107.3	CW
Ellenville	WRWD	1370	CW	Homer	WXHC	101.5	OLG
Ellenville	WFKP	99.3	AC/S	Honeoye Falls	WFXF	95.1	CH
Elmira	WENY	1230	TLK+	Hoosick Falls	WZEC	97.5	AC/S
Elmira	WELM	1410	SPTS	Hornell	WHHO	1320	OLG
Elmira	WCIH	90.3	RL/CC	Hornell	WLEA	1480	TLK+
Elmira	WENY	92.7	AC	Hornell	WSQA	88.7	CLA
Elmira	WLVY	94.3	CH	Hornell	WCKR	92.1	CW
Elmira	WECW	107.7	CH	Hornell	WKPQ	105.3	AC/H
Elmira Hts	WEHH	1600	AS	Horseheads	WWLZ	820	TLK+
Endicott	WENE	1430	SPTS	Horseheads	WLNL	1000	RL/CC
Endicott	WMRV	105.7	AC/H	Horseheads	WPGI	100.9	CW
Endwell	WBBI	107.5	RK/C	Houghton	WJSL	90.3	CLA
Essex	WCPV	101.3	RK/C	Hudson	WHUC	1230	AS
Fairport	WFKL	93.3	AH	Hudson	WHVP	91.1	RL/CC

Radio on the Road   New York

City	Station	Freq	Format	City	Station	Freq	Format
Hudson	WZCR	93.5	OLG	Loudonville	WVCR	88.3	CH
Hudson Falls	WENU	101.7	AC/S	Lowville	WLLG	99.3	CW
Huntington	WGSM	740	SIL	Malone	WICY	1490	OLG
Hyde Park	WHVW	950	VAR	Malone	WMHQ	90.1	RL
Hyde Park	WCZX	97.7	CH	Malone	WSLO	90.9	VAR
Irondequoit	WKGS	106.7	CH	Malone	WVNV	96.5	CW
Islip	WLIE	540	BIZ	Manlius	WAQX	95.7	RK/C
Ithaca	WHCU	870	N/TLK	Massena	WYBG	1050	TLK+
Ithaca	WTKO	1470	OLG	Massena	WMSA	1340	OLG
Ithaca	WSQG	90.9	CLA	Mechanicville	WABY	1160	AS
Ithaca	WICB	91.7	RK/C	Mechanicville	WABT	104.5	AC/H
Ithaca	WVBR	93.5	RK/C	Mexico	WVOA	103.9	RL
Ithaca	WYXL	97.3	AC	Middletown	WALL	1340	YTH
Ithaca	WQNY	103.7	CW	Middletown	WOSR	91.7	N/TLK
Jamestown	WJTN	1240	TLK+	Middletown	WRRV	92.7	RK/H
Jamestown	WKSN	1340	TLK+	Mineola	WTHE	1520	RL/G
Jamestown	WUBJ	88.1	N/TLK	Minetto	WKRH	106.5	RK/H
Jamestown	WNJA	89.7	N/TLK	Monroe	WLJP	89.3	RL/CC
Jamestown	WNJA	89.7	CLA	Montauk	WPKM	88.7	VAR
Jamestown	WNJA	89.7	CLA	Montauk	WMOS	104.7	CH
Jamestown	WNJA	89.7	N/TLK	Monticello	WSUL	98.3	AC/H
Jamestown	WCOT	90.9	RL/CC	Monticello	WJUX	99.7	RL
Jamestown	WWSE	93.3	AC/H	Montour Falls	WNGZ	104.9	RK/C
Jamestown	WHUG	101.9	CW	Morristown	WYSX	102.9	CH
Jeffersonville	WJFF	90.5	VAR	Mount Kisco	WVIP	1310	SP/RL
Jeffersonville	WDNB	102.1	N/TLK	Mount Kisco	WFAF	106.3	RK/C
Jeffersonville	WPDA	106.1	RK/C	Mt. Hope	WXHD	90.1	VAR
Johnson City	WLTB	101.7	AC	New City	WRKL	910	N/A
Johnstown	WIZR	930	AC	New Paltz	WBWZ	93.3	AC/H
Kingston	WGHQ	920	TLK+	New Rochelle	WVOX	1460	VAR
Kingston	WJGK	1200	N/A	New Rochelle	WRTN	93.5	AS
Kingston	WKNY	1490	AC	New York	WNYC	820	N/TLK
Kingston	WFGB	89.7	RL/CC	New York	WNYC	93.9	N/TLK
Kingston	WAMK	90.9	N/TLK	New York City	WMCA	570	RL
Kingston	WFRH	91.7	RL	New York City	WFAN	660	SPTS
Kingston	WKXP	94.3	CW	New York City	WOR	710	TLK+
Lake George	WCKM	98.5	OLG	New York City	WABC	770	TLK+
Lake Luzerne	WBAR	94.7	RL	New York City	WCBS	880	N/TLK
Lake Placid	WIRD	920	SPTS	New York City	WINS	1010	N/TLK
Lake Placid	WLPW	105.5	RK/C	New York City	WEPN	1050	SPTS
Lk Ronk'koma	WSHR	91.9	VAR	New York City	WBBR	1130	BIZ
Lake Success	WKTU	103.5	CH	New York City	WLIB	1190	TLK+
Lakewood	WKZA	106.9	AC/H	New York City	WADO	1280	SP/N
Lancaster	WXRL	1300	CW	New York City	WWRV	1330	SP/RL
Liberty	WVOS	1240	AC	New York City	WKDM	1380	N/A
Liberty	WGWR	88.1	RL/CC	New York City	WZRC	1480	N/A
Liberty	WVOS	95.9	AC	New York City	WQEW	1560	YTH
Little Falls	WLFH	1230	SPTS	New York City	WWRL	1600	N/A
Little Falls	WSKU	105.5	CH	New York City	WNYU	89.1	VAR
Lockport	WLVL	1340	TLK+	New York City	WKCR	89.9	VAR

129

City	Station	Freq	Format	City	Station	Freq	Format
New York City	WHCR	90.3	VAR	Oneonta	WSRK	103.9	AC/H
New York City	WFUV	90.7	AD/A	Ossining	WDFH	90.3	AD/A
New York City	WNYE	91.5	TLK+	Oswego	WSGO	1440	AS
New York City	WXRK	92.3	RK/C	Oswego	WNYO	88.9	RK/C
New York City	WPLJ	95.5	AC/H	Oswego	WRVO	89.9	N/TLK
New York City	WQXR	96.3	CLA	Oswego	WOLF	96.7	YTH
New York City	WQHT	97.1	CH	Oswego	WTKV	105.5	RK/C
New York City	WSKQ	97.9	SP/OM	Owego	WEBO	1330	TLK+
New York City	WRKS	98.7	URB	Palmyra	WZXV	99.7	RL
New York City	WBAI	99.5	VAR	Patchogue	WALK	1370	AS
New York City	WCBS	101.1	OLG	Patchogue	WLIM	1580	N/A
New York City	WQCD	101.9	JAZ	Patchogue	WALK	97.5	AC
New York City	WNEW	102.7	AC	Patchogue	WBLI	106.1	AC/H
New York City	WAXQ	104.3	RK/C	Patterson	WDBY	105.5	OLG
New York City	WWPR	105.1	URB	Pattersonville	WPGL	90.7	RL/CC
New York City	WLTW	106.7	AC/S	Peekskill	WLNA	1420	AS
New York City	WBLS	107.5	URB	Peekskill	WHUD	100.7	AC
Newark	WACK	1420	TLK+	Penn Yan	WYLF	850	AS
Newburgh	WGNY	1220	TLK+	Peru	WXLU	88.3	N/TLK
Newburgh	WGNY	103.1	AC/H	Phoenix	WZUN	102.1	AC
Newport Vil'ge	WBGK	99.7	CW	Plainview	WPOB	88.5	VAR
Niagara Falls	WHLD	1270	RL	Plattsburgh	WEAV	960	TLK+
Niagara Falls	WJJL	1440	OLG	Plattsburgh	WTWK	1070	TLK+
Niagara Falls	WKSE	98.5	CH	Plattsburgh	WIRY	1340	AC
North Creek	WXLG	89.9	VAR	Plattsburgh	WCEL	91.9	N/TLK
No. Syracuse	WTLA	1200	AS	Plattsburgh	WQAE	93.9	AD/A
No. Syracuse	WKRL	100.9	RK/C	Plattsburgh	WBTZ	99.9	RK/H
Norwich	WCHN	970	AS	Plattsburgh	WKOL	105.1	OLG
Norwich	WKXZ	93.9	AC/H	Port Henry	WVTK	92.1	CH
Norwich	WBKT	95.3	CW	Port Jervis	WDLC	1490	OLG
Norwood	WYSI	96.1	AC	Port Jervis	WRPJ	88.9	RL/CC
Noyack	WSUF	89.9	TLK+	Port Jervis	WTSX	96.7	OLG
Nyack	WNYK	88.7	RL/CC	Potsdam	WPDM	1470	AC
Ogdensburg	WSLB	1400	TLK+	Potsdam	WAIH	90.3	VAR
Ogdensburg	WBDB	92.7	CH	Potsdam	WTSC	91.1	VAR
Ogdensburg	WPAC	98.7	OLG	Potsdam	WSNN	99.3	AC
Olean	WOEN	1360	N/TLK	Poughkeepsie	WEOK	1390	YTH
Olean	WHDL	1450	OLG	Poughkeepsie	WKIP	1450	AS
Olean	WOLN	91.3	N/TLK	Poughkeepsie	WRHV	88.7	CLA
Olean	WPIG	95.7	CW	Poughkeepsie	WVKR	91.3	VAR
Olean	WMXO	101.5	AC/H	Poughkeepsie	WRNQ	92.1	AC/S
Olivebridge	WFSO	88.3	RL	Poughkeepsie	WPKF	96.1	CH
Oneida	WMCR	1600	AC	Poughkeepsie	WPDH	101.5	RK/C
Oneida	WMCR	106.3	AC	Poughkeepsie	WSPK	104.7	CH
Oneonta	WDOS	730	CW	Pulaski	WSCP	101.7	CW
Oneonta	WRHO	89.7	AD/A	Queensbury	WNYQ	105.7	CH
Oneonta	WONY	90.9	VAR	Ravena	WRCZ	94.5	RK/C
Oneonta	WSQC	91.7	CLA	Remsen	WADR	1480	SPTS
Oneonta	WSQC	91.7	CLA	Remsen	WOKR	93.5	CH
Oneonta	WZOZ	103.1	CH	Rensselaer	WTMM	1300	SPTS

City	Station	Freq	Format	City	Station	Freq	Format
Renssalaer	WQBK	103.9	RK/H	Sidney	WCDO	100.9	AC
Riverhead	WRIV	1390	AS	Smithtown	WFRS	88.9	RL
Riverhead	WFTU	1570	RK/H	Smithtown	WMJC	94.3	AC/H
Riverhead	WRCN	103.9	RK/C	Sodus	WUUF	103.5	CW
Rochester	WROC	950	TLK+	So Bristol Twn	WNVE	107.3	RK/H
Rochester	WLGZ	990	AS	So Glens Falls	WENU	1410	AS
Rochester	WHAM	1180	N/TLK	Southampton	WLIU	88.3	JAZ
Rochester	WHTK	1280	TLK+	Southampton	WRLI	91.3	CLA
Rochester	WXXI	1370	N/TLK	Southampton	WEHM	92.9	AD/A
Rochester	WHIC	1460	RL	Southampton	WHFM	95.3	RK/C
Rochester	WRUR	88.5	JAZ	Southold	WBEA	101.7	CH
Rochester	WBER	90.5	AD/A	Southport	WOKN	99.5	CW
Rochester	WXXI	91.5	CLA	Spencer	WCII	88.5	RL/CC
Rochester	WBEE	92.5	CW	Spring Valley	WRCR	1300	TLK+
Rochester	WCMF	96.5	RK/C	Springville	WSPQ	1330	TLK+
Rochester	WPXY	97.9	CH	St. Bonaventure	WSBU	88.3	AD/A
Rochester	WBZA	98.9	OLG	Staten Island	WSIA	88.9	VAR
Rochester	WVOR	100.5	AC/H	Stillwater	WQAR	101.3	AC
Rochester	WRMM	101.3	AC/S	Stony Brook	WUSB	90.1	AD/A
Rochester	WDKX	103.9	URB	Sylvan Beach	WBGJ	100.3	YTH
Rochester	WJZR	105.9	JAZ	Syosset	WKWZ	88.5	VAR
Rome	WRNY	1350	SPTS	Syracuse	WSYR	570	N/TLK
Rome	WYFY	1450	RL	Syracuse	WHEN	620	SPTS
Rome	WODZ	96.1	OLG	Syracuse	WNSS	1260	SPTS
Rome	WRBY	102.5	AC/H	Syracuse	WFBL	1390	TLK+
Rosedale	WFNP	88.7	AD/A	Syracuse	WOLF	1490	YTH
Rotterdam	WTRY	98.3	OLG	Syracuse	WAER	88.3	JAZ
Rouses Point	WKYJ	88.7	N/A	Syracuse	WJPZ	89.1	CH
Sag Harbor	WLNG	92.1	OLG	Syracuse	WRVD	90.3	VAR
Salamanca	WGGO	1590	AC/S	Syracuse	WCNY	91.3	CLA
Salamanca	WQRT	98.3	RK/C	Syracuse	WNTQ	93.1	CH
Sandy Creek	WSCP	1070	CW	Syracuse	WYYY	94.5	AC
Saranac Lake	WNBZ	1240	AS	Syracuse	WMHR	102.9	RL
Saranac Lake	WSLL	90.5	N/TLK	Syracuse	WLTI	105.9	AC/S
Saranac Lake	WYZY	106.3	AC	Syracuse	WWHT	107.9	CH
Saratoga Spr	WUAM	900	AS	Ticonderoga	WIPS	1250	OLG
Saratoga Spr	WSSK	89.7	RL/CC	Ticonderoga	WANC	103.9	N/TLK
Saratoga Spr	WSPN	91.1	JAZ	Troy	WOFX	980	SPTS
Saugerties	WBPM	92.9	OLG	Troy	WHAZ	1330	RL
Schenectady	WGY	810	N/TLK	Troy	WRPI	91.5	VAR
Schenectady	WVKZ	1240	OLG	Troy	WFLY	92.3	CH
Schenectady	WMHT	89.1	CLA	Trumansburg	WPIE	1160	SPTS
Schenectady	WRUC	89.7	VAR	Tupper Lake	WRGR	102.3	RK/C
Schenectady	WRVE	99.5	AC/H	Utica	WIBX	950	N/TLK
Schoharie	WMYY	97.3	RL	Utica	WRUN	1150	AS
Schuyler Falls	WAVX	90.9	RL/CC	Utica	WTLB	1310	AS
Scotia	WEGQ	93.7	CW	Utica	WUTQ	1550	SPTS
Seneca Falls	WSFW	1110	TLK+	Utica	WUNY	89.5	CLA
Seneca Falls	WSFW	99.3	RK/C	Utica	WPNR	90.7	AD/A
Sidney	WCDO	1490	AC	Utica	WRVN	91.9	N/TLK

City	Station	Freq	Format		City	Station	Freq	Format
Utica	WOUR	96.9	RK/H					
Utica	WLZW	98.7	AC					
Utica	WKVU	100.7	RL/CC					
Utica	WFRG	104.3	CW					
Utica	WRCK	107.3	RK/C					
Valhalla	WARY	88.1	VAR					
Vestal	WMXW	103.3	AC					
Voorheesville	WAJZ	96.3	URB					
Walton	WDLA	1270	AS					
Walton	WDLA	92.1	CW					
Warrensburg	WKBE	100.3	AC					
Warsaw	WCJW	1140	CW					
Warsaw	WCOU	88.3	RL/CC					
Warwick	WTBQ	1110	TLK+					
Waterloo	WNYR	98.5	AC/S					
Watertown	WTNY	790	TLK+					
Watertown	WATN	1240	TLK+					
Watertown	WNER	1410	SPTS					
Watertown	WSLJ	88.9	N/TLK					
Watertown	WWJS	90.1	RL/CC					
Watertown	WJNY	90.9	CLA					
Watertown	WRVJ	91.7	N/TLK					
Watertown	WCIZ	93.7	CH					
Watertown	WFRY	97.5	CW					
Watkins Glen	WTYX	1490	CW					
Waverly	WAVR	102.1	AC					
Webster	WFRW	88.1	RL					
Webster	WMHN	89.3	RL					
Webster	WDCZ	102.7	RL/CC					
Wellsville	WLSV	790	CW					
Wellsville	WJQZ	103.5	OLG					
Westhampton	WBON	98.5	RK/C					
Westport	WCLX	102.5	AD/A					
Wethersf'ld	WLKK	107.7	RK/C					
White Plains	WFAS	1230	AS					
White Plains	WFAS	103.9	AC					
Whitehall	WNYV	94.1	AC					
Whitesboro	WSKS	97.9	CH					
Willsboro	WXZO	96.7	TLK+					
Windham	WRIP	97.9	AC					
Woodstock	WDST	100.1	AD/A					
Wurtsboro	WZAD	97.3	CH					
Youngstown	WTOR	770	N/A					

Radio on the Road   North Carolina

City	Station	Freq	Format	City	Station	Freq	Format
Aberdeen	WQNX	1350	TLK+	Canton	WOXL	970	CH
Ahoskie	WRCS	970	RL/G	Carolina Bch	WMYT	1180	SP/RL
Ahoskie	WBKU	91.7	RL	Carolina Bch	WUIN	106.7	AD/A
Albemarle	WSPC	1010	TLK+	Chadbourn	WVOE	1590	RL/G
Albemarle	WZKY	1580	OLG	Chapel Hill	WCHL	1360	N/TLK
Asheboro	WZOO	710	RL/G	Chapel Hill	WLLQ	1530	SP/MEX
Asheboro	WKXR	1260	CW	Chapel Hill	WXYC	89.3	AD/A
Asheboro	WTJY	89.5	RL/G	Chapel Hill	WUNC	91.5	N/TLK
Asheboro	WKRR	92.3	RK/C	Charlotte	WFNZ	610	SPTS
Asheville	WWNC	570	TLK+	Charlotte	WYFQ	930	RL
Asheville	WSKY	1230	RL	Charlotte	WBT	1110	N/TLK
Asheville	WISE	1310	SPTS	Charlotte	WHVN	1240	RL
Asheville	WKJV	1380	RL/G	Charlotte	WGSP	1310	SP/MEX
Asheville	WCQS	88.1	CLA	Charlotte	WGFY	1480	YTH
Asheville	WLFA	91.3	RL/CC	Charlotte	WOGR	1540	RL/G
Atlantic	WTKF	107.3	TLK+	Charlotte	WFNA	1660	SPTS
Atlantic Bch	WBJD	91.5	N/TLK	Charlotte	WFAE	90.7	N/TLK
Aurora	WSTK	104.5	RL/G	Charlotte	WNKS	95.1	CH
Banner Elk	WZJS	100.7	RL/CC	Charlotte	WSOC	103.7	CW
Bath	WZPE	90.1	CLA	Charlotte	WKQC	104.7	AC
Bayboro	WRUP	97.9	CW	Charlotte	WLNK	107.9	AC
Beaufort	WXBE	88.3	RL	Cherryville	WCSL	1590	AC
Beech Mtn	WECR	102.3	AC/S	China Grove	WRNA	1140	RL/G
Belhaven	WQZL	101.1	CH	Claremont	WCXN	1170	SP/MEX
Belmont	WCGC	1270	RL	Clayton	WHPY	1590	RL
Benson	WPYB	1130	CW	Clinton	WRRZ	880	SP/MEX
Biltmore Forest	WOXL	96.5	CH	Clinton	WCLN	1170	OLG
Black Mtn	WFGW	1010	RL/G	Clinton	WCLN	107.3	RL/CC
Black Mtn	WZNN	1350	TLK+	Columbia	WERX	102.5	OLG
Black Mtn	WMIT	106.9	RL/CC	Columbia	WRSF	105.7	CW
Blowing Rock	WXIT	1200	TLK+	Concord	WEGO	1410	AS
Boiling Springs	WGWG	88.3	AD/A	Concord	WPEG	97.9	URB
Boone	WATA	1450	TLK+	Cullowhee	WWCU	90.5	CH
Boone	WASU	90.5	AD/A	Dallas	WZRH	960	TLK+
Brevard	WSQL	1240	VAR	Dallas	WSGE	91.7	AD/A
Bryson City	WBHN	1590	OLG	Davidson	WDAV	89.9	CLA
Buie's Creek	WCCE	90.1	RL	Dobson	WYZD	1560	RL/G
Burgaw	WVBS	1470	RL	Dunn	WCKB	780	RL/G
Burgaw	WKXB	99.9	URB	Dunn	WRCQ	103.5	RK/C
Burlington	WPCM	920	TLK+	Durham	WDNC	620	N/TLK
Burlington	WBAG	1150	AS	Durham	WTIK	1310	RL/SP
Burlington	WRSN	93.9	AC	Durham	WSRC	1410	RL/G
Burlington	WKXU	101.1	TLK+	Durham	WDUR	1490	SPTS
Burnsville	WKYK	940	CW	Durham	WXDU	88.7	AD/A
Buxton	WBUX	90.5	CLA	Durham	WNCU	90.7	JAZ
Buxton	WHDX	99.9	N/A	Durham	WDCG	105.1	CH
Buxton	WHDZ	101.5	N/A	Durham	WFXC	107.1	URB
Calabash	WYNA	104.9	AC/H	Eden	WCLW	1130	RL/G
Camp LeJeune	WSME	1120	CW	Eden	WLOE	1490	RL
Canton	WPTL	920	CW	Eden	WXRA	94.5	RK/C

City	Station	Freq	Format	City	Station	Freq	Format
Edenton	WZBO	1260	AS	Greensboro	WEAL	1510	RL/G
Edenton	WBXB	100.1	RL/G	Greensboro	WNAA	90.1	RL/G
Elizabeth City	WGAI	560	TLK+	Greensboro	WQFS	90.9	VAR
Elizabeth City	WCNC	1240	AS	Greensboro	WQMG	97.1	URB
Elizabeth City	WGPS	88.3	RL	Greensboro	WSMW	98.7	AH
Elizabeth City	WRVS	89.9	URB	Greensboro	WUAG	103.1	VAR
Elizabeth City	WKJX	96.7	AC/S	Greenville	WNCT	1070	TLK+
Elizabethtown	WBLA	1440	OLG	Greenville	WOOW	1340	RL/G
Elizabethtown	WGQR	105.7	OLG	Greenville	WZMB	91.3	AD/A
Elkin	WIFM	100.9	AC	Greenville	WNCT	107.9	OLG
Elon College	WSOE	89.3	VAR	Grifton	WXNR	99.5	RK/C
Erwin	WUAW	88.3	CH	Hamlet	WKDX	1250	RL/G
Fair Bluff	WOOR	105.3	OLG	Hamlet	WJSG	104.3	RL/G
Fairbluff	WZFB	1480	SPTS	Hark/Island	WLGP	100.3	RL
Fairmont	WFMO	860	RL/G	Harrisburg	WCCJ	92.7	URB
Fairmont	WSTS	100.9	RL/G	Hatteras	WWOC	94.5	CW
Fairview	WTZY	880	TLK+	Hatteras	WYND	97.1	AC/S
Farmville	WGHB	1250	TLK+	Havelock	WANG	1330	OLG
Farmville	WWGL	94.3	CW	Havelock	WKOO	105.1	OLG
Fayetteville	WFNC	640	TLK+	Henderson	WHNC	890	RL/G
Fayetteville	WFAY	1230	SPTS	Henderson	WIZS	1450	CW
Fayetteville	WAZZ	1490	AS	Henderson	WYFL	92.5	RL
Fayetteville	WIDU	1600	RL/G	Hendersonville	WHKP	1450	AC
Fayetteville	WFSS	91.9	JAZ	Hendersonville	WTZQ	1600	AS
Fayetteville	WQSM	98.1	AC/H	Hendersonville	WMYI	102.5	AC/S
Forest City	WWOL	780	RL/G	Hertford	WFMZ	104.9	CH
Forest City	WAGY	1320	CW	Hickory	WAIZ	630	OLG
Forest City	WPTP	93.3	RK/C	Hickory	WHKY	1290	TLK+
Franklin	WFSC	1050	AC/S	Hickory	WPIR	88.1	RL/G
Franklin	WPFJ	1480	RL/CC	Hickory	WFHE	90.3	N/TLK
Franklin	WFQS	91.3	CLA	Hickory	WXRC	95.7	CH
Franklin	WNCC	96.7	CW	Hickory	WLYT	102.9	AC/S
Fuquay-Var'a	WNNL	103.9	RL/G	High Point	WGOS	1070	SP/TK
Garner	WRTG	1000	SP/MEX	High Point	WMFR	1230	TLK+
Gaston	WLGQ	97.9	OLG	High Point	Wysr	1590	SPTS
Gastonia	WLTC	1370	RL/G	High Point	WHPU	90.3	AD/A
Gastonia	WGNC	1450	OLG	High Point	WHPE	95.5	RL
Gastonia	WBAV	101.9	URB	High Point	WMAG	99.5	AC/S
Gatesville	WQDK	99.3	CW	High Point	WVBZ	100.3	RK/C
Goldsboro	WFMC	730	RL/G	Highlands	WHLC	104.5	EZL
Goldsboro	WGBR	1150	TLK+	Hope Mills	WCCG	104.5	URB
Goldsboro	WSSG	1300	RL/G	Indian Trail	WPZS	100.9	N/TLK
Goldsboro	WKIX	96.9	SP/MEX	Jacksonville	WSRP	910	SP/MEX
Goldsboro	WKIX	102.3	CW	Jacksonville	WJNC	1240	TLK+
Graham	WSML	1200	TLK+	Jacksonville	WJCV	1290	RL/G
Granite Falls	WYCV	900	RL/G	Jacksonville	WQSL	92.3	CH
Greensboro	WPET	950	RL/G	Jacksonville	WILT	98.7	AC/S
Greensboro	WCOG	1320	YTH	Jacksonville	WXQR	105.5	RK/C
Greensboro	WKEW	1400	RL/G	Jefferson	WMMY	106.1	CW
Greensboro	WWBG	1470	SP/MEX	Kannapolis	WRKB	1460	RL/G

Radio on the Road

North Carolina

City	Station	Freq	Format	City	Station	Freq	Format
Kernersville	WTRU	830	RL	Mount Airy	WPAQ	740	CW
Kill Devil Hills	WCXL	104.1	AC/H	Mount Airy	WSYD	1300	RL/G
King	WKTE	1090	RL/G	Mount Olive	WDJS	1430	RL
Kings Mtn	WKMT	1220	TLK+	Moyock	WCDG	92.1	AC/S
Kinston	WRNS	960	RL/G	Murfreesboro	WWDR	1080	RL/G
Kinston	WELS	1010	RL/G	Murfreesboro	WDLZ	98.3	AC/S
Kinston	WLNR	1230	SP/MEX	Murphy	WCVP	600	CW
Kinston	WKNS	90.3	N/TLK	Murphy	WKRK	1320	CW
Kinston	WRNS	95.1	CW	Murphy	WCNG	102.7	AC/S
Kinston	WZBR	97.7	RL/G	Nags Head	WZPR	92.3	CH
Kinston	WELS	102.9	RL/G	Nashville	WZAX	99.7	AC
Laurinburg	WLNC	1300	AC	New Bern	WNOS	1450	AS
Laurinburg	WEWO	1460	RL/G	New Bern	WWNB	1490	RL/G
Laurinburg	WFLB	96.5	OLG	New Bern	WZNB	88.5	N/TLK
Leland	WAAV	980	TLK+	New Bern	WTEB	89.3	CLA
Leland	WKXS	94.1	URB	New Bern	WAAE	91.9	RL/CC
Lenoir	WKGX	1080	CW	New Bern	WIKS	101.9	URB
Lenoir	WJRI	1340	N/TLK	New Bern	WSFL	106.5	RK/C
Lenoir	WKVS	103.3	CW	New Hope	WAUG	750	RL/G
Lewisville	WSGH	1040	SP/MEX	Newland	WECR	1130	CW
Lexington	WLXN	1440	TLK+	Newport	WMGV	103.3	AC
Lexington	WTHZ	94.1	OLG	Newton	WNNC	1230	AC
Lillington	WLLN	1370	SP/MEX	Norlina	WZRN	90.5	AS
Lincolnton	WLON	1050	AC	Norlina	WJIJ	94.3	RL
Louisburg	WYRN	1480	CW	No Wilkesboro	WKBC	800	CW
Louisburg	WKXU	102.5	CW	No Wilkesboro	WKBC	97.3	AC/H
Lumberton	WAGR	1340	RL/G	Oak Island	WSFM	98.3	RK/S
Lumberton	WKML	95.7	CW	Ocean Isle Bch	WLQB	93.5	SP/MEX
Lumberton	WFNC	102.3	TLK+	Old Fort	WKSF	99.9	CW
Manteo	WUND	88.9	N/TLK	Oriental	WWEA	94.1	CW
Manteo	WURI	90.9	CLA	Oxford	WCBQ	1340	RL/G
Manteo	WOBX	98.1	RK/H	Pinehurst	WIOZ	550	AS
Manteo	WVOD	99.1	AD/A	Pinehurst	WBFY	90.3	RL
Marion	WBRM	1250	CW	Pinetops	WPWZ	95.5	URB
Mars Hill	WYQS	90.5	SIL	Pisgah Forest	WGCR	720	RL
Marshall	WHBK	1460	RL/G	Plymouth	WJPI	1470	RL/G
Mayodan	WMYN	1420	RL	Plymouth	WPNC	95.9	AC
Mebane	WGSB	1060	SP/MEX	Raeford	WMFA	1400	RL/G
Mint Hill	WNOW	1030	SP/MEX	Raeford	WRAE	88.7	RL
Mocksville	WDSL	1520	CW	Raleigh	WDTF	570	N/TLK
Monroe	WXNC	1060	SP/MEX	Raleigh	WPTF	680	TLK+
Monroe	WIXE	1190	CW	Raleigh	WRBZ	850	SPTS
Monroe	WDEX	1430	RL/G	Raleigh	WPJL	1240	RL
Mooresville	WHIP	1350	OLG	Raleigh	WCLY	1550	RL
Moreh'd City	WOTJ	90.7	RL/G	Raleigh	WKNC	88.1	AD/A
Moreh'd City	WRHT	96.3	CH	Raleigh	WSHA	88.9	JAZ
Morgantown	WCIS	760	RL/G	Raleigh	WCPE	89.7	CLA
Morgantown	WMNC	1430	CW	Raleigh	WQDR	94.7	CW
Morgantown	WMNC	92.1	CW	Raleigh	WBBB	96.1	RK/C
Kernersville	WTRU	830	RL	Raleigh	WRAL	101.5	AC

City	Station	Freq	Format	City	Station	Freq	Format
Raleigh	WWMY	102.9	OLG	Southport	WAZO	107.5	CH
Red Springs	WTEL	1160	RL/G	Sparta	WCOK	1060	CW
Reidsville	WREV	1220	SP/MEX	Spindale	WGMA	1520	RL/G
Reidsville	WJMH	102.1	JAZ	Spindale	WNCW	88.7	AD/A
Roanoke Raps	WCBT	1230	SPTS	Spring Lake	WCIE	1450	SP/OM
Roanoke Raps	WRTP	88.5	RL/CC	Spring Lake	WZRI	89.3	RL/CC
Roanoke Raps	WZRU	90.1	AS	Spruce Pine	WTOE	1470	AC/S
Roanoke Raps	WTGP	91.1	RL	St. Pauls	WKKE	1080	RL/G
Roanoke Raps	WPTM	102.3	CW	St. Pauls	WUKS	107.7	URB
Robbins	WLHC	103.1	JAZ	Statesville	WAME	550	AS
Robbisville	WCVP	95.9	CW	Statesville	WSIC	1400	TLK+
Rockingham	WLWL	770	OLG	Statesville	WKKT	96.9	CW
Rockingham	WAYN	900	AC	Statesville	WFMX	105.7	CW
Rockingham	WRSH	91.1	VAR	Swan Quarter	WHYC	88.5	VAR
Rocky Mount	WEED	1390	RL	Sylva	WRGC	680	AC
Rocky Mount	WRMT	1490	SPTS	Tabor City	WTAB	1370	RL/G
Rocky Mount	WRQM	90.9	N/TLK	Tarboro	WCPS	760	RL/G
Rocky Mount	WRSV	92.1	RL/G	Tarboro	WFXK	104.3	URB
Rocky Mount	WDWG	98.5	CW	Taylorsville	WACB	860	OLG
Rocky Mount	WRVA	100.7	AD/A	Taylorsville	WTLK	1570	RL/G
Rose Hill	WEGG	710	RL/G	Thomasville	WBLO	790	SPTS
Rose Hill	WZUP	104.7	SP/MEX	Thomasville	WIST	98.3	AS
Roxboro	WRXO	1430	AS	Topsail Beach	WWTB	103.9	N/TLK
Roxboro	WKRX	96.7	CW	Troy	WJRM	1390	RL/G
Rutherfordton	WCAB	590	CW	Tryon	WJFJ	1160	RL/G
Salisbury	WSAT	1280	AC/S	Valdese	WSVM	1490	OLG
Salisbury	WSTP	1490	N/TLK	Wadesboro	WADE	1340	RL/G
Salisbury	WOGR	93.3	RL/G	Wadesboro	WYFQ	93.5	RL
Salisbury	WEND	106.5	RK/S	Wake Forest	WDRU	1030	RL
Sanford	WWGP	1050	CW	Wallace	WZKB	94.3	RL/CC
Sanford	WXKL	1290	RL/G	Wanchese	WOBX	1530	RL/G
Sanford	WDCC	90.5	AD/A	Wanchese	WOBR	95.3	RK/C
Sanford	WFJA	105.5	OLG	Warrenton	WARR	1520	RL/G
Scotland Neck	WYAL	1280	RL/G	Washington	WDLX	930	TLK+
Selma	WBZB	1090	SPTS	Washington	WTOW	1320	RL/G
Semora	WKVE	106.7	RL/CC	Washington	WERO	93.3	CH
Shallotte	WVCB	1410	RL/G	Washington	WLGT	98.3	AC/S
Shallotte	WBNU	103.7	RK/C	Waxhaw	WNMX	106.1	AS
Shallotte	WLTT	106.3	N/TLK	Waynesville	WMXF	1400	AS
Shelby	WOHS	730	OLG	Waynesville	WQNS	104.9	RK/C
Shelby	WADA	1390	CW	Weldon	WSMY	1400	RL/G
Shelby	WIBT	96.1	CH	Wendell	WETC	540	SP/MEX
Siler City	WNCA	1570	AC	West Jefferson	WKSK	580	CW
Smithfield	WMPM	1270	RL/G	Whiteville	WENC	1220	RL/G
Snow Hill	WAGO	88.7	RL/G	Whiteville	WTXY	1540	TLK+
So Gastonia	WGAS	1420	RL/G	Whiteville	WZFX	99.1	URB
South'n Pines	WEEB	990	TLK+	Wilkesboro	WWWC	1240	RL/G
South'n Pines	WIOZ	102.5	AC/S	Wilkesboro	WSIF	94.7	VAR
South'n Pines	WKQB	106.9	CW	Williamston	WIAM	900	RL/G
Sout'rn Shores	WFMI	100.9	RL/G	Williamston	WRHD	103.7	CH

City	Station	Freq	Format		City	Station	Freq	Format
Wilmington	WMFD	630	SPTS					
Wilmington	WLSG	1340	RL/G					
Wilmington	WWIL	1490	RL/G					
Wilmington	WDVV	89.7	RL/CC					
Wilmington	WWIL	90.5	RL/CC					
Wilmington	WHQR	91.3	CLA					
Wilmington	WMNX	97.3	URB					
Wilmington	WWQQ	101.3	CW					
Wilmington	WGNI	102.7	AC					
Wilmington	WRQR	104.5	RK/C					
Wilson	WGTM	590	RL/G					
Wilson	WLLY	1350	RL/G					
Wilson	WVOT	1420	RL/G					
Wilson	WAJC	90.5	RL					
Wilson	WRDU	106.1	RK/C					
Windsor	WBTE	990	SIL					
Windsor	WURB	97.7	RL/G					
Windsor	WNBR	98.9	CW					
Wingate	WRCM	91.9	RL/CC					
Winston-Salem	WSJS	600	TLK+					
Winston-Salem	WPIP	880	RL/G					
Winston-Salem	WAAA	980	RL/G					
Winston-Salem	WPOL	1340	RL/G					
Winston-Salem	WTOB	1380	SP/MEX					
Winston-Salem	WSMX	1500	RL/G					
Winston-Salem	WBFJ	1550	RL/G					
Winston-Salem	WFDD	88.5	N/TLK					
Winston-Salem	WBFJ	89.3	RL/CC					
Winston-Salem	WSNC	90.5	JAZ					
Winston-Salem	WXRI	91.3	RL/G					
Winston-Salem	WMQX	93.1	OLG					
Winston-Salem	WTQR	104.1	CW					
Winston-Salem	WKZL	107.5	CH					
Wrightville Bch	WBNE	93.7	RK/C					
Yanceyville	WYNC	1540	RL/G					
Wilmington	WMFD	630	SPTS					
Wilmington	WLSG	1340	RL/G					
Wilmington	WWIL	1490	RL/G					
Wilmington	WDVV	89.7	RL/CC					
Wilmington	WWIL	90.5	RL/CC					
Wilmington	WHQR	91.3	CLA					
Wilmington	WMNX	97.3	URB					

City	Station	Freq	Format	City	Station	Freq	Format
Arthur	KVMI	96.7	SIL	Grand Forks	KJKJ	107.5	RK/C
Belcourt	KEYA	88.5	CW	Harvey	KHND	1470	AC
Beulah	KHOL	1410	AC	Harwood	KDJZ	100.7	RL/CC
Bismarck	KFYR	550	N/TLK	Hettinger	KNDC	1490	CW
Bismarck	KXMR	710	SPTS	Hope	KDAM	104.7	RK/C
Bismarck	KBMR	1130	CW	Jamestown	KSJB	600	CW
Bismarck	KBMK	88.3	N/A	Jamestown	KQDJ	1400	TLK+
Bismarck	KNRI	89.7	N/A	Jamestown	KPRJ	91.5	CLA
Bismarck	KCND	90.5	CLA	Jamestown	KSJZ	93.3	AC/H
Bismarck	KBFR	91.7	RL	Jamestown	KXGT	95.5	OLG
Bismarck	KYYY	92.9	CH	Kindred	KFAB	92.7	AH
Bismarck	KQDY	94.5	CW	Langdon	KNDK	1080	TLK+
Bismarck	KBYZ	96.5	RK/C	Langdon	KNDK	95.7	CW
Bismarck	KKCT	97.5	CH	Lincoln	KCJL	88.5	N/A
Bismarck	KACL	98.7	OLG	Lincoln	KVLQ	89.1	N/A
Bismarck	KSSS	101.5	RK/C	Lisbon	KQLX	890	FRM
Bottineau	KBTO	101.9	CW	Lisbon	KQLX	106.1	CW
Bowman	KPOK	1340	CW	Mandan	KLXX	1270	TLK+
Burlington	KWGO	102.9	N/A	Mandan	KNDR	104.7	RL/CC
Carrington	KDAK	1600	CW	Mayville	KMAV	1520	SPTS
Carrington	KYNU	98.3	CW	Mayville	KMAV	105.5	CW
Cavalier	KAOC	105.1	CW	Minot	KCJB	910	TLK+
Devil's Lake	KDLR	1240	CW	Minot	KHRT	1320	RL/G
Devil's Lake	KQZZ	96.7	RK/C	Minot	KRRZ	1390	OLG
Devil's Lake	KDVL	102.5	OLG	Minot	KMPR	88.9	CLA
Devil's Lake	KZZY	103.5	CW	Minot	KIZZ	93.7	CH
Dickinson	KDIX	1230	OLG	Minot	KYYX	97.1	CW
Dickinson	KLTC	1460	CW	Minot	KMXA	99.9	AC
Dickinson	KDPR	89.9	CLA	Minot	KZPR	105.3	RK/C
Dickinson	KZRX	92.1	CH	Minot	KHRT	106.9	RL/CC
Dickinson	KCAD	99.1	CW	Oakes	KDDR	1220	CW
Fargo	KFGO	790	TLK+	Rugby	KZZJ	1450	CW
Fargo	WDAY	970	N/TLK	Tioga	KTGO	1090	CW
Fargo	KFBN	88.7	RL/G	Valley City	KOVC	1490	CW
Fargo	KDSU	91.9	CLA	Valley City	KQDJ	101.1	AC/H
Fargo	WDAY	93.7	CH	Velva	KTZU	94.9	N/A
Fargo	KFNW	97.9	RL/CC	Wahpeton	KEGK	106.9	AC/H
Fargo	KKBX	101.9	RK/C	Walhalla	KYTZ	106.7	AC/H
Fargo	KPFX	107.9	RK/C	West Fargo	KFNW	1200	RL/CC
Fort Totten	KABU	90.7	VAR	West Fargo	KQWB	1550	N/TLK
Four Bears	KMHA	1340	CW	Williston	KEYZ	660	CW
Grafton	KXPO	1340	CW	Williston	KPPR	89.5	CLA
Grafton	KAUJ	100.9	OLG	Williston	KYYZ	96.1	CW
Grand Forks	KNOX	1310	N/TLK	Williston	KSPY	98.5	N/A
Grand Forks	KWTL	1370	RL	Williston	KDSR	101.1	CW
Grand Forks	KKXL	1440	SPTS				
Grand Forks	KUND	89.3	CLA				
Grand Forks	KFJM	90.7	AD/A				
Grand Forks	KKXL	92.9	CH				
Grand Forks	KNOX	94.7	CW				

City	Station	Freq	Format	City	Station	Freq	Format
Ada	WONB	94.9	AC	Caldwell	WWKC	104.9	CW
Akron	WHLO	640	N/TLK	Cambridge	WILE	1270	N/TLK
Akron	WARF	1350	TLK+	Cambridge	WOUC	89.1	AD/A
Akron	WAKR	1590	TLK+	Cambridge	WCMJ	96.7	AC
Akron	WZIP	88.1	CH	Campbell	WGFT	1330	RL/G
Akron	WAPS	91.3	AD/A	Canton	WCER	900	RL
Akron	WAKS	96.5	CH	Canton	WILB	1060	RL
Akron	WONE	97.5	RK/C	Canton	WHBC	1480	TLK+
Alliance	WDPN	1310	AC/S	Canton	WINW	1520	RL/G
Alliance	WRMU	91.1	JAZ	Canton	WHBC	94.1	AC
Alliance	WZKL	92.5	CH	Canton	WKDD	98.1	AC/H
Archbold	WBCY	89.5	RL/CC	Canton	WRQK	106.9	RK/C
Archbold	WMTR	96.1	AC	Castalia	WGGN	97.7	RL/CC
Ashland	WNCO	1340	AS	Cedarville	WCDR	90.3	RL/CC
Ashland	WRDL	88.9	AD/A	Celina	WCSM	1350	TLK+
Ashland	WNCO	101.3	CW	Celina	WKKI	94.3	AC
Ashtabula	WFUN	970	TLK+	Celina	WCSM	96.7	AC/S
Ashtabula	WREO	97.1	AC	Centerville	WCWT	101.5	AD/A
Ashtabula	WYBL	98.3	AC	Chillicothe	WCHI	1350	AS
Athens	WATH	970	OLG	Chillicothe	WBEX	1490	TLK+
Athens	WOUB	1340	VAR	Chillicothe		90.1	VAR
Athens	WOUB	91.3	AD/A	Chillicothe	WVXC	89.3	VAR
Athens	WJKW	95.9	RL/CC	Chillicothe	WOUH	91.9	AD/A
Athens	WXTQ	105.5	CH	Chillicothe	WLZT	93.3	AC/S
Bainbridge	WKHR	88.3	AS	Chillicothe	WKKJ	94.3	CW
Barnesville	WBNV	93.5	AC/S	Cincinnati	WKRC	550	TLK+
Batavia	WOBO	88.7	VAR	Cincinnati	WLW	700	TLK+
Batavia	WOBO	88.7	VAR	Cincinnati	WTSJ	1050	RL
Beach City	WOFN	88.7	RL	Cincinnati	WSAI	1360	SPTS
Beavercreek	WXEG	103.9	RK/C	Cincinnati	WCIN	1480	URB
Bellaire	WOMP	1290	TLK+	Cincinnati	WCKY	1530	TLK+
Bellaire	WOMP	100.5	CH	Cincinnati	WJVS	88.3	VAR
Bellefontaine	WBLL	1390	TLK+	Cincinnati	WGUC	90.9	CLA
Bellefontaine	WPKO	98.3	AC	Cincinnati	WVXU	91.7	VAR
Bellevue	WOHF	92.1	RK/C	Cincinnati	WOFX	92.5	RK/C
Belpre	WCVV	89.5	RL	Cincinnati	WAKW	93.3	RL/CC
Belpre	WLKP	91.9	RL/CC	Cincinnati	WVMX	94.1	AC/H
Belpre	WNUS	107.1	CW	Cincinnati	WRRM	98.5	AC
Berea	WBWC	88.3	AD/A	Cincinnati	WKRQ	101.9	AC/H
Bowling Grn	WJYM	730	RL	Cincinnati	WEBN	102.7	RK/H
Bowling Grn	WBGU	88.1	VAR	Cincinnati	WUBE	105.1	CW
Bowling Grn	WRQN	93.5	OLG	Circleville	WAZU	107.1	RK/C
Bryan	WQCT	1520	AS	Cleveland	WKNR	850	SPTS
Bryan	WGBE	90.9	CLA	Cleveland	WTAM	1100	N/TLK
Bryan	WBNO	100.9	CH	Cleveland	WHKW	1220	RL
Buchtel	WAIS	770	CW	Cleveland	WWMK	1260	YTH
Bucyrus	WBCO	1540	AS	Cleveland	WERE	1300	N/TLK
Bucyrus	WQEL	92.7	CH	Cleveland	WHK	1420	TLK+
Byesville	WILE	97.7	AS	Cleveland	WABQ	1540	RL/G
Cadiz	WCDK	106.3	RK/C	Cleveland	WCSB	89.3	VAR

City	Station	Freq	Format	City	Station	Freq	Format
Cleveland	WCPN	90.3	JAZ	Dayton	WQRP	89.5	RL/CC
Cleveland	WRUW	91.1	VAR	Dayton	WDPS	89.5	JAZ
Cleveland	WZAK	93.1	URB	Dayton	WIDR	98.1	VAR
Cleveland	WFHM	95.5	RL/CC	Dayton	WHKO	99.1	CW
Cleveland	WNCX	98.5	RK/C	Dayton	WTUE	104.7	RK
Cleveland	WGAR	99.5	CW	Dayton	WWSU	106.9	VAR
Cleveland	WMMS	100.7	RK/C	Dayton	WMMX	107.7	AC/H
Cleveland	WDOK	102.1	AC/S	De Graff	WDEQ	103.3	VAR
Cleveland	WCRF	103.3	RL	Defiance	WONW	1280	TLK+
Cleveland	WQAL	104.1	AC/H	Defiance	WGDE	91.9	CLA
Cleveland	WMJI	105.7	OLG	Defiance	WDFM	98.1	AC/H
Cleveland	WMVX	106.5	AC/H	Defiance	WZOM	105.9	CW
Cleveland	WENZ	107.9	URB	Delaware	WXOL	1550	SP/OM
Cleveland Hts	WJMO	1490	RL/G	Delaware	WJJE	89.1	RL
Cleveland Hts	WXTM	92.3	RK/C	Delaware	WJJE	89.1	RL
Clyde	WHVT	90.5	RL	Delaware	WSLN	98.7	VAR
Clyde	WMJK	100.9	RK/C	Delaware	WODB	107.9	OLG
Coal Grove	WBVB	97.1	OLG	Delhi Hills	WJYC	90.1	RL/CC
Columbus	WTVN	610	TLK+	Delphos	WBIE	91.5	RL/CC
Columbus	WOSU	820	N/TLK	Delphos	WDOH	107.1	AC/S
Columbus	WRFD	880	RL	Delta	WRWK	106.5	RK/H
Columbus	WMNI	920	AS	Dover	WJER	1450	AC
Columbus	WTPG	1230	TLK+	Dover	WJER	101.7	AC
Columbus	WBNS	1460	SPTS	East Liverpool	WOHI	1490	AS
Columbus	WVKO	1580	RL/G	East Liverpool	WOGF	104.3	CW
Columbus	WUFM	88.7	RL/CC	Eaton	WEDI	1130	CW
Columbus	WOSU	89.7	CLA	Eaton	WGTZ	92.9	CH
Columbus	WCBE	90.5	VAR	Edgewood	WZOO	102.5	AC/H
Columbus	WCOL	92.3	CW	Elyria	WEOL	930	N/TLK
Columbus	WSNY	94.7	AC	Elyria	WNWV	107.3	JAZ
Columbus	WLVQ	96.3	RK	Englewood	WDKF	94.5	CH
Columbus	WBNS	97.1	AC/H	Fairborn	WGNZ	1110	RL/G
Columbus	WNCI	97.9	CH	Fairfield	WCNW	1560	RL/G
Columbus	WBZX	99.7	RK/S	Fairfield	WMOJ	94.9	URB
Columbus	WCKX	107.5	URB	Findlay	WFIN	1330	TLK+
Columbus Grv	WLWD	93.9	CH	Findlay	WLFC	88.3	CH
Conneaut	WWOW	1360	N/TLK	Findlay	WKXA	100.5	CH
Conneaut	WGOJ	105.5	RL/G	Fort Shawnee	WZRX	107.5	RK/C
Cortland	WKTX	830	VAR	Fostoria	WFOB	1430	TLK+
Coshocton	WTNS	1560	CW	Fostoria	WBVI	96.7	AC/S
Coshocton	WOSE	91.1	CLA	Fredericktown	WWBK	98.3	CH
Coshocton	WOSE	91.1	CLA	Fremont	WFRO	99.1	AC
Coshocton	WTNS	99.3	AC	Gahanna	WCVO	104.9	RL/CC
Crestline	WYKL	98.7	RL/CC	Galion	WFXN	102.3	RK/C
Crooksville	WYBZ	107.3	OLG	Gallipolis	WJEH	990	AS
Cuyahoga F'ls	WCUE	1150	RL	Gallipolis	WRYV	101.5	CH
Dayton	WONE	980	SPTS	Gambier	WKCO	91.9	VAR
Dayton	WDAO	1210	URB	Geneva	WKKY	104.7	CW
Dayton	WHIO	1290	N/TLK	Georgetown	WAXZ	97.7	CH
Dayton	WING	1410	SPTS	Gibsonburg	WIMX	95.7	URB

City	Station	Freq	Format	City	Station	Freq	Format
Granville	WDUB	91.1	AD/A	Mansfield	WMAN	1400	N/TLK
Greenfield	WVNU	97.5	AC/S	Mansfield	WVMC	90.7	RL/CC
Greenville	WDPG	89.9	CLA	Mansfield	WOSV	91.7	CLA
Greenville	WDSJ	106.5	JAZ	Mansfield	WYHT	105.3	AC/H
Grove City	WWCD	101.1	RK/C	Mansfield	WVNO	106.1	AC
Hamilton	WMOH	1450	SPTS	Marietta	WLTP	910	TLK+
Hamilton	WHSS	89.5	RK/H	Marietta	WMOA	1490	AC/S
Hamilton	WGRR	103.5	OLG	Marietta	WMRT	88.3	CLA
Harrison	WNLT	104.3	RL/CC	Marietta	WCMO	98.5	CLA
Heath	WHTH	790	TLK+	Marietta	WRVB	102.1	CH
Hicksville	WFGA	106.7	AC	Marion	WMRN	1490	N/TLK
Hilliard	WFJX	105.7	RK/C	Marion	WOSB	91.1	CLA
Hillsboro	WSRW	1590	AS	Marion	WDIF	94.3	AC
Hillsboro	WSRW	106.7	CW	Marion	WMRN	106.9	CW
Holland	WPOS	102.3	RL/G	Marysville	WUCO	1270	CW
Hubbard	WBTJ	101.9	URB	Massilon	WTIG	990	SPTS
Huron	WKFM	96.1	CW	Maumee	WYSZ	89.3	RL/CC
Ironton	WIRO	1230	TLK+	McArthur	WYRO	98.7	RK/C
Ironton	WOUL	89.1	AD/A	McConnelsville	WJAW	100.9	SPTS
Ironton	WBKS	107.1	CH	Medina	WQMX	94.9	CW
Jackson	WCJO	97.7	CW	Miamisburg	WFCJ	93.7	RL
Jefferson	WCJV	90.9	RL/G	Middleport	WMPO	1390	SPTS
Johnstown	WVKO	103.1	SP/MEX	Middleport	WMPO	92.1	AC/H
Kent	WJMP	1520	TLK+	Middleport	WYVK	92.1	AC/H
Kent	WKSU	89.7	CLA	Middletown	WPFB	910	TLK+
Kent	WINR	100.1	TLK+	Middletown	WPFB	105.9	CW
Kenton	WKTN	95.3	AC	Milford	WKFS	107.1	CH
Kettering	WKET	98.3	RK/H	Millersburg	WVML	90.5	RL
Kettering	WLQT	99.9	AC/S	Millersburg	WKLM	95.3	AC
Lancaster	WLOH	1320	OLG	Montpelier	WLZZ	104.5	CW
Lancaster	WFCO	90.9	RL	Morrow	WLMH	89.1	VAR
Lancaster	WHOK	95.5	CW	Mount Vernon	WMVO	1300	AS
Lancaster	WJZA	103.5	JAZ	Mount Vernon	WNZR	90.9	RL/CC
Lebanon	WYGY	96.5	CW	Mount Vernon	WQIO	93.7	AC/S
Lima	WLJM	940	SPTS	Mt. Gilead	WVXG	95.1	RL/CC
Lima	WIMA	1150	N/TLK	Nad/Aolean	WNDH	103.1	AC
Lima	WYSM	89.3	RL/CC	Nelsonville	WSEO	107.7	CW
Lima	WYSM	89.3	CLA	New Boston	WIOI	1010	AS
Lima	WFGF	93.1	CW	New Concord	WMCO	90.7	VAR
Lima	WTGN	97.7	RL/G	New Lexington	WWJM	106.3	AC
Lima	WIMT	102.1	CW	N. Philadelphia	WKRJ	91.5	CLA
Lima	WUZZ	104.9	CH	N. Philadelphia	WNPQ	95.9	RL/CC
Logan	WLGN	1510	SPTS	Newark	WCLT	1430	N/TLK
Logan	WLGN	98.3	CW	Newark	WCLT	100.3	CW
London	WJYD	106.3	RL/G	Newark	WNKO	101.7	OLG
Lorain	WDLW	1380	OLG	Niles	WRTK	1540	URB
Lorain	WNZN	89.1	SP/MEX	Niles	WNCD	106.1	RK
Lorain	WCLV	104.9	CLA	Niles	WBBG	106.1	OLG
Loudonville	WBZW	107.7	CH	North Baltimore	WPFX	107.7	RK/C
Manchester	WAGX	101.3	CH	North Kingsville	WFXJ	107.5	RK/C

143

City	Station	Freq	Format	City	Station	Freq	Format
North Ridgeville	WJTB	1040	SP/RL	Springfield	WDHT	102.9	CH
Norwalk	WLKR	1510	OLG	St. Mary's	WMLX	103.3	AC
Norwalk	WNRK	90.7	CLA	Steubenville	WDIG	950	URB
Norwalk	WLKR	95.3	AC/S	Steubenville	WSTV	1340	TLK+
Oak Harbor	WJZE	97.3	URB	Steubenville	WBJV	88.9	RL
Oberlin	WOBL	1320	CW	Streetsboro	WSTB	91.5	AD/A
Oberlin	WOBC	91.5	VAR	Struthers	WKTL	90.7	VAR
Ontario	WRGM	1440	SPTS	Swanton	WJUC	107.3	URB
Ottawa	WBUK	106.3	OLG	Sylvania	WWWM	105.5	AC/H
Oxford	WMUB	88.5	JAZ	Thompson	WKSV	89.1	CLA
Oxford	WOXY	97.7	CH	Tiffin	WTTF	1600	AC
Painesville	WBKC	1460	NWS+	Tiffin	WHEI	88.9	AD/A
Parma	WCCD	1000	RL	Tiffin	WCKY	103.7	CW
Paulding	WKSD	99.7	OLG	Toledo	WCWA	1230	SPTS
Piketon	WXZQ	100.1	AC/H	Toledo	WSPD	1370	TLK+
Piqua	WPTW	1570	AC/S	Toledo	WLQR	1470	SPTS
Piqua	WDPT	95.7	OLG	Toledo	WTOD	1560	TLK+
Pleasant City	WBIK	92.1	RK/C	Toledo	WXUT	88.3	AD/A
Port Clinton	WXKR	94.5	RK/C	Toledo	WOTL	90.3	RL
Portsmouth	WNXT	1260	N/TLK	Toledo	WGTE	91.3	CLA
Portsmouth	WPAY	1400	TLK+	Toledo	WVKS	92.5	CH
Portsmouth	WOHP	88.3	RL/CC	Toledo	WKKO	99.9	CW
Portsmouth	WOSP	91.5	CLA	Toledo	WRVF	101.5	AC/S
Portsmouth	WNXT	99.3	AC	Toledo	WIOT	104.7	RK/C
Portsmouth	WPAY	104.1	CW	Troy	WOKL	96.9	RL/CC
Portsmouth	WZZZ	107.5	RK/C	Uhrichsville	WBTC	1540	TLK+
Proctorville	WMEJ	91.9	RL/G	Uhrichsville	WTUZ	99.9	CW
Reading	WMKV	89.3	AS	Union City	WTGR	97.5	CW
Richwood	WJZK	104.3	JAZ	University Hts	WJCU	88.7	AD/A
Ripley	WAOL	99.5	CH	Upper Arlington	WXMG	98.9	OLG
Rossford	WDMN	1520	RL/G	Up. Sandusky	WXML	90.1	RL/CC
Rushville	WLRY	88.5	RL/CC	Up. Sandusky	WYNT	95.9	OLG
Salem	WSOM	600	AS	Urbana	WKSW	101.7	CW
Salem	WQXK	105.1	CW	Van Wert	WERT	1220	AS
Sandusky	WLEC	1450	AS	Van Wert	WBYR	98.9	RK/C
Sandusky	WVMS	89.5	RL	Wapakoneta	WZOQ	92.1	CH
Sandusky	WCPZ	102.7	AC/H	Warren	WHKZ	1440	RL
Shadyside	WVKF	95.7	CH	Warren	WANR	1570	RL/CC
Shelby	WAUI	88.3	RL	Washington Ct	WCHO	1250	AS
Shelby	WSWR	100.1	OLG	Washington Ct	WCHO	105.5	CW
Sidney	WMVR	1080	AC/S	Wauseon	WYSA	88.5	RL/CC
Sidney	WMVR	105.5	AC/S	Wauseon	WNKL	96.9	RL/CC
South Vienna	WVSO	88.3	RL	Waverly	WXIC	660	RL/G
South Webster	WSNA	94.9	RL/CC	Waverly	WXIZ	100.9	CW
So. Zanesville	WCVZ	92.7	RL/CC	Wellston	WYPC	1330	AS
Spencerville	WBCJ	88.1	RL/CC	Wellston	WKOV	96.7	AC
Springfield	WIZE	1340	SPTS	West Carrollton	WDPR	88.1	CLA
Springfield	WULM	1600	OLG	West Carrollton	WROU	92.1	URB
Springfield	WUSO	89.1	AD/A	West Chester	WLHS	89.9	VAR
Springfield	WEEC	100.7	RL/G	West Union	WVKW	89.5	VAR

Radio on the Road Ohio

City	Station	Freq	Format	City	Station	Freq	Format
West Union	WRAC	103.1	RL/G				
Westerville	WOBN	101.5	AD/A				
Westerville	WTDA	103.9	AH				
Wilberforce	WCSU	88.9	RL/G				
Willard	WLRD	96.9	RL/G				
Willoughby	WELW	1330	TLK+				
Wilmington	WKFI	1090	CW				
Wilmington	WSWO	102.3	RL/CC				
Wooster	WKVX	960	OLG				
Wooster	WRKW	89.3	CLA				
Wooster	WCWS	90.9	VAR				
Wooster	WQKT	104.5	CW				
Xenia	WBZI	1500	CW				
Xenia	WZLR	95.3	RK/C				
Yellow Springs	WYSO	91.3	AD/A				
Youngstown	WKBN	570	TLK+				
Youngstown	WBBW	1240	SPTS				
Youngstown	WNIO	1390	AS				
Youngstown	WASN	1500	RL/G				
Youngstown	WYSU	88.5	CLA				
Youngstown	WYTN	91.7	RL				
Youngstown	WNCD	93.3	RK/C				
Youngstown	WMXY	98.9	AC/H				
Youngstown	WHOT	101.1	CH				
Zanesville	WHIZ	1240	AS				
Zanesville	WOUZ	90.1	AD/A				
Zanesville	WJIC	91.7	RL				
Zanesville	WCVZ	92.7	RL				
Zanesville	WHIZ	102.5	AC/H				

City	Station	Freq	Format	City	Station	Freq	Format
Ada	KADA	1230	SPTS	Cushing	KUSH	1600	TLK+
Ada	KTGS	89.9	RL/G	Del City	KOCY	1560	YTH
Ada	KAKO	91.3	RL	Dickson	KTRX	92.7	RK/C
Ada	KADA	99.3	CW	Duncan	KKEN	1350	TLK+
Ada	KQUJ	102.3	RL	Duncan	KKEN	102.3	CW
Altus	KWHW	1450	CW	Durant	KSEO	750	SPTS
Altus	KOCU	90.1	CLA	Durant	KAYC	91.1	RL
Altus	KKVO	90.9	RL/CC	Durant	KSSU	91.9	CH
Altus	KRKZ	93.5	RK/C	Durant	KLAK	97.5	AC
Altus	KEYB	107.9	CW	Durant	KLBC	107.1	CW
Alva	KPAK	97.5	AD/A	Edmond	KCSC	90.1	CLA
Alvar	KALV	1430	OLG	Edmond	KOKF	90.9	RL/CC
Alvar	KNID	99.7	CW	Edmond	KKWD	97.9	CH
Alvar	KTTL	105.7	AC	El Reno	KZUE	1460	SP/MEX
Anadarko	KVSP	103.5	URB	Elk City	KADS	1240	SPTS
Ardmore	KVSO	1240	TLK+	Elk City	KXOO	94.3	RL/CC
Ardmore	KQPD	91.1	RL	Elk City	KECO	96.5	CW
Ardmore	KKAJ	95.7	CW	Elk City	KTIJ	98.5	AC
Ardmore	KLCU	97.5	CLA	Enid	KGWA	960	N/TLK
Ardmore	KACO	98.5	OLG	Enid	KCRC	1390	SPTS
Atoka	KEOR	1110	RL/G	Enid	KFXY	1640	SPTS
Atoka	KHKC	102.1	CW	Enid	KKRD	91.1	RL/CC
Bartlesville	KWON	1400	TLK+	Enid	KQOB	96.9	RK/C
Bartlesville	KWRI	89.1	RL/CC	Enid	KOFM	103.1	CW
Bartlesville	KYFM	100.1	AC	Eufaula	KTNT	102.5	CW
Bennington	KFYZ	98.3	CW	Frederick	KTAT	1570	AC
Bethany	WWLS	104.9	SPTS	Frederick	KSYE	91.5	RL/CC
Bixby	KJMM	105.3	URB	Frederick	KYBE	95.9	CW
Blackwell	KOKB	1580	TLK+	Glenpool	KTSO	94.1	CH
Bristow	KREK	104.9	CW	Goodwell	KPSU	91.7	VAR
Broken Arrow	KNYD	90.5	RL/G	Grandfield	KWKL	89.9	RL/CC
Broken Arrow	KIZS	92.1	AC/H	Grove	KGVE	99.3	CW
Broken Bow	KKBI	106.1	CW	Guthrie	KMFS	1490	RL
Byng	KYKC	100.1	CW	Guymon	KGYN	1210	CW
Cache	KARU	88.9	N/A	Guymon	KKBS	92.7	RK/C
Carnedie	KJCC	89.5	RL	Healdton	KICM	93.7	CW
Chickasha	KWCO	1560	CH	Heavener	KPRV	92.5	CW
Chickasha	KTUZ	105.5	SPN	Henryetta	KVAZ	91.5	RL/G
Claremore	KRVT	1270	AS	Henryetta	KXBL	99.5	CW
Claremore	KRSC	91.3	AD/A	Hobart	KTJS	1420	CW
Claremore	KTFR	100.7	RL/CC	Hobart	KQTZ	105.9	AC/H
Clinton	KCLI	1320	SP/MEX	Holdenville	KTLS	1370	AC/H
Clinton	KYCU	89.1	CLA	Hollis	KKRE	92.5	OLG
Clinton	KQMX	95.5	AC	Hugo	KIHN	1340	CW
Clinton	KKFC	105.5	CW	Hugo	KITX	96.5	CW
Coalgate	KKFC	105.5	CW	Idabel	KBEL	1240	SPTS
Collinsville	KTBT	101.5	CH	Idabel	KXRT	90.9	RL
Comanche	KDDQ	96.7	RK/C	Idabel	KBEL	96.7	CW
Cordell	KCDL	99.3	RK/C	Idabel	KQIB	102.9	AC/H
Coweta	KDIM	88.1	RL	Ketchum	KOSN	107.5	CLA

City	Station	Freq	Format	City	Station	Freq	Format
Kingfisher	KINB	105.3	SP/MEX	Oklahoma City	KATT	100.5	RK/S
Lahoma	KXLS	95.7	AC/H	Oklahoma City	KTST	101.9	CW
Langston	KALU	89.3	URB	Oklahoma City	KJYO	102.7	CHR
Lawton	KKRX	1050	URB	Oklahoma City	KMGL	104.1	AC
Lawton	KXCA	1380	SPTS	Oklahoma City	KRXO	107.7	RK/C
Lawton	KCCU	89.3	CLA	Okmulgee	KOKL	1240	CW
Lawton	KVRS	90.3	RL	Owasso	KQLL	106.1	OLG
Lawton	KJRF	91.1	RL	Pauls Valley	KVLH	1470	OLG
Lawton	KZCD	94.1	RK/C	Pawhuska	KPGM	1500	RL/G
Lawton	KMGZ	95.3	CH	Pawhuska	KBVL	103.9	AC/H
Lawton	KJMZ	98.1	URB	Perry	KOKP	1020	TLK+
Lawton	KBZQ	99.5	AC	Perry	KOSB	105.1	CH
Lawton	KLAW	101.3	CW	Pocola	KKRI	88.1	RL/CC
Lawton	KVRW	107.3	OLG	Ponca City	WBBZ	1230	OLG
Lindsay	KBLP	105.1	CW	Ponca City	KLVV	88.7	RL
Locust Grove	KEMX	94.5	RL/CC	Ponca City	KJTH	89.7	RL/CC
Lone Grove	KYNZ	106.7	AC/H	Ponca City	KLOR	99.3	OLG
Madill	KMAD	1550	CW	Ponca City	KPNC	100.9	CW
Mangum	KHIM	97.7	TLK+	Ponca City	KIXR	104.7	AC/H
Marlow	KFXI	92.1	CW	Poteau	KPRV	1280	AC/S
McAlester	KNED	1150	CW	Poteau	KARG	91.7	RL
McAlester	KTMC	1400	AS	Poteau	KZBB	97.9	CHR
McAlester	KBCW	91.9	CLA	Poteau	KOMS	107.3	CW
McAlester	KMCO	101.3	CW	Pryor	KMUR	1570	RL
McAlester	KTMC	105.1	RK/C	Pryor	KMYZ	104.5	RK/H
Miami	KVIS	910	RL/G	Roland	KREU	92.3	SP/MEX
Miami	KGLC	100.9	RL/CC	Sallisaw	KKUZ	1560	SILNT
Midwest City	KTLV	1220	RL/G	Sallisaw	KKBD	95.9	RK/C
Moore	WWLS	640	SPTS	Sand Springs	KTFX	1340	RL/G
Moore	KMSI	88.1	RL	Sand Springs	KRTQ	102.3	RK/S
Muskogee	KBIX	1490	OLG	Sapulpa	KYAL	1550	SPTS
Muskogee	KBIX	1490	SPTS	Sapulpa	KXOJ	100.9	RL/CC
Muskogee	KMMY	97.1	CH	Seminole	KXTH	89.1	RL/CC
Muskogee	KHTT	106.9	CHR	Seminole	KIRC	105.9	CW
Newcastle	KKNG	93.3	CW	Shawnee	KGFF	1450	AS
Norman	KREF	1400	SPTS	Shawnee	KQCV	95.1	RL
Norman	KGRU	106.3	JAZ	Snyder	KJCM	100.3	AC
Nowata	KRIG	94.3	CW	Spencer	KROU	105.7	JAZ
Okarche	KTUZ	106.7	SP/MEX	Sperry	KMUS	1380	YTH
Oklahoma City	KQCV	800	RL	Stigler	KTKL	88.5	RL/CC
Oklahoma City	KTLR	890	TLK+	Stillwater	KSPI	780	TLK+
Oklahoma City	WKY	930	N/TLK	Stillwater	KOSU	91.7	CLA
Oklahoma City	KTOK	1000	N/TLK	Stillwater	KSPI	93.7	AC/H
Oklahoma City	KRMP	1140	URB	Stillwater	KVRO	98.1	OLG
Oklahoma City	KOKC	1520	TLK+	Stillwater	KGFY	105.5	CW
Oklahoma City	KYLV	88.9	RL/CC	Stuart	KLRB	89.3	RL
Oklahoma City	KOMA	92.5	OLG	Sulphur	KFXT	90.7	RL/G
Oklahoma City	KHBZ	94.7	RK/H	Sulphur	KXLT	90.7	CW
Oklahoma City	KXXY	96.1	CW	Taft	KXCR	100.3	RL/CC
Oklahoma City	KYIS	98.9	AC/H	Tahlequah	KTLQ	1350	RL/G

Radio on the Road

City	Station	Freq	Format		City	Station	Freq	Format
Tahlequah	KEOK	101.7	CW					
Tishomingo	KAZC	88.3	RL/G					
Tonkawa	KAYE	90.7	RK					
Tulsa	KRMG	740	N/TLK					
Tulsa	KCFO	970	RL					
Tulsa	KGTO	1050	URB					
Tulsa	KFAQ	1170	TLK+					
Tulsa	KAKC	1300	BIZ					
Tulsa	KTBZ	1430	SPTS					
Tulsa	KWTU	88.7	CLA					
Tulsa	KWGS	89.5	N/TLK					
Tulsa	KBEZ	92.9	AC/S					
Tulsa	KWEN	95.5	CW					
Tulsa	KRAV	96.5	AC/H					
Tulsa	KMOD	97.5	RK					
Tulsa	KVOO	98.5	CW					
Tulsa	KJSR	103.3	CH					
Vinita	KITO	1470	CW					
Vinita	KITO	96.1	CW					
Wagoner	KXTD	1530	SP/MEX					
Warner	KTFX	102.1	CW					
Watonga	KIMY	93.5	RL/G					
Weatherford	KWEY	1590	CW					
Weatherford	KAYM	90.5	RL					
Weatherford	KWEY	97.3	CW					
Wewoka	KWSH	1260	CW					
Wewoka	KSLE	104.7	OLG					
Wilburton	KSEC	103.7	AC/S					
Woodward	KSIW	1450	OLG					
Woodward	KJOV	90.7	RL/CC					
Woodward	KMZE	92.1	AC					
Woodward	KWFX	93.5	CW					
Woodward	KAZY	95.9	AC					
Woodward	KWOX	101.1	CW					
Woodward	KWDQ	102.3	RK/C					

City	Station	Freq	Format	City	Station	Freq	Format
Albany	KWIL	790	RL/G	Coos Bay	KHSN	1230	SPTS
Albany	KWIL	791	RL/G	Coos Bay	KMHS	1420	CHR
Albany	KWIL	792	RL/G	Coos Bay	KSBA	88.5	JAZ
Albany	KWIL	793	RL/G	Coos Bay	KJCH	90.9	RL
Albany	KWIL	794	RL/G	Coos Bay	KDCQ	93.5	OLG
Albany	KWIL	795	RL/G	Coos Bay	KYTT	98.7	RL/CC
Albany	KWIL	796	RL/G	Coos Bay	KYSG	105.9	JAZ
Albany	KWIL	797	RL/G	Coquille	KWRO	630	TLK+
Albany	KWIL	798	RL/G	Coquille	KSHR	97.3	CW
Albany	KWIL	799	RL/G	Corvallis	KOAC	550	N/TLK
Albany	KWIL	800	RL/G	Corvallis	KEJO	1240	TLK+
Albany	KWIL	801	RL/G	Corvallis	KLOO	1340	TLK+
Albany	KWIL	802	RL/G	Corvallis	KBVR	88.7	VAR
Astoria	KLOY	88.7	N/A	Corvallis	KFLY	101.5	RK/C
Astoria	KWYA	89.7	RL/CC	Corvallis	KLOO	106.3	CH
Astoria	KORM	90.5	N/A	Cottage Grove	KNND	1400	CW
Astoria	KMUN	91.9	VAR	Cottage Grove	KCGR	100.5	AC
Astoria	KAST	92.9	AC/S	Creswell	KUJZ	95.3	CW
Baker	KBKR	1490	TLK+	Dallas	KWIP	880	SP/MEX
Baker	KDJC	88.1	VAR	Depoe Bay	KPPT	100.7	CH
Baker	KANL	90.7	RL	Eagle Point	KEPO	92.1	SIL
Baker	KOBK	91.5	VAR	Eagle Point	KZZE	106.3	RK
Baker	KKBC	95.3	OLG	Enterprise	KWVR	1340	TLK+
Baker	KCMB	104.7	CW	Enterprise	KWVR	92.1	CW
Bandon	KBDN	96.5	RK/C	Eugene	KUGN	590	N/TLK
Banks	KVMX	107.5	OLG	Eugene	KKNX	840	OLG
Bay City	KIXT	96.3	N/A	Eugene	KPNW	1120	TLK+
Beaverton	KKCW	103.3	AC	Eugene	KRVM	1280	N/TLK
Bend	KICE	940	SPTS	Eugene	KSCR	1320	SPTS
Bend	KBND	1110	TLK+	Eugene	KLZS	1450	AS
Bend	KVLB	90.5	RL/CC	Eugene	KOPT	1600	TLK+
Bend	KOAB	91.3	N/TLK	Eugene	KWVA	88.1	VAR
Bend	KXIX	94.1	CH	Eugene	KLCC	88.7	AD/A
Bend	KNLR	97.5	RL/CC	Eugene	KWAX	91.1	CLA
Bend	KTWS	98.3	RK/C	Eugene	KRVM	91.9	AD/A
Bend	KMTK	99.7	CW	Eugene	KMGE	94.5	AC
Bend	KMGX	100.7	AC/S	Eugene	KZEL	96.1	RK/C
Bend	KQAK	105.7	OLG	Eugene	KNRQ	97.9	RK/H
Bonanza	KYSF	102.9	CHR	Eugene	KODZ	99.1	OLG
Brookings	KURY	910	AS	Florence	KCST	1250	AS
Brookings	KMWR	90.7	RL/CC	Florence	KLFO	88.1	AD/A
Brookings	KURY	95.3	CW	Florence	KWVZ	91.7	CLA
Brownsville	KEHK	102.3	AC	Florence	KDUK	104.7	CH
Brownsville	KEHK	102.3	AC/H	Florence	KCST	106.9	AC/S
Burns	KZZR	1230	CW	Garibaldi	KDEP	105.5	AC
Burns	KQHC	92.7	CH	Gleneden Bch	KSHL	97.5	CW
Cannon Beach	KCBZ	96.5	AC/H	Gold Beach	KGBR	92.7	CW
Canyon City	KJDY	94.5	CW	Gold Hill	KRWQ	100.3	CW
Cave Junction	KCNA	102.7	OLG	Grants Pass	KAJO	1270	TLK+
Cherryville	KLVP	88.7	RL/CC	Grants Pass	KAD/AK	91.1	RL

Oregon

City	Station	Freq	Format	City	Station	Freq	Format
Grants Pass	KROG	96.9	RK/C	Medford	KDOV	91.7	RL/G
Gresham	KMUZ	1230	SP/OM	Medford	KTMT	93.7	AH
Gresham	KMHD	89.1	JAZ	Medford	KBOY	95.7	RK/C
Harbeck	KLDR	98.3	CHR	Medford	KLDZ	103.5	OLG
Hermiston	KOHU	1360	CW	Milton Free'ter	KHTO	97.9	CH
Hermiston	KQFM	99.3	CH	Milton Free'ter	KLRF	88.5	RL/G
Hillsboro	KUIK	1360	TLK+	Milwaukie	KZNY	1010	SP/OM
Hood River	KIHR	1340	CW	Molalla	KRSK	105.1	AC/H
Hood River	KQHR	90.1	CLA	Monmouth	KSND	95.1	AD/A
Hood River	KCGB	105.5	AC/H	Myrtle Point	KOOZ	94.1	CLA
John Day	KJDY	1400	CW	Newport	KNPT	1310	TLK+
John Day	KGNR	91.9	RL/CC	Newport	KLCO	90.5	AD/A
Jordan Valley	KIDH	90.9	N/A	Newport	KNCU	92.7	CW
Junction City	KXOR	660	SP/MEX	Newport	KYTE	102.7	AC/H
Keizer	KYKN	1430	N/TLK	North Bend	KBBR	1340	N/TLK
Klamath Falls	KKJX	960	SP/MEX	North Bend	KOOS	94.9	CH
Klamath Falls	KAGO	1150	TLK+	North Bend	KACW	107.3	AC/H
Klamath Falls	KFLS	1450	N/TLK	North Powder	KEFS	89.5	N/A
Klamath Falls	KLMF	88.5	CLA	Nyssa	KARO	98.7	RL/CC
Klamath Falls	KKLJ	88.9	RL/CC	Oakridge	KMKR	88.5	AD/A
Klamath Falls	KTEC	89.5	AD/A	Oakridge	KAVE	92.1	AD/A
Klamath Falls	KSKF	90.9	JAZ	Ontario	KSRV	1380	CH
Klamath Falls	KLAD	92.5	CW	Ontario	KSRV	96.1	CW
Klamath Falls	KAGO	99.5	RK/C	Oregon City	KGDD	1520	SP/MEX
Klamath Falls	KKRB	106.9	AC/S	Pendleton	KTIX	1240	SPTS
La Grande	KLBM	1450	TLK+	Pendleton	KUMA	1290	TLK+
La Grande	KTVR	90.3	N/TLK	Pendleton	KRBM	90.9	N/TLK
La Grande	KEOL	91.7	AD/A	Pendleton	KWHT	103.5	CW
La Grande	KUBQ	98.7	CH	Pendleton	KUMA	107.7	AC
La Grande	KWRL	100.1	AC	Phoenix	KCMX	880	N/TLK
La Pine	KKLP	90.1	N/A	Phoenix	KAPL	1300	RL
Lake Oswego	KKSL	1290	RL/G	Phoenix	KAKT	105.1	CW
Lake Oswego	KDZR	1640	YTH	Portland	KPOJ	620	TLK+
Lake Oswego	KLTH	106.7	AC	Portland	KXL	750	N/TLK
Lakeview	KQIK	1230	CW	Portland	KPDQ	800	RL
Lakeview	KOAP	88.7	N/TLK	Portland	KCMD	970	TLK+
Lakeview	KQIK	93.5	AC/S	Portland	KFXX	1080	SPTS
Lakeview	KLCR	95.3	RK/C	Portland	KXMG	1150	SP/OM
Lebanon	KSHO	920	AS	Portland	KEX	1190	TLK+
Lebanon	KGAL	1580	N/TLK	Portland	KKPZ	1330	SP/RL
LEBANON	KXPC	103.7	CW	Portland	KBNP	1410	BIZ
LEBANON	KXPC	103.7	CW	Portland	KBPS	1450	VAR
Lincoln City	KBCH	1400	AS	Portland	KBVM	88.3	RL
Lincoln City	KCRF	96.7	RK/C	Portland	KBPS	89.9	CLA
Malin	KBUG	100.5	RL/G	Portland	KBOO	90.7	VAR
McMinnvile	KLYC	1260	OLG	Portland	KOPB	91.5	N/TLK
McMinnvile	KSLC	90.3	CHR	Portland	KGON	92.3	RK/C
Medford	KRTA	610	SP/OM	Portland	KPDQ	93.7	RL
Medford	KEZX	730	AS	Portland	KXJM	95.5	URB
Medford	KMED	1440	N/TLK	Portland	KYCH	97.1	AH

152

City	Station	Freq	Format	City	Station	Freq	Format
Portland	KRRC	97.9	AD/A	Tillamook	KTCB	89.5	VAR
Portland	KUPL	98.5	CW	Tillamook	KTMK	91.1	VAR
Portland	KWJJ	99.5	CW	Tillamook	KTIL	94.1	AS
Portland	KKRZ	100.3	CHR	Toledo	KCUP	1230	AS
Portland	KUFO	101.1	RK/C	Tri-City	KKMX	104.3	AC/H
Portland	KINK	101.9	AD/A	Troutdale	KPAM	860	N/TLK
Prineville	KRCO	690	CW	Umatilla	KLWJ	1090	RL/CC
Prineville	KLTW	95.1	AC/S	Veneta	KEUG	105.5	AH
Redmond	KRDM	1240	TLK+	Waldport	KORC	820	AS
Redmond	KWRX	88.5	CLA	Warm Springs	KWSO	91.9	AC
Redmond	KLRR	101.7	AD/A	Warm Springs	KWLZ	96.5	RK/C
Redmond	KSJJ	102.9	CW	Welches	KZRI	90.3	RL/CC
Reedsport	KDUN	1030	CW	West Klamath	KRAM	1070	AS
Reedsport	KLFR	89.1	VAR	Weston	KMMG	101.9	SP/OM
Reedsport	KSYD	92.1	AD/A	Winchester	KLOV	89.3	RL/CC
Reedsport	KJMX	99.5	OLG	Winston	KGRV	700	RL/CC
Rogue River	KRRM	94.7	CW	Woodburn	KWBY	940	SP/MEX
Roseburg	KTBR	950	N/TLK				
Roseburg	KQEN	1240	N/TLK				
Roseburg	KRNR	1490	CW				
Roseburg	KMPQ	88.1	AD/A				
Roseburg	KSRS	91.5	CLA				
Roseburg	KRSB	103.1	CW				
Salem	KCCS	1220	RL				
Salem	KSLM	1390	OLG				
Salem	KBZY	1490	OLG				
Salem	KAJC	90.1	RL				
Salem	KWBX	90.3	RL/CC				
Scappoose	KFIS	104.1	RL/CC				
Seaside	KSWB	840	OLG				
Seaside	KCYS	98.1	CW				
Seaside	KCRX	102.3	RK/C				
Selma	KJKL	88.7	RL/CC				
Sisters	KWPK	104.1	AC				
Springfield	KORE	1050	RL				
Springfield	KQFE	88.9	RL				
Springfield	KKNU	93.1	CW				
St. Helens	KOHI	1600	CW				
Stanfield	KLKY	96.1	SP/MEX				
Stayton	KCKX	1460	CW				
Sutherlin	KAVJ	101.1	OLG				
Sweet Home	KFIR	720	CW				
Sweet Home	KLVU	107.1	RL/CC				
Talent	KSJK	1230	N/TLK				
The Dalles	KACI	1300	TLK+				
The Dalles	KODL	1440	AS				
The Dalles	KACI	97.7	OLG				
The Dalles	KMCQ	104.5	AC/H				
Tigard	KLVP	1040	RL/CC				
Tillamook	KMBD	1590	TLK+				

City	Station	Freq	Format	City	Station	Freq	Format
Allentown	WDIY	88.1	CLA	Brookville	WYTR	103.3	OLG
Allentown	WAEB	790	TLK+	Brookville	WMKX	105.5	CH
Allentown	WTKZ	1320	SPTS	Brownsville	WASP	1130	OLG
Allentown	WKAP	1470	OLG	Burgettstown	WOGH	103.5	CW
Allentown	WHOL	1600	SP/OM	Burnham	WVNW	96.7	CW
Allentown	WJCS	89.3	RL	Butler	WISR	680	TLK+
Allentown	WMUH	91.7	VAR	Butler	WBUT	1050	OLG
Allentown	WLEV	100.7	AC/S	Butler	WLER	97.7	AC
Allentown	WAEB	104.1	CHR	California	WCAL	91.9	RK/C
Altoona	WRTA	1240	TLK+	Cambridge Spr	WXXO	104.5	AC/H
Altoona	WFBG	1290	TLK+	Canonsburg	WWCS	540	YTH
Altoona	WVAM	1430	SPTS	Canton	WHGL	100.3	CW
Altoona	WFGY	98.1	CW	Carbondale	WCDL	1440	CW
Altoona	WWOT	100.1	CHR	Carbondale	WCIG	91.3	RL
Ambridge	WMBA	1460	TLK+	Carbondale	WNAK	94.3	AS
Annville	WWSM	1510	CW	Carlisle	WHYL	960	TLK+
Apollo	WAVL	910	RL/CC	Carlisle	WIOO	1000	CW
Avis	WQBR	99.9	CW	Carlisle	WDCV	88.3	VAR
Avoca	WFEZ	103.1	AC/S	Carlisle	WCAT	102.3	CW
Barnesboro	WNCC	950	SIL	Carnegie	WZUM	1590	RL
Barnesboro	WHPA	93.5	OLG	Cashtown	WFKJ	890	RL/G
Beaver Falls	WBVP	1230	TLK+	Central City	WCCL	101.7	OLG
Beaver Falls	WGEV	88.3	RL/CC	Chambersburg	WCHA	800	AS
Beaver Falls	WITX	90.9	SIL	Chambersburg	WHGT	1590	SIL
Beaver Falls	WAMO	106.7	URB	Chambersburg	WZXQ	88.3	N/A
Beaver Springs	WLZS	106.1	OLG	Chambersburg	WIKZ	95.1	AC/H
Bedford	WBFD	1310	AC/S	Charleroi	WFGI	940	CW
Bedford	WHJD	1600	RL/G	Chester	WVCH	740	RL/CC
Bedford	WAYC	100.9	AC	Chester	WPWA	1590	RL/G
Bedford	WBVE	107.5	RK/C	Chester	WDNR	89.5	VAR
Bellefonte	WBLF	970	N/TLK	Clarendon	WKNB	104.3	CW
Bellefonte	WZWW	95.3	AC	Clarion	WWCH	1300	CW
Bellwood	WALY	103.9	OLG	Clarion	WWCH	1300	CW
Benton	WGGI	95.9	CW	Clarion	WCUC	91.7	AD/A
Berwick	WFBS	1280	OLG	Clarion	WCCR	92.7	AC
Berwick	WKAB	103.5	CH	Clearfield	WCPA	900	OLG
Bethlehem	WGPA	1100	VAR	Clearfield	WQYX	93.5	AC/H
Bethlehem	WLVR	91.3	VAR	Coatesville	WCOJ	1420	TLK+
Bethlehem	WZZO	95.1	RK	Columbia	WVZN	1580	SP/MEX
Blairsville	WLCY	106.3	AC/S	Connellsville	WPNT	1340	OLG
Bloomsburg	WHLM	930	AC	Cooperstown	WUUZ	107.7	CH
Bloomsburg	WBUQ	91.1	AD/A	Corry	WWCB	1370	AC
Bloomsburg	WFYY	106.5	AC/H	Coudersport	WFRM	600	CW
Boalsburg	WBUS	93.7	RK/C	Coudersport	WFRM	96.7	AC
Boyertown	WBYN	107.5	RL/CC	Covington	WDKC	101.5	CW
Braddock	WURP	1550	TLK+	Cresson	WBXQ	94.3	RK/C
Braddock	WRRK	96.9	RK/C	Curwensville	WOKW	102.9	AC
Bradford	WESB	1490	AC	Dallas	WSJR	93.7	CW
Bradford	WBRR	100.1	CH	Danville	WPGM	1570	RL/CC
Bristol	WLBS	91.7	AS	Danville	WPGM	96.7	RL/CC

Pennsylvania

City	Station	Freq	Format	City	Station	Freq	Format
Doylestown	WISP	1570	RL	Glen Mills	WZZE	97.3	CHR
Du Bois	WCED	1420	TLK+	Grantham	WVMM	90.7	RL/CC
Du Bois	WOWQ	102.1	CW	Greencastle	WQCM	94.3	RK/C
Du Bois	WDBA	107.3	RL/CC	Greensburg	WJJJ	107.1	URB
E. Stroudsburg	WESS	90.3	VAR	Greenville	WGRP	940	AC
E Stroudsburg	WOGI	90.3	CW	Greenville	WTGP	88.1	VAR
Easton	WEEX	1230	SPTS	Greenville	WEXC	107.1	OLG
Easton	WEST	1400	AC/S	Grove City	WSAJ	1340	CLA
Easton	WCTO	96.1	CW	Grove City	WSAJ	91.1	CLA
Easton	WODE	99.9	CH	Grove City	WWGY	95.1	CW
Easton	WJRH	104.9	AD/A	Hanover	WHVR	1280	CW
Ebensburg	WRDD	1580	BIZ	Harrisburg	WHP	580	TLK+
Ebensburg	WYOT	99.1	CH	Harrisburg	WKBO	1230	RL/CC
Edinboro	WFSE	88.9	AD/A	Harrisburg	WTCY	1400	URB
Edinboro	WXTA	97.9	CW	Harrisburg	WWKL	1460	SPTS
Elizabethtown	WPDC	1600	SPTS	Harrisburg	WXPH	88.1	AD/A
Elizabethtown	WWEC	88.3	VAR	Harrisburg	WITF	89.5	CLA
Elizabethville	WYGL	100.5	CW	Harrisburg	WRBT	94.9	CW
Ellwood City	WKST	92.1	OLG	Harrisburg	WRVV	97.3	CH
Emporium	WLEM	1250	CW	Harrisburg	WHKF	99.3	CHR
Emporium	WQKY	98.9	OLG	Harrisburg	WNNK	104.1	AC/H
Emporium	WQKY	98.9	AC	Havertown	WHHS	107.9	VAR
Ephrata	WRTL	90.7	CLA	Hawley	WBYH	89.1	RL/CC
Ephrata	WIOV	105.1	CW	Hawley	WYCY	105.3	OLG
Erie	WRIE	1260	AS	Hazleton	WAZL	1490	CH
Erie	WFNN	1330	SPTS	Hazleton	WBSX	97.9	RK/C
Erie	WJET	1400	TLK+	Hershey	WCPP	106.7	CH
Erie	WPSE	1450	BIZ	Hollidaysburg	WRKY	104.9	RK/C
Erie	WEFR	88.1	RL	Homer City	WCCS	1160	AC
Erie	WMCE	88.5	CLA	Honesdale	WPSN	1590	OLG
Erie	WERG	89.9	RK/H	Honesdale	WDNH	95.3	AC/H
Erie	WQLN	91.3	CLA	Hughesville	WRKK	1200	TLK+
Erie	WFGO	94.7	OLG	Huntingdon	WHUN	1150	TLK+
Erie	WXKC	99.9	AC	Huntingdon	WKVR	92.3	AD/A
Erie	WQHZ	102.3	RK/C	Huntingdon	WLAK	103.5	AC/H
Erie	WRTS	103.7	CHR	Huntingdon	WWLY	106.3	OLG
Everett	WZSK	1040	N/TLK	Indiana	WDAD	1450	OLG
Everett	WSKE	104.3	CW	Indiana	WIUP	90.1	VAR
Fairview (Erie)	WUSE	93.9	CW	Indiana	WQMU	92.5	AC/H
Farrell	WLOA	1470	RL	Irwin	WKHB	620	TLK+
Folsom	WRSD	94.9	VAR	Jackson Twn	WRTY	91.1	CLA
Forest City	WQFN	100.1	OLG	Jeanette	WKFB	770	OLG
Franklin	WFRA	1450	N/TLK	Jenkintown	WPPZ	103.9	RL/G
Franklin	WAWN	89.5	RL	Jersey Shore	WJSA	1600	RL/CC
Franklin	WOXX	99.3	AC/H	Jersey Shore	WJSA	96.3	RL/CC
Freeland	WKRZ	98.5	CH	Johnsonburg	WJNG	100.5	CH
Galeton	WCOG	100.7	RL/CC	Johnstown	WNTJ	850	N/TLK
Gettysburg	WGET	1320	AC	Johnstown	WCRO	1230	AS
Gettysburg	WZBT	91.1	AD/A	Johnstown	WPRR	1490	SPTS
Gettysburg	WGTY	107.7	CW	Johnstown	WFRJ	88.9	RL

City	Station	Freq	Format	City	Station	Freq	Format
Johnstown	WQEJ	89.7	CLA	Mckeesport	WPTT	1360	TLK+
Johnstown	WRKW	92.1	RK/C	Meadville	WMGW	1490	TLK+
Johnstown	WYKE	95.5	CW	Meadville	WARC	90.3	VAR
Johnstown	WYKE	96.5	AC	Meadville	WVME	91.9	RL
Kane	WPSB	90.1	CLA	Meadville	WZPR	100.3	CW
Kane	WLMI	103.9	CW	Mechanicsburg	WTPA	93.5	RK/C
Kittanning	WTYM	1380	OLG	Media	WPHI	100.3	CH
Lancaster	WLAN	1390	AS	Mercer	WLLF	96.7	AC/S
Lancaster	WLPA	1490	SPTS	Mercer	WWIZ	103.9	RK/C
Lancaster	WFNM	89.1	AD/A	Mercersburg	WPPT	92.1	AC
Lancaster	WJTL	90.3	RL/CC	Mexico	WJUN	1220	SPTS
Lancaster	WLCH	91.3	SP/TK	Mexico	WJUN	92.5	CW
Lancaster	WLAN	96.9	CHR	Meyersdale	WQZS	93.3	OLG
Lancaster	WROZ	101.3	AC/S	Middletown	WZXM	88.7	N/A
Lansdale	WNPV	1440	N/TLK	Middletown	WMSS	91.1	AD/A
Lansford	WLSH	1410	AC/S	Mifflinburg	WWBE	98.3	CW
Laporte	WCOZ	103.9	AC	Mifflintown	WQJU	107.1	RL
Latrobe	WCNS	1480	OLG	Mill Hall	WVRT	97.7	CH
Latrobe	WQTW	1570	AC/H	Millersburg	WQLV	98.9	AC
Lebanon	WADV	940	RL/G	Millersville	WIXQ	91.7	VAR
Lebanon	WLBR	1270	TLK+	Millvale	WAMO	860	URB
Lebanon	WQIC	100.1	AC/S	Milton	WMLP	1380	TLK+
Lehighton	WYNS	1160	SPTS	Milton	WVLY	100.9	AC/S
Levittown	WBCB	1490	AS	Monroeville	WPGR	1510	RL/G
Lewisburg	WVBU	90.5	RK/C	Montrose	WPEL	1250	RL/G
Lewisburg	WGRC	91.3	RL/CC	Montrose	WPEL	96.5	RL
Lewisburg	WXCR	103.7	RK/C	Mount Carmel	WSPI	99.7	CH
Lewistown	WIEZ	670	TLK+	Mount Union	WXOT	99.5	CH
Lewistown	WKVA	920	OLG	Mountaintop	WBHT	97.1	CHR
Lewistown	WJRC	90.9	RL/CC	Mt. Pocono	WPLY	960	OLG
Lewistown	WMRF	95.7	AC/H	Muncy	WBZD	93.3	OLG
Lewistown	WCHX	105.5	RK/C	Murrysville	WRWJ	88.1	RL/G
Lincoln Univ.	WWLU	88.7	URB	N. Cambria	WPCL	97.3	RL/CC
Linesville	WVCC	101.7	OLG	Nanticoke	WNAK	730	AS
Lock Haven	WBPZ	1230	OLG	Nanticoke	WSFX	89.1	AD/A
Lock Haven	WSNU	92.1	AC/H	Nanticoke	WQFM	92.1	OLG
Loretto	WWGE	1400	TLK+	New Berlin	WBGM	88.1	RL/CC
Mansfield	WNTE	89.5	VAR	New Castle	WKST	1200	TLK+
Mansfield	WNBQ	92.3	AC/H	New Castle	WJST	1280	OLG
Markleysburg	WLOG	89.1	RL	New Castle	WVMN	90.1	RL
Martinsburg	WJSM	1110	RL/G	New Kensington	WGBN	1150	RL/G
Martinsburg	WJSM	92.7	RL/G	New Kensington	WZPT	100.7	AC/H
Masontown	WYFU	88.5	RL	New Wilmington	WWNW	88.9	AC/H
Masontown	WRIJ	106.9	RL/G	Norristown	WNAP	1110	RL/G
McConnellsburg	WFYL	1530	TLK+	North East	WYNE	1530	CLA
McConnellsburg	WFYL	1530	TLK+	North East	WRKT	100.9	RK
McConnellsburg	WWCF	88.7	N/A	Northumberland	WEGH	107.3	CH
McConnellsburg	WEEO	103.7	CH	Oil City	WKQW	1120	TLK+
McConnellsburg	WEEQ	103.7	CH	Oil City	WOYL	1340	N/TLK
McKeesport	WEDO	810	VAR	Oil City	WKQW	96.3	OLG

City	Station	Freq	Format	City	Station	Freq	Format
Oil City	WGYI	98.5	CW	Pittsburgh	WLTJ	92.9	AC/S
Oliver	WOGG	94.9	CW	Pittsburgh	WRKZ	93.7	RK/C
Olyphant	WQOR	750	RL	Pittsburgh	WWSW	94.5	OLG
Olyphant	WBHD	95.7	CH	Pittsburgh	WKST	96.1	CH
Palmyra	WWKL	92.1	CH	Pittsburgh	WSHH	99.7	AC/S
Patton	WBRX	94.3	RK/C	Pittsburgh	WORD	101.5	RL/CC
Pen Argyl	WWPJ	89.5	CLA	Pittsburgh	WDVE	102.5	RK
Philadelphia	WFIL	560	RL	Pittsburgh	WPGB	104.7	TLK+
Philadelphia	WIP	610	SPTS	Pittsburgh	WXDX	105.9	RK
Philadelphia	WWDB	860	BIZ	Pittsburgh	WDSY	107.9	CW
Philadelphia	WURD	900	TLK+	Pittston	WITK	1550	OLG
Philadelphia	WPEN	950	OLG	Pittston	WDMT	102.3	AD/A
Philadelphia	WNTP	990	TLK+	Plains	WYCK	1340	OLG
Philadelphia	KYW	1060	N/TLK	Pleasant Gap	WQWK	98.7	RK/C
Philadelphia	WPHT	1210	TLK+	Pocono Pines	WPZX	105.9	RK/C
Philadelphia	WHAT	1340	TLK+	Port Allegany	WHKS	94.9	AC/S
Philadelphia	WDAS	1480	RL/G	Port Matilda	WKVB	107.9	RL/CC
Philadelphia	WNWR	1540	N/A	Portage	WLKJ	1470	RL/CC
Philadelphia	WPEB	88.1	VAR	Pottstown	WPAZ	1370	OLG
Philadelphia	WXPN	88.5	AD/A	Pottsville	WPPA	1360	AC
Philadelphia	WRTI	90.1	CLA	Pottsville	WPAM	1450	RK/C
Philadelphia	WHYY	90.9	N/TLK	Pottsville	WAVT	101.9	CH
Philadelphia	WKDU	91.7	AD/A	Punxsutawney	WECZ	1540	TLK+
Philadelphia	WXTU	92.5	CW	Punxsutawney	WPXZ	104.1	AC
Philadelphia	WMMR	93.3	RK	Radnor Twnshp	WYBF	89.1	VAR
Philadelphia	WYSP	94.1	TLK+	Reading	WEEU	830	TLK+
Philadelphia	WEJM	95.7	AH	Reading	WIOV	1240	SPTS
Philadelphia	WRDW	96.5	CH	Reading	WRAW	1340	OLG
Philadelphia	WOGL	98.1	OLG	Reading	WXAC	91.3	AD/A
Philadelphia	WUSL	98.9	URB	Reading	WRFY	102.5	AC/H
Philadelphia	WBEB	101.1	AC	Red Lion	WTHM	1440	SIL
Philadelphia	WIOQ	102.1	CHR	Red Lion	WSOX	96.1	OLG
Philadelphia	WMGK	102.9	RK/C	Renovo	WZYY	106.9	RK/C
Philadelphia	WSNI	104.5	AC	Reynoldsville	WDSN	106.5	AC
Philadelphia	WDAS	105.3	URB	Ridgebury	WREQ	96.9	RL
Philadelphia	WJJZ	106.1	JAZ	Riverside	WLGL	92.3	CW
Phillipsburg	WPHB	1260	CW	Roaring Spring	WKMC	1370	AS
Phillipsburg	WUBZ	105.9	RK	Russell	WQFX	103.1	RK/C
Phoenixville	WPHE	690	SP/RL	Saegertown	WHUZ	94.3	CH
Pittsburgh	WPIT	730	RL	Salladasburg	WBYL	95.5	CW
Pittsburgh	WBGG	970	SPTS	Sayre	WATS	960	AC
Pittsburgh	KDKA	1020	N/TLK	Schnecksville	WXLV	90.3	AD/A
Pittsburgh	WWNL	1080	RL	Scottdale	WLSW	103.9	AC/H
Pittsburgh	WEAE	1250	SPTS	Scranton	WEJL	630	SPTS
Pittsburgh	WJAS	1320	AS	Scranton	WGBI	910	N/TLK
Pittsburgh	KQV	1410	NWS+	Scranton	WICK	1400	OLG
Pittsburgh	WRCT	88.3	AD/A	Scranton	WVIA	89.9	N/TLK
Pittsburgh	WYEP	91.3	AD/A	Scranton	WVMW	91.5	RK/H
Pittsburgh	WPTS	92.1	JAZ	Scranton	WUSR	99.5	VAR
Pittsburgh	WPTS	92.1	AD/A	Scranton	WGGY	101.3	CW

City	Station	Freq	Format	City	Station	Freq	Format
Scranton	WWDL	104.9	AC	Tunkhannock	WEMR	1460	AC
Scranton	WEZX	106.9	RK/C	Tunkhannock	WBZR	107.7	CW
Scranton/W-B	WARM	590	N/TLK	Tyrone	WTRN	1340	AC/S
Selinsgrove	WYGL	1240	CW	Tyrone	WGMR	101.1	CH
Selinsgrove	WQSU	88.9	AD/A	Union City	WCTL	106.3	RL/CC
Sellersville	WBYO	88.9	RL/CC	Uniontown	WMBS	590	AS
Shamokin	WISL	1480	TLK+	Uniontown	WPKL	99.3	OLG
Shamokin	WBLJ	95.3	CW	University Park	WOWY	97.1	OLG
Sharon	WPIC	790	TLK+	Villanova	WXVU	89.1	AD/A
Sharon	WYFM	102.9	RK/C	Warminster	WRDV	89.3	AS
Sharpsville	WAKZ	95.9	CH	Warren	WNAE	1310	AC
Shippensburg	WEEO	1480	SPTS	Warren	WRRN	92.3	OLG
Shippensburg	WSYC	88.7	AD/A	Warwick	WZZD	88.1	RL/CC
Shiremanstown	WWII	720	RL/CC	Washington	WKZV	1110	CW
Slippery Rock	WRSK	88.1	AD/A	Washington	WJPA	1450	OLG
Smethport	WQRM	106.3	AC	Washington	WNJR	91.7	VAR
Somerset	WNTN	990	N/TLK	Washington	WJPA	95.3	OLG
Somerset	WBHV	1330	RL/G	Waynesboro	WCBG	1380	SPTS
Somerset	WLKH	97.7	RL/CC	Waynesboro	WFYN	101.5	RK/C
South Waverly	WPHD	96.1	OLG	Waynesburg	WANB	1580	CW
So Williamsport	WZXR	99.3	RK	Waynesburg	WCYJ	88.7	AC/H
Spangler	WPCL	97.3	RL/CC	Waynesburg	WANB	103.1	CW
St. Mary's	WKBI	1400	AS	Wellsboro	WNBT	1490	AS
St. Mary's	WKBI	93.9	AC	Wellsboro	WNBT	104.5	AC/H
Starview	WSJW	92.7	JAZ	West Chester	WCHE	1520	TLK+
State College	WRSC	1390	N/TLK	West Chester	WCUR	91.7	VAR
State College	WMAJ	1450	SPTS	West Hazleton	WKZN	1300	N/TLK
State College	WRXV	89.1	RL/CC	Whitneyville	WLIH	107.1	RL/G
State College	WTLR	89.9	RL/CC	Wilkes-Barre	WILK	980	N/TLK
State College	WTLR	89.9	RL	Wilkes-Barre	WBAX	1240	SPTS
State College	WKPS	90.7	AD/A	Wilkes-Barre	WRKC	88.5	VAR
State College	WPSU	91.5	CLA	Wilkes-Barre	WCLH	90.7	VAR
State College	WLTS	94.5	AC/S	Wilkes-Barre	WMGS	92.9	AC/S
State College	WJHT	103.1	CHR	Wilkinsburg	WPYT	1190	TLK+
Stroudsburg	WVPO	840	OLG	Williamsport	WLYC	1050	SPTS
Stroudsburg	WBYX	88.7	RL/CC	Williamsport	WWPA	1340	N/TLK
Stroudsburg	WSBG	93.5	CH	Williamsport	WRAK	1400	TLK+
Summerdale	WJAZ	91.7	JAZ	Williamsport	WPTC	88.1	RK/H
Sunbury	WKOK	1070	N/TLK	Williamsport	WVYA	89.7	VAR
Sunbury	WQKX	94.1	CHR	Williamsport	WCRG	90.7	RL/CC
Susquehanna	WCDW	100.5	OLG	Williamsport	WRLC	91.7	AD/A
Swarthmore	WSRN	91.5	VAR	Williamsport	WKSB	102.7	AC/H
Sweet Valley	WRGN	88.1	RL/CC	Williamsport	WILQ	105.1	CW
Tamaqua	WMGH	105.5	AC	Williamsport	WRVH	107.9	AC/S
Telford	WBMR	91.7	RL	Wyomissing	WYTL	91.7	RL/CC
Tioga	WMTT	94.7	RK/C	York	WSBA	910	TLK+
Titusville	WTIV	1230	TLK+	York	WQXA	1250	CW
Tobyhanna	WKRF	107.9	CHR	York	WOYK	1350	SPTS
Towanda	WTTC	1550	OLG	York	WVYC	99.7	AD/A
Towanda	WTTC	95.3	OLG	York	WARM	103.3	AC
Trout Run	WCIT	90.1	RL/CC	York	WQXA	105.7	RK/C
Troy	WTZN	1310	SPTS	York-Hanover	WYCR	98.5	CH
				Youngsville	WTMV	88.5	RL/CC

Radio on the Road   Rhode Island

City	Station	Freq	Format		City	Station	Freq	Format
Block Island	WCRI	95.9	CLA					
Block Island	WADK	99.3	AS					
Bristol	WQRI	88.3	AD/A					
Coventry	WCVY	91.5	AD/A					
East Providence	WPMZ	1110	SP/OM					
Greenville	WALE	990	TLK+					
Greenville	WALE	990	SP/MEX					
Hope Valley	WCNX	1180	N/TLK					
Kingston	WRIU	90.3	VAR					
Middletown	WKKB	100.3	SP/OM					
Narrag'ett Pier	WAKX	102.7	N/A					
Newport	WADK	1540	N/TLK					
Pawtuckett	WDDZ	550	YTH					
Portsmouth	WJHD	90.7	VAR					
Providence	WPRO	630	TLK+					
Providence	WSKO	790	SPTS					
Providence	WHJJ	920	TLK+					
Providence	WRIB	1220	SP/RL					
Providence	WRNI	1290	N/TLK					
Providence	WELH	88.1	VAR					
Providence	WDOM	91.3	VAR					
Providence	WPRO	92.3	CHR					
Providence	WHJY	94.1	RK					
Providence	WBRU	95.5	RK/H					
Providence	WWBB	101.5	OLG					
Providence	WWLI	105.1	AC					
Smithfield	WJMF	88.7	VAR					
Wakefield	WZRA	99.7	OLG					
Wakefield	WSKO	99.7	SPTS					
Warwick	WARV	1590	RL					
West Warwick	WLKW	1450	AS					
Westerly	WXNI	1230	N/TLK					
Westerly	WBLQ	88.1	TLK/S					
Westerly	WBLQ	88.1	TLK+					
Westerly	WEEI	103.7	SPTS					
Wickford	WKFD	1370	SILNT					
Woonsocket	WOON	1240	OLG					
Woonsocket	WNRI	1380	TLK+					
Woonsocket	WWKX	106.3	CHR					

City	Station	Freq	Format	City	Station	Freq	Format
Abbeville	WABV	1590	SILNT	Charleston	WEZL	103.5	CW
Abbeville	WZLA	92.9	OLG	Cheraw	WCRE	1420	OLG
Aiken	WLJK	89.1	JAZ	Cheraw	WJMX	103.3	CHR
Aiken	WKSP	96.3	URB	Chester	WGCD	1490	RL/G
Aiken	WKXC	99.5	CW	Chester	WBT	99.3	N/TLK
Allendale	WDOG	1460	CW	Chesterfield	WRFE	89.3	N/A
Allendale	WDOG	93.5	CW	Chesterfield	WVSZ	107.3	CW
Anderson	WAIM	1230	TLK+	Clearwater	WSLT	98.3	AC/S
Anderson	WANS	1280	RL/G	Clemson	WAHT	1560	OLG
Anderson	WROQ	101.1	RK/C	Clemson	WSBF	88.1	VAR
Anderson	WJMZ	107.3	URB	Clemson	WCCP	104.9	SPTS
Andrews	WGTN	100.7	AC/H	Clinton	WPCC	1410	SPTS
Atlantic Bch	WMIR	1200	RL/G	Columbia	WVOC	560	N/TLK
Atlantic Bch	WSEA	100.3	CHR	Columbia	WCEO	840	SP/MEX
Bamberg	WRIT	790	RL/G	Columbia	WOIC	1230	TLK+
Bamberg	WWBD	95.7	CH	Columbia	WISW	1320	TLK+
Batesburg	WBLR	1430	SP/RL	Columbia	WCOS	1400	SPTS
Batesburg	WZMJ	93.1	SPTS	Columbia	WQXL	1470	RL/CC
Beaufort	WVGB	1490	RL/G	Columbia	WMHK	89.7	RL/CC
Beaufort	WAGP	88.7	RL	Columbia	WUSC	90.5	VAR
Beaufort	WJWJ	89.9	JAZ	Columbia	WLTR	91.3	CLA
Beaufort	WYKZ	98.7	AC/S	Columbia	WARQ	93.5	RK/H
Belton	WHPB	1390	N/A	Columbia	WCOS	97.5	CW
Belton	WEPC	88.5	RL/CC	Columbia	WOMG	103.1	OLG
Belvedere	WAFJ	88.3	RL/CC	Columbia	WNOK	104.7	CHR
Bennettsville	WBSC	1550	OLG	Conway	WIQB	1050	SPTS
Bishopville	WAGS	1380	CW	Conway	WPJS	1330	RL/G
Blackville	WIIZ	97.9	URB	Conway	WHMC	90.1	CLA
Bluffton	WGZR	106.9	CW	Conway	WJXY	93.9	SPTS
Blythewood	WBAJ	890	RL	Cross Hill	WHZQ	94.1	CW
Bowman	WSPX	94.5	RL/G	Darlington	WPFM	1350	RL/G
Branchville	WGFG	105.1	OLG	Darlington	WDAR	105.5	AC/S
Briarcliffe Acr	WQSD	107.1	CH	Dillon	WDSC	800	RL/G
Buck013	WGTR	107.9	CW	Dillon	WWHW	91.1	N/A
Camden	WQIS	1130	SPTS	Dillon	WEGX	92.9	CW
Camden	WCAM	1590	AS	Dorchester	WTMZ	910	SPTS
Camden	WPUB	102.7	OLG	Easley	WELP	1360	RL
Cayce	WGCV	620	RL/G	Easley	WOLI	103.9	RL/CC
Cayce	WYFV	88.7	RL	Elloree	WORG	100.3	AC
Cayce	WLTY	96.7	AC/S	Florence	WYNN	540	RL/G
Charleston	WLTQ	730	TLK+	Florence	WJMX	970	TLK+
Charleston	WTMA	1250	N/TLK	Florence	WOLS	1230	SPTS
Charleston	WQSC	1340	TLK+	Florence	WLPG	91.7	RL
Charleston	WXTC	1390	RL/G	Florence	WYNN	106.3	URB
Charleston	WQNT	1450	NWS+	Folly Beach	WYBB	98.1	RK/C
Charleston	WFCH	88.5	RL	Forest Acres	WWNQ	94.3	CW
Charleston	WSCI	89.3	CLA	Fountain Inn	WFIS	1600	N/TLK
Charleston	WSSX	95.1	CHR	Gaffney	WFGN	1180	RL/G
Charleston	WSUY	96.9	AC/S	Gaffney	WEAC	1500	RL/G
Charleston	WALC	100.5	CH	Gaffney	WYFG	91.1	RL

City	Station	Freq	Format	City	Station	Freq	Format
Gaffney	WAGI	105.3	RL/G	Lake City	WHYM	1260	RL/G
Garden City	WWXM	97.7	CH	Lake City	WWFN	100.1	SPTS
Georgetown	WGTN	1400	N/TLK	Lamar	WSIM	93.7	OLG
Georgetown	WLMC	1470	RL/G	Lancaster	WAGL	1560	SP/MEX
Georgetown	WWGS	1580	N/A	Lancaster	WRHM	107.1	CW
Georgetown	WXJY	93.7	SPTS	Latta	WCMG	94.3	URB
Georgetown	WSYN	106.5	OLG	Laurens	WLBG	860	RL/G
Goose Creek	WSCC	94.3	TLK+	Lexington	WLGO	1170	SP/OM
Gray Court	WSSL	100.5	CW	Lexington	WLXC	98.5	URB
Greenville	WLFJ	660	RL	Loris	WLSC	1240	CW
Greenville	WMUU	1260	RL	Loris	WVCO	94.9	OLG
Greenville	WYRD	1330	TLK+	Manning	WYMB	920	SPTS
Greenville	WGVL	1440	SP/TK	Marion	WHLZ	92.5	CW
Greenville	WPCI	1490	URB	Mauldin	WBZT`96.7	98.9	RK/C
Greenville	WLFJ	89.3	RL/CC	McClellanville	WAZS	98.9	SP/MEX
Greenville	WEPR	90.1	CLA	McClellanville	WAZS	98.9	SP/MEX
Greenville	WTBI	91.7	RL	Moncks Crner	WQTK	950	SPTS
Greenville	WESC	92.5	CW	Moncks Crner	WCSQ	92.5	AC/H
Greenville	WFBC	93.7	CHR	Mt Pleasant	WZJY	1480	RL
Greenville	WMUU	94.5	RL	Mt Pleasant	WRFQ	104.5	CH
Greenville	WPLS	96.7	AD/A	Mullins	WJAY	1280	RL/G
Greenwood	WCZZ	1090	OLG	Murrell's Inlet	WMBJ	88.3	RL/CC
Greenwood	WLMA	1350	TLK+	Murrell's Inlet	WYEZ	94.5	EZL
Greenwood	WCRS	1450	AS	Myrtle Beach	WQJM	1450	N/TLK
Greenwood	WZSN	103.5	AC/S	Myrtle Beach	WMYB	92.1	AC
Greer	WPJM	800	RL/G	Myrtle Beach	WKZQ	101.7	RK
Greer	WCKI	1300	RL	Myrtle Beach	WYAV	104.1	RK/C
Greer	WOLT	103.3	RL/CC	New Ellenton	WGOR	102.7	OLG
Hampton	WHGS	1270	N/A	Newberry	WKDK	1240	AC
Hampton	WBHC	92.1	AC	Newberry	WKMG	1520	URB
Hanahan	WAVF	96.1	RK/H	Newberry	WGVC	106.3	URB
Hardeeville	WLVH	101.1	URB	North Augusta	WPCH	1380	CW
Hartsville	WHSC	1450	SPTS	North Augusta	WKZK	1600	RL/G
Hartsville	WTNI	1490	CW	N. Charleston	WYFH	90.7	RL
Hartsville	WBZF	98.5	RL/G	N. Charleston	WXLY	102.5	OLG
Hemingway	WLGI	90.9	VAR	N Myrtle Bch	WNMB	900	OLG
Hilton Head Is	WFXH	1130	SPTS	N Myrtle Bch	WKVC	88.9	RL/CC
Hilton Head Is	WFXH	106.1	RK/H	N Myrtle Bch	WEZV	105.9	EZL
Holly Hill	WJBS	1440	RL/G	Orangeburg	WPJK	1580	RL/G
Homeland Pk	WRIX	1020	RL/G	Orangeburg	WSSB	90.3	JAZ
Honea Path	WRIX	103.1	TLK+	Orangeburg	WQKI	102.9	URB
Irmo	WWNU	92.1	CW	Orangeburg	WHXT	103.9	URB
Johnsonville	WPDT	105.1	RL/G	Orangeburg	WTCB	106.7	AC
Johnston	WKSX	92.7	OLG	Pageland	WRML	102.3	RL/G
Kershaw	WKSC	1300	OLG	Pamplico	WMXT	102.1	RK/C
Kingstree	WDKD	1310	OLG	Parris Island	WGZO	103.1	CH
Kingstree	WGSS	94.1	RL/G	Pawleys Isl	WDAI	98.5	URB
Kingstree	WWKT	99.3	URB	Pickens	WTBI	1540	RL
Ladson	WKCL	91.5	RL/CC	Port Royal	WXST	99.7	URB
Ladson	WJNI	106.3	RL/G	Port Royal	WLOW	107.9	AS

City	Station	Freq	Format		City	Station	Freq	Format
Ravenel	WMGL	101.7	URB					
Richburg	WRBK	90.3	OLG					
Ridgeland	WNFO	1430	SP/MEX					
Ridgeland	WVVV	104.9	AD/A					
Ridgeville	WPAL	100.9	URB					
Rock Hill	WAVO	1150	RL					
Rock Hill	WRHI	1340	TLK+					
Rock Hill	WNSC	88.9	JAZ					
Saluda	WJES	92.1	OLG					
Sans Souci	WCSZ	1070	RL/G					
Scranton	WWRK	102.9	RK/C					
Seneca	WSNW	1150	OLG					
Seneca	WHZT	98.1	CHR					
Socastee	WRNN	99.5	N/TLK					
So Congaree	WFMV	95.3	RL/G					
Spartanburg	WSPA	910	N/TLK					
Spartanburg	WORD	950	TLK+					
Spartanburg	WSPG	1400	TLK+					
Spartanburg	WASC	1530	URB					
Spartanburg	WSPA	98.9	AC/S					
St. Andrews	WMFX	102.3	RK/C					
St. George	WQIZ	810	RL					
St. George	WNKT	107.5	CW					
St. Mathews	WQKI	710	RL/G					
St. Mathews	WIGL	93.9	AC					
St. Stephen	WTUA	106.1	RL/G					
Summerton	WLJI	98.3	RL/G					
Summerville	WAZS	980	SP/MEX					
Summerville	WWWZ	93.3	URB					
Sumter	WDXY	1240	TLK+					
Sumter	WQMC	1290	RL/G					
Sumter	WSSC	1340	RL					
Sumter	WRJA	88.1	JAZ					
Sumter	WICI	94.7	AC/H					
Sumter	WWDM	101.3	URB					
Surfside Bch	WYAK	103.1	CW					
Surfside Bch	WYAK	103.1	CW					
Trav'lers Rest	WDAB	1580	SP/MEX					
Union	WBCU	1460	CW					
Walhalla	WWOF	1000	TLK+					
Walhalla	WGOG	96.3	CW					
Walterboro	WALD	1080	RL					
Walterboro	WALI	93.7	CW					
Wedgefield	WIBZ	95.5	OLG					
West Columbia	WXBT	100.1	URB					
Williston	WAAW	94.7	RL/G					
Woodruff	WDRF	1510	CW					
York	WBZK	980	SP/MEX					

City	Station	Freq	Format	City	Station	Freq	Format
Aberdeen	KSDN	930	SPTS	Mitchell	KMIT	105.9	CW
Aberdeen	KGIM	1420	TLK+	Mitchell	KQRN	107.3	AC/H
Aberdeen	KKAA	1560	RL	Mobridge	KOLY	1300	AS
Aberdeen	KSDN	94.1	RK/C	Mobridge	KOLY	99.5	AC
Aberdeen	KLRJ	94.9	RL/CC	Pierre	KGFX	1060	CW
Aberdeen	KBFO	106.7	AC	Pierre	KCCR	1240	CH
Aberdeen	KBFO	106.7	AC/H	Pierre	KSQP	1450	TLK+
Aberdeen	KBFO	106.7	EZL	Pierre	KGFX	92.7	AC/H
Belle Fourche	KBFS	1450	CW	Pierre	KLXS	95.3	AC
Belle Fourche	KZZI	95.9	CW	Porcupine	KILI	90.1	N/A
Belle Fourche	KFMH	102.1	OLG	Rapid City	KKLS	920	OLG
Brookings	KBRK	1430	AS	Rapid City	KIMM	1150	CW
Brookings	KESD	88.3	CLA	Rapid City	KTOQ	1340	TLK+
Brookings	KSDJ	90.7	RK/H	Rapid City	KOTA	1380	N/TLK
Brookings	KBRK	93.7	AC	Rapid City	KTPT	88.3	RL/CC
Canton	KYBB	102.7	CH	Rapid City	KBHE	89.3	CLA
Canton	KYBB	102.7	RK/C	Rapid City	KQFR	89.9	RL
Clear Lake	KDBX	107.1	AP	Rapid City	KASD	90.3	N/A
Clear Lake	KDBX	107.1	AC/H	Rapid City	KWRC	90.9	N/A
Custer	KFCR	1490	AC	Rapid City	KQRQ	92.3	CH
Custer	KAWK	105.1	AC	Rapid City	KKMK	93.9	AC
Deadwood	KDSJ	980	OLG	Rapid City	KLMP	97.9	RL
Deadwood	KSQY	95.1	AD/A	Rapid City	KOUT	98.7	CW
Dell Rapids	KSQB	95.7	CH	Rapid City	KFXS	100.3	RK/C
Dell Rapids	KSQB	95.7	AC	Rapid City	KIQK	104.1	CW
Faith	KPSD	97.1	CLA	Rapid City	KZLK	106.3	AC
Flandreau	KKHG	107.9	CW	Rapid City	KZLK	106.3	AC
Flandreau	KWSF	107.9	CW	Redfield	KQKD	1380	RL
Freeman	KVCF	90.5	RL	Redfield	KNBZ	97.7	AC/H
Gregory	KVCX	101.5	RL	Redfield	KNBZ	97.7	OLG
Hot Springs	KZMX	580	CW	Redfield	KGIM	103.7	CW
Hot Springs	KZMX	96.7	CW	Reliance	KTSD	91.1	CLA
Huron	KOKK	1210	CW	Reliance	KPLO	94.5	CW
Huron	KIJV	1340	AC/S	Salem	KIKN	100.5	CW
Huron	KZNC	99.1	CW	Sioux Falls	KXRB	1000	CW
Huron	KZKK	105.1	AC/H	Sioux Falls	KSOO	1140	TLK+
Lead	KCYT	94.3	N/A	Sioux Falls	KWSN	1230	SPTS
Lemmon	KBJM	1400	CW	Sioux Falls	KNWC	1270	RL/CC
Little Eagle	KLND	89.5	VAR	Sioux Falls	KELO	1320	TLK+
Little Eagle	KLND	89.5	VAR	Sioux Falls	KSQB	1520	AS
Lowry	KQSD	91.9	CLA	Sioux Falls	KRSD	88.1	CLA
Lowry	KMLO	100.7	CW	Sioux Falls	KAUR	89.1	JAZ
Lowry	KMLO	100.7	CW	Sioux Falls	KCSD	90.9	CLA
Madison	KJAM	1390	AC	Sioux Falls	KELO	92.5	AC/S
Madison	KJAM	103.1	CW	Sioux Falls	KCFS	94.5	RL/CC
Martin	KZSD	102.5	CLA	Sioux Falls	KNWC	96.5	RL/CC
Milbank	KMSD	1510	OLG	Sioux Falls	KMXC	97.3	AC/H
Milbank	KKSD	104.3	OLG	Sioux Falls	KTWB	101.9	CW
Mission	KLGH	100.7	N/A	Sioux Falls	KRRO	103.7	RK
Mitchell	KORN	1490	TLK+	Sioux Falls	KKLS	104.7	CHR

City	Station	Freq	Format	City	Station	Freq	Format
Sisseton	KSWS	89.3	CW				
Sisseton	KBWS	102.9	CW				
Spearfish	KBHU	89.1	AD/A				
Spearfish	KDDX	101.1	RK				
Spearfish	KSLT	107.3	RL/CC				
Sturgis	KBHB	810	FRM				
Sturgis	KRCS	93.1	CHR				
Vermillion	KVTK	1570	SPTS				
Vermillion	KUSD	89.7	CLA				
Vermillion	KAOR	91.1	AD/A				
Vermillion	KVHT	106.3	AC/H				
Volga	KJJQ	910	TLK+				
Volga	KKQQ	102.3	CW				
Watertown	KWAT	950	FRM				
Watertown	KSDR	1480	TLK+				
Watertown	KJBB	89.1	RL/G				
Watertown	KSDR	92.9	CW				
Watertown	KIXX	96.1	AC/H				
Watertown	KDLO	96.9	CW				
Wessington Springs	KGGK	98.3	OLG				
Wessington Springs	KUQL	98.3	OLG				
Winner	KWYR	1260	CW				
Winner	KWYR	93.7	AC				
Yankton	WNAX	570	N/TLK				
Yankton	KYNT	1450	AC/S				
Yankton	KKYA	93.1	CW				
Yankton	WNAX	104.1	CW				

City	Station	Freq	Format	City	Station	Freq	Format
Alamo	WCTA	810	N/TLK	Church Hill	WMCH	1260	RL/G
Alamo	WWGM	93.1	RL/G	Clarksville	WJQI	540	RL/CC
Alcoa	WBCR	1470	TLK+	Clarksville	WJZM	1400	TLK+
Alcoa	WYLV	89.1	RL/CC	Clarksville	WCTZ	1550	RL/G
Algood	WATX	1590	RL/CC	Clarksville	WAYQ	88.3	RL/CC
Ardmore	WSLV	1110	CW	Clarksville	WAPX	91.9	AD/A
Ashland City	WQSV	790	VAR	Cleveland	WBAC	1340	TLK+
Athens	WYXI	1390	TLK+	Cleveland	WCLE	1570	RL/G
Athens	WLAR	1450	CW	Cleveland	WALV	95.3	AC/H
Athens	WJSQ	101.7	CW	Cleveland	WUSY	100.7	CW
Atwood	WTBK	93.7	RL/CC	Clifton	WLVS	106.5	CW
Bartlett	WMPS	1210	AD/A	Clinton	WYSH	1380	CW
Bartlett	WMFS	92.9	RK/H	Clinton	WDVX	89.9	CW
Baxter	WBXE	93.7	RK/C	Clinton	WYFC	95.3	RL
Belle Meade	WNFN	106.7	SPTS	Coalmont	WSGM	104.7	RL/G
Benton	WBIN	1540	RL/CC	Collegedale	WSMC	90.5	CLA
Benton	WBIN	1540	RL	Collierville	WCRV	640	RL
Benton	WOCE	93.1	SP/MEX	Collinwood	WMSR	94.9	CH
Berry Hill	WVOL	1470	URB	Colonial Hts	WPWT	870	TLK+
Blountville	WGOC	640	CW	Colonial Hts	WRZK	1059	RK
Bolivar	WBOL	1560	OLG	Colubia	WHRS	91.7	CLA
Bolivar	WOJG	94.7	RL/G	Columbia	WMRB	910	RL/G
Bolivar	WMOD	96.7	CW	Columbia	WMCP	1280	CW
Brentwood	WNSR	560	SPTS	Columbia	WKRM	1340	AC
Bristol	WOPI	1490	AS	Columbia	WAYM	88.7	RL/CC
Bristol	WBCV	1550	RL	Columbia	WKOM	101.7	OLG
Bristol	WHCB	91.5	RL	Cookeville	WPTN	780	TLK+
Bristol	WXBQ	96.9	CW	Cookeville	WHUB	1400	CW
Brownsville	WNWS	1520	SP/MEX	Cookeville	WTTU	88.5	AD/A
Brownsville	WQNN	88.3	N/A	Cookeville	WWOG	90.9	RL/G
Brownsville	WTBG	95.3	RL/G	Cookeville	WGSQ	94.7	CW
Bulls Gap	WBGQ	100.7	AC	Cookeville	WGIC	98.5	CH
Calhoun	WCLE	104.1	AC	Copperhill	WLSB	1400	CW
Camden	WFWL	1220	CW	Covington	WKBL	1250	CW
Camden	WRJB	98.3	AC	Covington	WKBQ	93.5	AC/S
Carthage	WRKM	1350	SPTS	Cowan	WZYX	1440	OLG
Carthage	WUCZ	104.1	CW	Crossville	WAEW	1330	TLK+
Celina	WVFB	101.5	CW	Crossville	WCSV	1490	SPTS
Centerville	WNKX	1570	CW	Crossville	WMKW	89.3	RL/CC
Centerville	WNKX	96.7	CW	Crossville	WPBX	99.3	AC
Chattanooga	WGOW	1150	TLK+	Crossville	WOWF	102.5	CW
Chattanooga	WNOO	1260	RL/G	Dayton	WDNT	1280	TLK+
Chattanooga	WDOD	1310	AS	Dayton	WDNT	104.9	OLG
Chattanooga	WDEF	1370	TLK+	Dickson	WDKN	1260	CW
Chattanooga	WLMR	1450	RL	Dickson	WNRZ	91.5	RL/CC
Chattanooga	WJOC	1490	RL/G	Dickson	WQZQ	102.5	CHR
Chattanooga	WUTC	88.1	JAZ	Donelson	WAMB	1160	AS
Chattanooga	WMBW	88.9	RL/CC	Dresden	WCDZ	95.1	OLG
Chattanooga	WDYN	89.7	RL	Dunlap	WSDQ	1190	CW
Chattanooga	WDEF	92.3	AC/S	Dyer	WLSQ	94.3	CH
Chattanooga	WDOD	96.5	AD/A	Dyersburg	WTRO	1450	OLG
Chattanooga	WSKZ	106.5	RK/C	Dyersburg	WKNQ	90.7	N/TLK

City	Station	Freq	Format	City	Station	Freq	Format
Dyersburg	WASL	100.1	AC/H	Jackson	WJAK	1460	RL/G
East Ridge	WOGT	107.9	OLG	Jackson	WAMP	88.1	RL/CC
Elizabethton	WBEJ	1240	CW	Jackson	WKNP	90.1	CLA
Elizabethton	WHHQ	1520	RL	Jackson	WNWS	101.5	TLK+
Elizabethton	WUMC	90.5	VAR	Jackson	WMXX	103.1	OLG
Elizabethton	WTZR	99.3	RK/H	Jackson	WTNV	104.1	CW
Englewood	WENR	1090	AS	Jamestown	WCLC	1260	RL/G
Erwin	WEMB	1420	RL/G	Jamestown	WDEB	1500	CW
Erwin	WXIS	103.9	CH	Jamestown	WDEB	103.9	CW
Etowah	WCPH	1220	AS	Jamestown	WCLC	105.1	RL/G
Etowah	WLLJ	103.1	RL/CC	Jasper	WWAM	820	RL/G
Fairview	WPFD	850	CW	Jefferson City	WJFC	1480	CW
Farragut	WMTY	670	TLK+	Jefferson City	WEnrx	99.3	SPTS
Fayetteville	WEKR	1240	CW	Jellico	WJJT	1540	RL/G
Fayetteville	WYTM	105.5	CW	Jellico	WEKX	102.7	AC
Franklin	WAKM	950	CW	Johnson City	WETS	89.5	VAR
Franklin	WHEW	1380	SP/MEX	Johnson City	WETB	790	RL/G
Franklin	WRLT	100.1	AD/A	Johnson City	WJCW	910	N/TLK
Gallatin	WHIN	1010	CW	Johnson City	WQUT	101.5	RK/C
Gallatin	WYXE	1130	SP/RL	Jonesborough	WKTP	1590	AS
Gallatin	WMRO	1560	OLG	Karns	WMVU	93.1	AC/H
Gallatin	WVCP	88.5	AC/S	Kingsport	WHGG	1090	RL/CC
Gallatin	WGFX	104.5	SPTS	Kingsport	WKIN	1320	TLK+
Gatlinburg	WSEV	105.5	AC	Kingsport	WKPT	1400	AS
Germantown	WPLX	1170	RL/CC	Kingsport	WCQR	88.3	RL/CC
Germantown	WOWW	1430	YTH	Kingsport	WCSK	90.3	VAR
Germantown	WMBZ	94.1	AC	Kingsport	WTFM	98.5	AC
Germantown	WHBQ	107.5	CHR	Kingsport	WKOS	104.9	AC
Goodlettsville	WRQQ	97.1	OLG	Kingston	WBBX	1410	RL/G
Graysville	WAYB	95.7	RL	Kingston	WKTS	90.1	AC
Greeneville	WGRV	1340	CW	Kingston Sprs	WFFI	93.7	RL/CC
Greeneville	WSMG	1450	OLG	Knoxville	WRJZ	620	RL
Greeneville	WAEZ	94.9	AC/H	Knoxville	WETR	760	TLK+
Harriman	WBZH	92.7	AC	Knoxville	WKVL	850	TLK+
Harrogate	WRWB	740	TLK+	Knoxville	WKXV	900	RL/G
Harrogate	WLMU	91.3	AC	Knoxville	WNOX	990	SPTS
Harrogate	WXJB	96.5	CW	Knoxville	WVLZ	1180	SPTS
Hartsville	WTNK	1090	CW	Knoxville	WIFA	1240	RL
Henderson	WFHC	91.5	VAR	Knoxville	WKGN	1340	URB
Henderson	WFKX	95.7	URB	Knoxville	WITA	1490	RL
Henderson	WHHM	107.7	AC/S	Knoxville	Wnpz	1580	RL/G
Hendersonville	WQQK	92.1	URB	Knoxville	WUTK	90.3	AD/A
Henry	WMUF	104.7	CW	Knoxville	WKCS	91.1	OLG
Hohenwald	WMLR	1230	CW	Knoxville	WUOT	91.9	CLA
Hohenwald	WAUO	90.7	RL	Knoxville	WJXB	97.5	AC/S
Humboldt	WIRJ	740	N/TLK	Knoxville	WIMZ	103.5	RK/C
Humboldt	WHMT	1190	SPTS	Knoxville	WQIX	104.5	CH
Humboldt	WZDQ	102.3	RK/C	Knoxville	WIVK	107.7	CW
Humboldt	WLSZ	105.3	CH	La Follette	WGLH	960	RL/G
Huntingdon	WDAP	1530	CW	La Follette	WLAF	1450	RL/G
Huntingdon	WVHR	100.9	CW	La Follette	WQLA	104.9	CW
Jackson	WDXI	1310	BIZ	Lafayette	WEEN	1460	RL/G
Jackson	WTJS	1390	TLK+	Jackson	WJAK	1460	RL/G

Radio on the Road   Tennessee

City	Station	Freq	Format	City	Station	Freq	Format
Lafayette	WLCT	102.1	RL	Memphis	WWTQ	680	TLK+
Lakeland	WMQM	1600	RL	Memphis	WMC	790	SPTS
Lavergne	WBUZ	102.9	RK/H	Memphis	KWAM	990	TLK+
Lawrenceburg	WWLX	590	CW	Memphis	WGSF	1030	SP/MEX
Lawrenceburg	WDXE	1370	CW	Memphis	WDIA	1070	URB
Lawrenceburg	WZXX	88.5	RL	Memphis	WLOK	1340	RL/G
Lawrenceburg	WAWI	89.7	RL	Memphis	WBBP	1480	RL/G
Lawrenceburg	WLLX	97.5	CW	Memphis	WQOX	88.5	URB
Lawrenceburg	WDXE	106.7	AC	Memphis	WYPL	89.3	READ
Lebanon	WCOR	900	CW	Memphis	WEVL	89.9	VAR
Lebanon	WCKD	1490	N/A	Memphis	WKNO	91.1	CLA
Lebanon	WFMQ	91.5	VAR	Memphis	WUMR	91.7	JAZ
Lebanon	WANT	98.9	CW	Memphis	WHRK	97.1	URB
Lebanon	WRVW	107.5	CHR	Memphis	WMC	99.7	AC/H
Lenoir City	WLIL	730	CW	Memphis	WEGR	102.7	RK
Lenoir City	WBLC	1360	RL	Memphis	WEVR	104.5	AC
Lenoir City	WKZX	93.5	CW	Memphis	WRVR	104.5	AC
Lewisburg	WAXO	1220	CW	Memphis	WGKX	105.9	CW
Lewisburg	WJJM	1490	CW	Middleton	WYDL	100.7	CH
Lewisburg	WJJM	94.3	CW	Milan	WYNU	92.3	CH
Lexington	WDXL	1490	RL/G	Millington	WLRM	1380	RL/G
Lexington	WIGH	88.7	RL	Millington	WXMX	98.1	RK/C
Lexington	WZLT	99.3	AC	Minor Hill	WEUZ	92.1	URB
Livingston	WLIV	920	CW	Monterey	WLIV	104.7	CW
Livingston	WLQK	95.9	AC/S	Monterey	WKXD	106.9	CHR
Lobelville	WFGZ	94.5	RL/CC	Morristown	WCRK	1150	AC/S
Lookout Mtn	WFLI	1070	RL/G	Morristown	WMTN	1300	OLG
Loudon	WLOD	1140	TLK+	Morristown	WMXK	95.9	AC
Loudon	WNML	99.1	SPTS	Mt Pleasant	WXRQ	1460	RL/G
Loudon	WKVL	105.3	AD/A	Mountain City	WMCT	1390	CW
Madison	WKDA	1430	N/TLK	Munford	WMPW	98.9	CH
Madisonville	WRKQ	1250	TLK+	Murfreesboro	WMGC	810	SP/TK
Madisonville	WYGO	99.5	AC	Murfreesboro	WGNS	1450	N/TLK
Manchester	WMSR	1320	CH	Murfreesboro	WMTS	88.3	VAR
Manchester	WWTN	99.7	TLK+	Murfreesboro	WMOT	89.5	JAZ
Manchester	WFTZ	101.5	AC	Murfreesboro	WFCM	91.7	RL
Martin	WCMT	1410	TLK+	Murfreesboro	WCJK	96.3	AH
Martin	WUTM	90.3	CH	Nashville	WSM	650	CW
Maryville	WKCE	1120	SPTS	Nashville	WENO	760	RL
Maryville	WGAP	1400	TLK+	Nashville	WMDB	880	RL/G
Maryville	WTXM	95.7	OLG	Nashville	WYFN	980	RL
Maynardsville	WOEZ	88.3	EZL	Nashville	WKDA	1200	SP/MEX
Mckenzie	WHDM	1440	OLG	Nashville	WNSG	1240	RL/G
Mckenzie	WAJJ	89.3	RL/CC	Nashville	WNQM	1300	SP/RL
Mckenzie	WWYN	106.9	CW	Nashville	WNAH	1360	RL/G
Mckinnon	WTPR	101.5	OLG	Nashville	WLAC	1510	N/TLK
Mcminnville	WBMC	960	RL/G	Nashville	WFSK	88.1	JAZ
Mcminnville	WAKI	1230	TLK+	Nashville	WNAZ	89.1	RL/CC
Mcminnville	WCPI	91.3	VAR	Nashville	WPLN	90.3	CLA
Mcminnville	WTRZ	103.9	CW	Nashville	WRVU	91.1	VAR
Memphis	WHBQ	560	SPTS	Nashville	WJXA	92.9	AC/S
Memphis	WREC	600	TLK+	Nashville	WSM	95.5	CW
Memphis	WREC	600	N/TLK	Nashville	WSIX	97.9	CW

171

City	Station	Freq	Format	City	Station	Freq	Format
Nashville	WKDF	103.3	CW	Smithville	WJLE	101.7	CW
Nashville	WNRQ	105.9	RK/C	Smyrna	WFCM	710	RL/CC
New Johns'vlle	WAYW	89.9	RL/CC	Smyrna	WFFH	94.1	RL/CC
Newport	WNPC	1060	CW	Sneedville	WSDC	88.5	RL/G
Newport	WLIK	1270	RL	Soddy- Daisy	WGOW	102.3	TLK+
Newport	WNPC	92.9	CW	Soddy-Daisey	WSDT	1240	RL/G
Norris	WRMX	106.7	OLG	Soddy-Daisy	WSDT	1240	RL/G
Oak Ridge	WATO	1290	TLK+	Somerville	WSTN	1410	RL
Oak Ridge	WNFZ	94.3	RK	South Fulton	WCMT	101.3	AC
Oak Ridge	WOKI	100.3	TLK+	So Pittsburg	WEPG	910	CW
Olive Hill	WDNX	89.1	RL	So Pittsburg	WLOV	97.3	OLG
Oliver Springs	WOKI	98.7	AD/A	Sparta	WTZX	860	CW
Oneida	WOCV	1310	AC	Sparta	WSMT	1050	RL/G
Oneida	WBNT	105.5	CW	Sparta	WRKK	105.5	RK/C
Paris	WTPR	710	OLG	Spencer	WZYZ	90.1	RL/G
Paris	WMUF	1000	CW	Spencer	WKZP	107.3	AC/H
Paris	WLZK	94.1	AC	Spring City	WXQK	970	TLK+
Paris	WAKQ	105.5	CHR	Spring City	WAYA	93.9	AC/H
Parkers Cross.	WBFG	96.5	SPTS	Springfield	WSGI	1100	CW
Parsons	WKJQ	1550	RL/G	Springfield	WDBL	1590	TLK+
Parsons	WKJQ	97.3	CW	St. Joseph	WJOR	101.5	RL/G
Pegram	WQZQ	102.5	CH	Static	WSBI	1210	OLG
Pikeville	WUAT	1110	CW	Summertown	WUTZ	88.3	VAR
Pikeville	WUAT	1110	CW	Surgoinsville	WEYE	104.3	CH
Portland	WQKR	1270	OLG	Sweetwater	WDEH	800	RL/G
Powell	WQBB	1040	SPTS	Sweetwater	WLOD	98.3	OLG
Pulaski	WKSR	1420	OLG	Tazewell	WNTT	1250	CW
Pulaski	WKSR	98.3	CW	Tazewell	WCTU	94.1	CW
Red Bank	WAWL	91.5	AD/A	Trenton	WTNE	1500	AS
Red Bank	WJTT	94.3	URB	Trenton	WTNE	97.7	CW
Ripley	WTRB	1570	CW	Tullahoma	WJIG	740	RL/G
Ripley	WAUV	89.7	RL	Tullahoma	WAUT	88.5	RL
Ripley	WKVZ	94.9	RL/CC	Tullahoma	WTML	91.5	CLA
Rockwood	WOFE	580	CW	Tullahoma	WHRP	93.3	URB
Rockwood	WOFE	105.7	CW	Tusculum	WIKQ	103.1	CW
Rogersville	WRGS	1370	CW	Union City	WENK	1240	OLG
Rogersville	WJDT	106.5	CW	Union City	WTNN	88.9	N/A
Savannah	WORM	1010	OLG	Union City	WYVY	104.9	CW
Savannah	WAZD	88.1	RL	Union City	WQAK	105.7	RK
Savannah	WKWX	93.5	CW	Wartburg	WECO	940	RL/G
Savannah	WORM	101.7	CW	Wartburg	WEC0	101.3	CW
Selmer	WDTM	1150	RL/G	Waverly	WQMV	1060	SIL
Selmer	WSIB	93.9	RL/CC	Waverly	WVRY	105.1	RL/G
Selmer	WXOQ	105.5	CW	Waynesboro	WTNR	930	OLG
Sevierville	WSEV	930	SPTS	White Bluff	WQSE	1030	RL/G
Sevierville	WMYU	102.1	CHR	Winchester	WCDT	1340	CW
Sewanee	WUTS	91.3	AD/A	Woodbury	WBRY	1540	CW
Seymour	WJBZ	96.3	RL/G	Woodbury	WBOZ	104.9	RL/G
Shelbyville	WZNG	1400	TLK+				
Shelbyville	WLIJ	1580	CW				
Shelbyville	WBIA	88.3	RL/G				
Signal Mountain	WKXJ	98.1	CHR				
Smithville	WJLE	1480	CW				

Radio on the Road   Texas

City	Station	Freq	Format	City	Station	Freq	Format
Abilene	KSLI	1280	TLK+	Atlanta	KALT	1610	RL
Abilene	KWKC	1340	N/TLK	Atlanta	KNRB	100.1	RL
Abilene	KYYW	1470	CW	Austin	KLBJ	590	TLK+
Abilene	KZQQ	1560	AS	Austin	KVET	1300	SPTS
Abilene	KGNZ	88.1	RL/CC	Austin	KFON	1490	SP/MEX
Abilene	KACU	89.7	CLA	Austin	KAZI	88.7	URB
Abilene	KAGT	90.5	RL/G	Austin	KMFA	89.5	CLA
Abilene	KAQD	91.3	RL	Austin	KUT	90.5	N/TLK
Abilene	KULL	92.5	AC	Austin	KVRX	91.7	VAR
Abilene	KFGL	100.7	RK/C	Austin	KLBJ	93.7	RK/C
Abilene	KEAN	105.1	CW	Austin	KKMJ	95.5	AC
Abilene	KKHR	106.3	SP/TO	Austin	KVET	98.1	CW
Abilene	KEYJ	107.9	RK	Austin	KASE	100.7	CW
Alamo	KJAV	104.9	OLG	Austin	KPEZ	102.3	AD/A
Alamo Hts	KDRY	1100	RL	Azle	KTCY	101.7	SP/OM
Alice	KOPY	1070	CW	Baird	KORQ	95.1	CH
Alice	KOPY	92.1	SP/TO	Balch Springs	KSKY	660	TLK+
Alice	KNDA	102.9	CHR	Ballinger	KRUN	1400	CW
Allen	KESN	103.3	SPTS	Ballinger	KKCN	103.1	CW
Alpine	KVLF	1240	AS	Bandera	KEEP	98.3	CW
Alpine	KALP	92.7	CW	Bastrop	KHIB	88.5	RL
Alvin	KTEK	1110	RL	Bastrop	KGSR	107.1	AD/A
Alvin	KACC	89.7	RK	Bay City	KFRT	88.1	RL
Amarillo	KGNC	710	N/TLK	Bay City	KZBJ	89.5	N/A
Amarillo	KIXZ	940	N/TLK	Bay City	KXGJ	101.7	SP/MEX
Amarillo	KTNZ	1010	SP/RL	Bay City	KMKS	102.5	CW
Amarillo	KZIP	1310	SP/TK	Baytown	KWWJ	1360	RL/G
Amarillo	KDJW	1360	CW	Beaumont	KLVI	560	TLK+
Amarillo	KPUR	1440	SPTS	Beaumont	KZZB	990	RL/G
Amarillo	KJRT	88.3	RL	Beaumont	KRCM	1380	N/TLK
Amarillo	KXLV	89.1	RL/CC	Beaumont	KIKR	1450	SPTS
Amarillo	KACV	89.9	AD/A	Beaumont	KLBT	88.1	N/A
Amarillo	KAVW	90.7	RL	Beaumont	KTXB	89.7	RL
Amarillo	KYFA	91.9	RL/CC	Beaumont	KVLU	91.3	CLA
Amarillo	KQIZ	93.1	CHR	Beaumont	KQXY	94.1	CHR
Amarillo	KMXJ	94.1	AC/H	Beaumont	KYKR	95.1	CW
Amarillo	KMML	96.9	CW	Beaumont	KFNC	97.5	N/TLK
Amarillo	KGNC	97.9	CW	Beaumont	KTCX	102.5	URB
Amarillo	KPRF	98.7	CHR	Beaumont	KQQK	107.9	SP/OM
Amarillo	KBZD	99.7	SP/OM	Beeville	KIBL	1490	SP/RL
Amarillo	KXGL	100.9	CH	Beeville	KVFM	91.3	SP/RL
Amarillo	KATP	101.9	CW	Beeville	KTKO	105.7	CW
Amarillo	KRGN	103.1	RL/CC	Beeville	KRXB	107.1	RK/C
Amarillo	KJJP	105.7	CLA	Bellaire	KILE	1560	SP/MEX
Andrews	KACT	1360	CW	Bells	KMKT	93.1	CW
Andrews	KACT	105.5	CW	Bellville	KNUZ	1090	AC/S
Anson	KTLT	98.1	RL/CC	Belton	KTON	940	CW
Arlington	KLTY	94.9	RL/CC	Belton	KOOC	106.3	CH
Athens	KLVQ	1410	RL/G	Benavides	KXTM	107.7	SP/MEX
Atlanta	KPYN	900	RL	Benbrook	KDXX	107.1	SP/OM

City	Station	Freq	Format	City	Station	Freq	Format
Big Lake	KPDB	98.3	SIL	Bryan	KKYS	104.7	AC
Big Lake	KWTR	104.1	CW	Buda	KROX	101.5	RK/H
Big Sandy	KTAA	90.7	RL	Burkburnett	KYYI	104.7	RK/C
Big Spring	KBYG	1400	OLG	Burleson	KTFW	1460	CW
Big Spring	KBST	1490	TLK+	Burnet	KRHC	1340	CW
Big Spring	KBCX	91.5	RL	Burnet	KBEY	92.5	CW
Big Spring	KBTS	94.3	RK/C	Burnet	KHLB	106.9	CW
Big Spring	KBST	95.7	CW	Bushland	KTXP	91.5	CLA
Bishop	KFLZ	106.9	SP/MEX	Byrne	KLRW	88.5	RL/CC
Bloomington	KHVT	91.5	RL	Caldwell	KLTR	107.3	AC/S
Bloomington	KLUB	106.9	RK/C	Callisburg	KPFC	91.9	CH
Boerne	KBRN	1500	SP/RL	Cameron	KMIL	1330	CW
Bonham	KFYN	1420	CW	Cameron	KNVR	94.3	SIL
Borger	KQTY	1490	CW	Cameron	KXCS	103.9	RK/H
Borger	KASV	88.7	RL	Camp Wood	KAYG	99.1	SP/RL
Borger	KAXH	91.5	RL	Campbell	KRVA	107.1	CW
Borger	KQFX	104.3	SP/MEX	Canton	KVCI	1510	RL/CC
Borger	KQTY	106.7	CW	Canyon	KZRK	1550	SPTS
Bowie	KNTX	1410	OLG	Canyon	KWTS	91.1	AD/A
Brady	KNEL	1490	OLG	Canyon	KPUR	107.1	OLG
Brady	KNEL	95.3	CW	Canyon	KZRK	107.9	RK
Breckenridge	KROO	1430	OLG	Carrrzo Spgs	KBEN	1450	RL
Breckenridge	KLXK	93.5	CW	Carrizo Sprgs	KZCO	92.1	SP/RL
Brenham	KWHI	1280	CW	Carrollton	KJON	850	SP/TO
Brenham	KULF	94.1	AC/S	Carthage	KGAS	1590	RL/G
Brenham	KTTX	106.1	CW	Carthage	KTUX	98.9	RK
Bridgeport	KBFR	90.5	RL	Carthage	KGAS	104.3	CW
Bridgeport	KBOC	98.3	CW	Cedar Park	KDHT	93.3	CHR
Brookshire	KCHN	1050	N/A	Center	KDET	930	CW
Brownfield	KKUB	1300	SP/RL	Center	KQBB	100.5	CW
Brownfield	KPBB	88.5	SP/RL	Centerville	KUZN	105.9	SILNT
Brownfield	KCWV	90.7	N/A	Childress	KCTX	1510	AS
Brownfield	KLZK	104.3	AC/S	Childress	KCTX	96.1	CW
Brownsville	KVNS	1700	TLK+	Clarendon	KEFH	99.3	OLG
Brownsville	KBNR	88.3	SP/RL	Clarksville	KCAR	1350	CW
Brownsville	KKPS	99.5	SP/TO	Clarksville	KGAP	98.5	OLG
Brownsville	KTEX	100.3	CW	Claude	KARX	95.7	RK/C
Brownwood	KXYL	1240	SP/TO	Cleburne	KCLE	1120	CW
Brownwood	KBWD	1380	AC	Cleveland	KTHT	97.1	CW
Brownwood	KPBE	89.3	SP/RL	Clifton	KWOW	103.3	SP/MEX
Brownwood	KBUB	90.3	RL	Cockrell Hill	KRVA	1600	N/A
Brownwood	KHPU	91.7	SIL	Coleman	KSTA	1000	CW
Brownwood	KXYL	96.9	TLK+	Coleman	KXCT	102.3	RK/C
Brownwood	KPSM	99.3	RL/CC	College Sta	KZNE	1150	SPTS
Brownwood	KOXE	101.3	CW	College Sta	WTAW	1620	TLK+
Bryan	KTAM	1240	SP/MEX	College Sta	KEOS	89.1	VAR
Bryan	KAGC	1510	RL/CC	College Sta	KAMU	90.9	CLA
Bryan	KORA	98.3	CW	Colorado City	KVMC	1320	CW
Bryan	KNFX	99.5	RK/C	Colorado City	KAUM	107.1	CW
Bryan	KKYS	104.7	AC	Columbus	KULM	98.3	CW

City	Station	Freq	Format	City	Station	Freq	Format
Comanche	KYOX	94.3	CW	Dallas	KBFB	97.9	CH
Comfort	KCOR	95.1	SP/OM	Dallas	KLUV	98.7	OLG
Commerce	KETR	88.9	VAR	Dallas	KJKK	100.3	AH
Commerce	KYJC	91.3	RL	Dallas	WRR	101.1	CLA
Conroe	KJOJ	880	SP/MEX	Dallas	KDMX	102.9	AC/H
Conroe	KYOK	1140	RL/G	Dallas	KKDA	104.5	URB
Conroe	KAFR	88.3	RL/CC	Dallas	KLLI	105.3	TLK+
Conroe	KHPT	106.9	OLG	Decatur	KDKR	91.3	RL
Copperas Cve	KSSM	103.1	URB	Decatur	KRNB	105.7	URB
Corpus Christi	KCTA	1030	RL	Del Rio	KTJK	1230	SP/TO
Corpus Christi	KCCT	1150	TLK+	Del Rio	KWMC	1490	OLG
Corpus Christi	KSIX	1230	SPTS	Del Rio	KDLK	94.3	CW
Corpus Christi	KKTX	1360	TLK+	Del Rio	KTDR	96.3	AC/H
Corpus Christi	KUNO	1400	SP/MEX	Del Valle	KIXL	970	RL
Corpus Christi	KEYS	1440	TLK+	Denison	KYNG	950	N/A
Corpus Christi	KKLM	88.7	RL/CC	Denton	KFZO	99.1	SP/OM
Corpus Christi	KBNJ	91.7	RL/CC	Denton	KHKS	106.1	CHR
Corpus Christi	KMXR	93.9	OLG	Devine	KRPT	92.5	TLK+
Corpus Christi	KBSO	94.7	CW	Diboll	KSML	1260	SPTS
Corpus Christi	KZFM	95.5	CH	Diboll	KAFX	95.5	CH
Corpus Christi	KLTG	96.5	AC/H	Dilley	KLMO	98.9	SP/TO
Corpus Christi	KRYS	99.1	CW	Dimmitt	KDHN	1470	CW
Corpus Christi	KEDT	90.3	CLA	Dimmitt	KNNK	100.5	RL/G
Corsicana	KAND	1340	CW	Doss	KGLF	88.1	RL/CC
Crane	KXOI	810	SP/RL	Dripping Spgs	KXXS	104.9	SP/OM
Crane	KMMZ	101.3	SP/MEX	Dublin	KSTV	93.1	CW
Creedmore	KZNX	1530	SPTS	Dumas	KDDD	800	CW
Crockett	KIVY	1290	TLK+	Dumas	KDDD	95.3	OLG
Crockett	KIVY	1290	AS	Eagle Pass	KEPS	1270	SP/TO
Crockett	KIVY	92.7	AS	Eagle Pass	KEPI	88.7	RL/CC
Crockett	KIVY	92.7	CW	Eagle Pass	KEPX	89.5	SP/RL
Crockett	KBHT	93.5	CW	Eagle Pass	KINL	92.7	AC
Crystal Beach	KPTI	105.3	CH	Eastland	KEAS	1590	CW
Crystal City	KHER	94.3	SP/TO	Eastland	KEAS	97.7	CW
Cuero	KTLZ	89.9	RL	Edinburg	KURV	710	N/TLK
Cypress	KYND	1520	SP/MEX	Edinburg	KOIR	88.5	SP/RL
Daingerfield	KNGR	1560	RL/G	Edinburg	KBFM	104.1	CH
Dalhart	KXIT	1240	CW	Edinburg	KVLY	107.9	AC
Dalhart	KXIT	96.3	RK/C	Edna	KTMR	1130	SP/TO
Dallas	KLIF	570	TLK+	Edna	KGUL	96.1	CW
Dallas	KGGR	1040	RL/BG	El Campo	KULP	1390	CW
Dallas	KRLD	1080	N/TLK	El Campo	KIOX	96.9	SP/OM
Dallas	KTRA	1190	OLG	El Paso	KROD	600	N/TLK
Dallas	KTCK	1310	SPTS	El Paso	KTSM	690	N/TLK
Dallas	KNIT	1480	RL/G	El Paso	KAMA	750	SP/OM
Dallas	KNON	89.3	VAR	El Paso	KBNA	920	SP/OM
Dallas	KERA	90.1	N/TLK	El Paso	XEJ	970	SPN
Dallas	KCBI	90.9	RL/CC	El Paso	KXPL	1060	SP/N
Dallas	KVTT	91.7	RL	El Paso	KSVE	1150	SP/TK
Dallas	KZPS	92.5	RK/C	El Paso	KVIV	1340	SP/RL

City	Station	Freq	Format	City	Station	Freq	Format
El Paso	KHEY	1380	SPTS	Fort Worth	KSCS	96.3	CW
El Paso	KHEY	1380	SPTS	Fort Worth	KEGL	97.1	AC/S
El Paso	KELP	1590	RL	Fort Worth	KPLX	99.5	CW
El Paso	KHRO	1650	RK/H	Fort Worth	KOAI	102.1	JAZ
El Paso	KTEP	88.5	CLA	Fort Worth	KDGE	102.1	RK/C
El Paso	KXCR	89.5	RL/CC	Franklin	KZTR	101.9	AC
El Paso	KVER	91.1	SP/RL	Frankston	KOYE	96.7	SP/MEX
El Paso	KOFX	92.3	SP/TK	Frankston	KTXV	890	SP/MEX
El Paso	KSII	93.1	AC/H	Fredericksburg	KNAF	910	CW
El Paso	KINT	93.9	SP/OM	Fredericksburg	KNAF	910	TLK+
El Paso	KYSE	94.7	SP/OM	Fredericksburg	KNAF	105.7	CW
El Paso	KLAQ	95.5	RK	Freeport	KBRZ	1460	RL/SP
El Paso	KHEY	96.3	CW	Freeport	KJOJ	103.3	SP/MEX
El Paso	KBNA	97.5	SP/OM	Freer	KPBN	90.7	SP/RL
El Paso	XHPX	98.3	SP/MEX	Freer	KBRA	95.9	SIL
El Paso	KTSM	99.9	AC/S	Friona	KGRW	94.7	SP/MEX
El Paso	XHH	100.7	SP/MEX	Frisco	KXEB	910	TLK+
El Paso	KPRR	102.1	CH	Gainesville	KGAF	1580	CW
El Paso	XHTO	104.3	SP/MEX	Gainesville	KSOC	94.5	URB
El Paso	XHGU	105.9	SP/MEX	Galveston	KHCB	1400	SP/RL
El Paso	XHNC	107.5	SP/MEX	Galveston	KGBC	1540	RL
Electra	KOLI	94.9	CW	Galveston	KOVE	106.5	SP/MEX
Elgin	KKLB	92.5	SP/MEX	Ganado	KZAM	104.7	CW
Fabens	KPAS	103.1	SP/RL	Garland	KAAM	770	AS
Fairfield	KNES	92.1	CW	Gatesville	KVLZ	98.3	RL/CC
Falfurrias	KLDS	1260	SP/RL	Georgetown	KHFI	96.7	CHR
Falfurrias	KDFM	103.3	SP/RL	Georgetown	KINV	107.7	SP/MEX
Falfurrias	KPSO	106.3	SP/TO	Giddings	KANJ	91.5	RL
Fannett	KZFT	90.5	RL	Gilmer	KOFY	1060	SP/MEX
Farmersville	KFCD	990	TLK+	Gilmer	KFRO	95.3	AC/S
Farmersville	KXEZ	92.1	OLG	Gladewater	KEES	1430	TLK+
Farwell	KMUL	830	SP/MEX	Glen Rose	KTFW	92.1	CW
Farwell	KIJN	1060	SP/RL	Goliad	KHMC	95.9	SP/TO
Farwell	KIJN	92.3	RL/CC	Gonzales	KCTI	1450	CW
Farwell	KICA	98.3	RK/C	Gonzales	KQQT	106.3	RL/CC
Ferris	KDFT	540	SP/RL	Graham	KSWA	1330	CW
Floresville	KWCB	89.7	CW	Graham	KWKQ	94.7	CH
Floresville	KTFM	94.1	OLG	Granbury	KPAR	1420	CW
Flower Mound	KTYS	96.7	CW	Grand Prairie	KKDA	730	OLG
Floydada	KFLP	900	FRM	GREENVILLE	KGVL	1400	CW
Floydada	KFLP	106.1	CW	Greenville	KIKT	93.5	CW
Fort Stockton	KFST	860	AC/S	Gregory	KPUS	104.5	RK/C
Fort Stockton	KFST	94.3	CW	Groves	KCOL	92.5	OLG
Fort Worth	WBAP	820	N/TLK	Hallettsville	KHLT	1520	CW
Fort Worth	KFJZ	870	SP/TO	Hallettsville	KTXM	99.9	CW
Fort Worth	KHVN	970	RL/G	Haltom City	KDBM	93.3	RK/C
Fort Worth	KFLC	1270	SP/TK	Hamilton	KCLW	900	CW
Fort Worth	KKGM	1630	RL/G	Hamlin	KCDD	103.7	CH
Fort Worth	KTCU	88.7	VAR	Harker Hts	KUSG	105.5	CW
Fort Worth	KLNO	94.1	SP/MEX	Harlingen	KGBT	1530	SP/MEX

City	Station	Freq	Format	City	Station	Freq	Format
Harlingen	KMBH	88.9	CLA	Houston/Galveston	KRBE	104.1	CH
Harlingen	KFRQ	94.5	RK	Houston/Galveston	KHCB	105.7	RL
Harlingen	KBTQ	96.1	URB	Howe	KHYI	95.3	CW
Haskell	KVRP	97.1	CW	Hudson	KLSN	96.3	CW
Hearne	KVJM	103.1	URB	Humble	KGOL	1180	SP/MEX
Hebbronville	KAZF	91.9	RL/CC	Humble	KSBJ	89.3	RL/CC
Hebbronville	KEKO	101.7	SP/RL	Huntington	KYBI	101.9	SP/MEX
Helotes	KONO	101.1	OLG	Huntsville	KHCH	1400	RL
Hemphill	KPBL	1240	SIL	Huntsville	KHVL	1490	OLG
Hemphill	KTHP	103.9	CW	Huntsville	KSHU	90.5	CLA
Hempstead	KEZB	105.3	AC/S	Huntsville	KSAM	101.7	CW
Henderson	KWRD	1470	CW	Hurst	KMNY	1350	BIZ
Hereford	KPAN	860	CW	Hutto	KQKZ	92.1	RL/CC
Hereford	KJNZ	103.5	SP/MEX	Idalou	KRBL	105.7	CW
Hereford	KPAN	106.3	CW	Ingleside	KRPX	107.3	RK/C
Highland Park	KBIS	1150	SIL	Ingram	KTXI	90.1	CLA
Highland Park	KVIL	103.7	AC	Jacksboro	KJKB	101.7	SILNT
Highland Village	KWRD	100.7	RL	Jacksonville	KEBE	1400	CW
Hillsboro	KHBR	1560	CW	Jacksonville	KBJS	90.3	RL
Hillsboro	KBRQ	102.5	RK/C	Jacksonville	KLJT	102.3	AC/S
Hondo	KCWM	1460	CW	Jacksonville	KOOI	106.5	AC/S
Hondo	KMFR	105.9	RK/C	Jasper	KCOX	1350	URB
Hondo	KMFR	105.9	RK/C	Jasper	KTXJ	102.7	CW
Hooks	KPWW	95.9	CHR	Jasper	KJAS	107.3	AC/S
Hornsby	KOOP	91.7	VAR	Jefferson	KHJC	91.9	RL
Houston	KUHF	88.7	CLA	Jefferson	KJTX	104.5	RL/G
Houston/Galv'ston	KILT	610	SPTS	Johnson City	KFAN	107.9	AD/A
Houston/Galv'ston	KTRH	740	N/TLK	Jourdanton	KLEY	95.7	SP/MEX
Houston/Galv'ston	KBME	790	SPTS	Junction	KMBL	1450	AC/S
Houston/Galv'ston	KEYH	850	SP/MEX	Junction	KOOK	93.5	CW
Houston/Galv'ston	KPRC	950	TLK+	Karnes City	KTXX	103.1	CH
Houston/Galv'ston	KLAT	1010	SP/TK	Keene	KJCR	88.3	RL/CC
Houston/Galv'ston	KENR	1070	TLK+	Kenedy	KAML	990	CW
Houston/Galv'ston	KQUE	1230	SP/MEX	Kenedy	KTNR	92.1	SP/RL
Houston/Galv'ston	KXYZ	1320	BIZ	Kerens	KRVF	106.9	CW
Houston/Galv'ston	KCOH	1430	URB	Kermit	KERB	600	SP/RL
Houston/Galv'ston	KMIC	1590	YTH	Kermit	KERB	106.3	SP/RL
Houston/Galv'ston	KPFT	90.1	VAR	Kerrville	KERV	1230	AC/S
Houston/Galv'ston	KTSU	90.9	JAZ	Kerrville	KKER	88.7	RL
Houston/Galv'ston	KTRU	91.7	VAR	Kerrville	KKER	91.1	RL
Houston/Galv'ston	KKRW	93.7	RK/C	Kerrville	KHKV	91.1	SP/RL
Houston/Galv'ston	KTBZ	94.5	RK/C	Kerrville	KRNH	92.3	CW
Houston/Galv'ston	KHJZ	95.7	JAZ	Kerrville	KRVL	94.3	CW
Houston/Galv'ston	KHMX	96.5	AC/H	Kilgore	KBGE	1240	RK/C
Houston/Galv'ston	KBXX	97.9	CH	Kilgore	KTPB	88.7	CLA
Houston/Galv'ston	KODA	99.1	AC/S	Kilgore	KKTX	96.1	RK/C
Houston/Galv'ston	KILT	100.3	CW	Killeen	KRMY	1050	RL/G
Houston/Galv'ston	KLOL	101.1	SP/OM	Killeen	KNCT	91.3	EZL
Houston/Galv'ston	KMJQ	102.1	URB	Killeen	KIIZ	92.3	URB
Houston/Galv'ston	KLTN	102.9	SP/MEX	Kingsville	KINE	1330	SP/RL

City	Station	Freq	Format	City	Station	Freq	Format
Kingsville	KTAI	91.1	RK	Lubbock	KXTQ	93.7	SP/TO
Kingsville	KKBA	92.7	AC	Lubbock	KFMX	94.5	RK
Kingsville	KFTX	97.5	CW	Lubbock	KLLL	96.3	CW
Krum	KNOR	93.7	URB	Lubbock	KQBR	99.5	CW
La Grange	KVLG	1570	OLG	Lubbock	KONE	101.1	RK/C
La Grange	KBUK	104.9	CW	Lubbock	KZII	102.5	CHR
Lake Jackson	KYBJ	91.1	RL/CC	Lubbock	KEJS	106.5	SP/TO
Lake Jackson	KLDE	107.5	OLG	Lufkin	KRBA	1340	RL/G
Lamesa	KPET	690	CW	Lufkin	KLDN	88.9	CLA
Lamesa	KBKN	91.3	RL	Lufkin	KSWP	90.9	RL/CC
Lamesa	KTXC	104.7	SP/MEX	Lufkin	KAVX	91.9	RL
Lampasas	KCYL	1450	CW	Lufkin	KYBI	100.1	AC
Laredo	KVOZ	890	SP/RL	Lufkin	KYKS	105.1	CW
Laredo	KLAR	1300	SP/RL	Luling	KAMX	94.7	AC
Laredo	KLNT	1490	TLK+	Lytle	KZLV	91.3	RL/CC
Laredo	KHOY	88.1	EZL	Madisonville	KMVL	1220	AS
Laredo	KBNL	89.9	SP/RL	Madisonville	KAGG	96.1	CW
Laredo	KJBZ	92.7	SP/TO	Madisonville	KMVL	100.5	CW
Laredo	KQUR	94.9	AC/H	Malakoff	KCKL	95.9	CW
Laredo	KRRG	98.1	CHR	Manor	KELG	1440	SP/MEX
Laredo	KNEX	106.1	CH	Marble Falls	KBMD	88.5	RL
Leakey	KBLT	104.3	RL/CC	Marion	KBIB	1000	SP/RL
Leander	KHHL	98.9	SP/MEX	Markham	KZRC	92.5	AD/A
Levelland	KLVT	1230	RL/G	Marlin	KLRK	92.9	AC/H
Levelland	KLVT	105.5	CW	Marshall	KCUL	1410	OLG
Lewisville	KESS	107.9	SP/MEX	Marshall	KMHT	1450	AS
Liberty	KSHN	99.9	AC	Marshall	KBWC	91.1	URB
Linden	KLBL	99.3	CW	Marshall	KCUL	92.3	SP/MEX
Littlefield	KZZN	1490	CW	Marshall	KMHT	103.9	CW
Livingston	KETX	1440	SIL	Mason	KOTY	95.7	N/A
Livingston	KETX	92.3	CW	Mason	KHLE	102.5	CW
Llano	KQBT	96.3	AS	McAllen	KRIO	910	SP/RL
Llano	KITY	102.9	OLG	McAllen	KHID	88.1	CLA
Lockhart	KFIT	1060	RL/G	McAllen	KVMV	96.9	RL/CC
Lometa	KACQ	101.9	CW	McAllen	KGBT	98.5	SP/MEX
Longview	KFRO	1370	N/TLK	McCamey	KPBM	95.3	SP/RL
Longview	KYKX	105.7	CW	McCook	KCAS	91.5	RL
Lorenzo	KKCL	98.1	OLG	McKinney	KNTU	88.1	JAZ
Los Ybanez	KYMI	98.5	SP/RL	McQueeney	KLTO	97.7	SP/TO
Lubbock	KRFE	580	AS	Memphis	KLSR	105.3	CW
Lubbock	KFYO	790	N/TLK	Mercedes	KHKZ	106.3	AC/H
Lubbock	KJTV	950	N/TLK	Merkel	KHXS	102.7	RK/C
Lubbock	KKAM	1340	SPTS	Mertzon	KMEO	91.9	N/A
Lubbock	KLFB	1420	SP/RL	Mesquite	KEOM	88.5	CH
Lubbock	KBZO	1460	SP/MEX	Mexia	KRQX	1590	CW
Lubbock	KDAV	1590	OLG	Mexia	KYCX	104.9	CW
Lubbock	KTXT	88.1	AD/A	Midland	KPOE	880	N/A
Lubbock	KOHM	89.1	CLA	Midland	KPBJ	90.1	N/A
Lubbock	KAMY	90.1	RL	Midland/Odessa	KCRS	550	TLK+
Lubbock	KKLU	90.9	RL/CC	Midland/Odessa	KWEL	1070	TLK+

City	Station	Freq	Format	City	Station	Freq	Format
Midland/Odessa	KJBC	1150	RL	Palacios	KKOS	99.7	CW
Midland/Odessa	KMND	1510	SPTS	Palestine	KNET	1450	CW
Midland/Odessa	KNFM	92.3	CW	Palestine	KYFP	89.1	RL
Midland/Odessa	KBAT	93.3	CHR	Palestine	KYYK	98.3	CW
Midland/Odessa	KQRX	95.1	RK/H	Pampa	KGRO	1230	AC/S
Midland/Odessa	KCRS	103.3	CH	Pampa	KAXH	90.9	RL
Midland/Odessa	KCHX	106.7	AC	Pampa	KOMX	100.3	CW
Mineola	KMOO	99.9	CW	Pampa	KOMX	100.3	CW
Mineral Wells	KJSA	1140	CW	Paris	KPJC	1250	JAZ
Mineral Wells	KFWR	95.9	CW	Paris	KPLT	1490	CW
Mirando City	KBDR	100.5	SP/MEX	Paris	KHCP	89.3	RL/CC
Mission	KIRT	1580	SP/MEX	Paris	KBCV	89.3	RL
Mission	KQXX	105.5	OLG	Paris	KOYN	93.9	CW
Missouri City	KPTY	104.9	CH	Paris	KBUS	101.1	RK/C
Monahans	KLBO	1330	OLG	Paris	KPLT	107.7	AC
Monahans	KGEE	99.9	RL/CC	Pasadena	KIKK	650	N/TLK
Monahans	KFZX	102.1	RK	Pasadena	KLVL	1480	SP/MEX
Mount Pleasant	KIMP	960	TLK+	Pasadena	KFTG	88.1	SP/RL
Muenster	KXGM	106.5	SP/OM	Pasadena	KKBQ	92.9	CW
Muleshoe	KMUL	103.1	CW	Pearsall	KVWG	1280	RL/CC
Nacogdoches	KSFA	860	TLK+	Pearsall	KVWG	95.3	CW
Nacogdoches	KSAU	90.1	JAZ	Pearsall	KRIO	104.1	SP/OM
Nacogdoches	KJCS	103.3	CW	Pecan Grove	KREH	900	SP/MEX
Nacogdoches	KTBQ	107.7	RK/C	Pecos	KIUN	1400	SP/OM
Navasota	KWBC	1550	N/TLK	Pecos	KKLY	97.3	CW
Navasota	KHTZ	92.5	CW	Pecos	KPTX	98.3	AC
Nederland	KBED	1510	RL/G	Perryton	KEYE	1400	CW
New Boston	KNBO	1530	RL/G	Perryton	KEYE	96.1	CH
New Boston	KEWL	95.1	OLG	Pflugersville	KOKE	1600	TLK+
New Boston	KZRB	103.5	URB	Pharr	KVJY	840	CW
New Braunfels	KGNB	1420	N/TLK	Pilot Point	Kzmp	104.9	SP/OM
New Braunfels	KNBT	92.1	CW	Pittsburg	KKXI	91.7	SIL
New Ulm	KNRG	92.3	CH	Pittsburg	KSCN	96.9	CW
Nolanville	KLFX	107.3	RK	Pittsburg	Kdve	103.1	TLK+
Odem	KLHB	98.3	SP/TO	Plains	KPHS	90.3	VAR
Odessa	KENT	920	RL/CC	Plainview	Kvop	1090	TLK+
Odessa	KOZA	1230	SP/RL	Plainview	Krew	1400	OLG
Odessa	KRIL	1410	SPTS	Plainview	KPMB	88.5	SP/RL
Odessa	KLVW	88.7	RL/CC	Plainview	KBAH	90.5	RL
Odessa	KBMM	89.5	RL/CC	Plainview	KWLD	91.5	RL/CC
Odessa	KFLB	90.5	RL/CC	Plainview	KSTQ	97.3	SP/MEX
Odessa	KOCV	91.3	CLA	Plainview	KRIA	103.9	SP/TO
Odessa	KMRK	96.1	CH	Plainview	KKYN	106.9	CW
Odessa	KMCM	96.9	OLG	Plano	KMKI	620	YTH
Odessa	KODM	97.9	AC	Pleasanton	KFNI	1380	SP/RL
Odessa	KHKX	99.1	CW	Point Comfort	KAJI	94.1	CW
Odessa	KQLM	107.9	SP/OM	Port Arthur	KDEI	1250	RL
Orange	KOGT	1600	CW	Port Arthur	KOLE	1340	N/TLK
Orange	KKMY	104.5	AC	Port Arthur	KQBU	93.3	SP/MEX
Orange	KIOC	106.1	RK	Port Arthur	KTJM	98.5	SP/MEX
Ore City	KAZE	106.9	CH	Port Isabel	KVPA	101.1	SP/TO
Overton	KPXI	100.7	RL	Port Lavaca	KITE	93.3	OLG
Ozona	KYXX	94.3	CW	Port Neches	KUHD	1150	SP/RL

City	Station	Freq	Format	City	Station	Freq	Format
Portland	KMJR	105.5	SP/MEX	San Antonio	KPAC	88.3	CLA
Post	KPOS	107.3	RL/CC	San Antonio	KSTX	89.1	N/TLK
Prairie View	KSGR	91.3	RL/G	San Antonio	KSYM	90.1	AD/A
Premont	KMFM	100.7	SP/RL	San Antonio	KYFS	90.9	RL
Quanah	KREL	1150	CW	San Antonio	KRTU	91.7	JAZ
Quanah	KIXC	100.9	OLG	San Antonio	KROM	92.9	SP/MEX
Ralls	KCLR	1530	SP/RL	San Antonio	KXXM	96.1	CHR
Raymondville	KSOX	1240	SPTS	San Antonio	KAJA	97.3	CW
Raymondville	KBUC	102.1	CW	San Antonio	KISS	99.5	RK
Raymondville	KBIC	105.7	SP/RL	San Antonio	KCYY	100.3	CW
Refugio	KTKY	106.1	SILNT	San Antonio	KQXT	101.9	AC
Rio Grande City	KQBO	107.5	SP/MEX	San Antonio	KSRX	102.7	RK/C
Robinson	KHCK	107.9	SP/MEX	San Antonio	KZEP	104.5	RK/C
Robstown	KROB	1510	SP/TO	San Antonio	KXTN	107.5	SP/TO
Robstown	KLUX	89.5	EZL	San Augustine	KQSI	92.5	CW
Robstown	KSAB	99.9	SP/TO	San Diego	KUKA	105.9	SP/TO
Robstown	KMIQ	105.1	SP/TO	San Juan	KUBR	1210	SP/RL
Rockdale	KRXT	98.5	CW	San Marcos	KUOL	1470	SIL
Rockport	KKPN	102.3	AC/H	San Marcos	KTSW	89.9	AD/A
Rollingwood	KJCE	1370	UR/AC	San Marcos	KBPA	103.5	AH
Rollingwood	KJCE	1370	TLK+	San Saba	KBAL	1410	AS
Roma	KBMI	97.7	SIL	San Saba	KBAL	97.1	CW
Rosenberg	KRTX	980	SP/TO	Sanger	KVRK	89.7	RL/CC
Round Rock	KNLE	88.1	RL/CC	Sanger	KDTK	104.1	SPTS
Round Rock	KFMK	105.9	URB	Santa Fe	KJIC	90.5	RL/G
Rudolph	KTER	90.7	RL	Schertz	KBBT	98.5	CH
Rusk	KTLU	1580	OLG	Seabrook	KROI	92.1	SP/MEX
Rusk	KWRW	97.7	OLG	Seadrift	KMAT	105.1	RL
San Angelo	KGKL	960	TLK+	Seguin	KWED	1580	CW
San Angelo	KKSA	1260	N/TLK	Seguin	KSMG	105.3	AC/H
San Angelo	KCRN	1340	RL/CC	Seminole	KIKZ	1250	CW
San Angelo	KHBQ	89.3	N/A	Seminole	KSEM	106.3	CW
San Angelo	KDCD	92.9	CW	Seymour	KSEY	1230	SP/TO
San Angelo	KCRN	93.9	RL/CC	Seymour	KSEY	94.3	TLK+
San Angelo	KIXY	94.7	CHR	Shamrock	KBKH	92.9	OLG
San Angelo	KGKL	97.5	CW	Sherman	KJIM	1500	AS
San Angelo	KELI	98.7	AC/S	Sherman	KKLF	1700	TLK+
San Angelo	KYZZ	100.1	OLG	Silsbee	KSET	1300	SPTS
San Angelo	KWFR	101.9	RK/C	Silsbee	KAYD	101.7	CW
San Angelo	KWFR	105.7	RK/C	Sinton	KDAE	1590	SP/RL
San Angelo	KMDX	106.1	AC	Sinton	KNCN	101.3	RK
San Angelo	KSJT	107.5	SP/MEX	Sinton	KOUL	103.7	CW
San Antonio	KTSA	550	N/TLK	Slaton	KJAK	92.7	RL
San Antonio	KSLR	630	RL	Snyder	KSNY	1450	RL/CC
San Antonio	KKYX	680	CW	Snyder	KLYD	98.9	RK/H
San Antonio	KTKR	760	SPTS	Somerset	KSJL	810	URB
San Antonio	KONO	860	OLG	Sonora	KHOS	980	SP/RL
San Antonio	KENS	1160	YTH	Sonora	KHOS	92.1	CW
San Antonio	WOAI	1200	N/TLK	So Padre Island	KESO	92.7	SP/MEX
San Antonio	KZDC	1250	SP/MEX	So Padre Island	KZSP	95.3	OLG
San Antonio	KXTN	1310	TLK+	Spearman	KRDF	98.3	CW
San Antonio	KCOR	1350	SP/OM	Springtown	KMQX	89.1	OLG
San Antonio	KCHL	1420	RL/G	Stamford	KVRP	1400	CW

City	Station	Freq	Format	City	Station	Freq	Format
Stanton	KKJW	105.9	CW	Victoria	KVRT	90.7	CLA
Stephenville	KSTV	1510	SP/MEX	Victoria	KQVT	92.3	AC/H
Stephenville	KQXS	89.1	OLG	Victoria	KVIC	95.1	CH
Sterling City	KNRX	96.5	RK/H	Victoria	KTXN	98.7	CW
Sulphur Springs	KSST	1230	AS	Victoria	KEPG	100.9	URB
Sulphur Springs	KSCH	95.9	CW	Victoria	KIXS	107.9	CW
Sweetwater	KXOX	1240	CW	Waco	KBBW	1010	RL
Sweetwater	KXOX	96.7	CW	Waco	KWTX	1230	N/TLK
Tahoka	KMMX	100.3	AC/H	Waco	KQRL	1580	SPTS
Tahoka	KAMZ	103.5	SP/MEX	Waco	KRZI	1660	SPTS
Taylor	KWNX	1260	SPTS	Waco	KVLW	88.1	RL/CC
Taylor	KXBT	104.3	CH	Waco	KBCT	94.5	CW
Temple	KTEM	1400	TLK+	Waco	KBGO	95.7	OLG
Temple	KBDE	89.9	RL	Waco	KWTX	97.5	CHR
Temple	KLTD	101.7	RK/C	Waco	WACO	99.9	CW
Terrell	KPYK	1570	AS	Waco	KWBU	103.3	CLA
Terrell Hills	KLUP	930	TLK+	Wake Village	KHTA	92.5	RL
Terrell Hills	KELZ	106.7	CHR	Waxahachie	KBEC	1390	CW
Texarkana	KCMC	740	SPTS	Weatherford	KZEE	1220	RL
Texarkana	KTFS	940	TLK+	Weatherford	KMQX	88.5	OLG
Texarkana	KEWL	1400	OLG	Weatherford	KYQX	89.5	AC/S
Texarkana	KTAL	98.1	RK/C	Wells	KVLL	94.7	AC/S
Texarkana	KKYR	102.5	CW	Weslaco	KRGE	1290	SP/RL
Texas City	KYST	920	SP/TK	West Lake Hills	KTXZ	1560	SP/TO
Thorndale	KJAZ	99.3	JAZ	Wharton	KANI	1500	RL
Three Rivers	KEMA	94.5	CW	Wheeler	KPDR	90.5	RL
Tomball	KSEV	700	TLK+	Whitehouse	KISX	107.3	CHR
Tulia	KTUE	1260	SP/RL	Whitesboro	KMAD	102.5	RK/C
Tulia	KBTE	104.9	CH	Wichita Falls	KWFS	1290	TLK+
Tye	KWFA	1030	N/A	Wichita Falls	KMCU	88.7	CLA
Tye	KBCY	99.7	CW	Wichita Falls	KMOC	89.5	RL/CC
Tyler	KTBB	600	TLK+	Wichita Falls	KTEO	90.5	RL
Tyler	KZEY	690	URB	Wichita Falls	KNIN	92.9	CHR
Tyler	KGLD	1330	RL/G	Wichita Falls	KLUR	99.9	CW
Tyler	KYZS	1490	SPTS	Wichita Falls	KWFS	102.3	CW
Tyler	KVNE	89.5	RL/CC	Wichita Falls	KQXC	103.9	CH
Tyler	KGLY	91.3	RL/G	Wichita Falls	KBZS	106.3	RK/C
Tyler	KDOK	92.1	OLG	Willis	KVST	99.7	CW
Tyler	KTYL	93.1	AC/H	Willis	KIOL	103.7	RK/C
Tyler	KNUE	101.5	CW	Winfield	KALK	97.7	AC/H
Tyler	KKUS	104.1	CW	Winnie	KKHT	100.7	RL
Universal City	KSAH	720	SP/MEX	Winnsboro	KWNS	104.7	RL/G
University Park	KTNO	1440	SP/RL	Winona	KBLZ	102.7	CH
University Park	KZMP	1540	SP/RL	Winters	KNCE	96.1	OLG
Uvalde	KVOU	1400	AS	Wolforth	KAIQ	95.5	SP/OM
Uvalde	KBNU	93.7	RL/CC	Woodville	KWUD	1490	CW
Uvalde	KUVA	102.3	SP/TO	Wylie	KHSE	700	N/A
Uvalde	KVOU	104.9	CW	Yoakum	KYKM	92.5	CW
Vernon	KVWC	1490	CW	Zapata	KBAW	93.5	SP/RL
Vernon	KVWC	103.1	OLG				
Victoria	KVNN	1340	TLK+				
Victoria	KNAL	1410	AS				
Victoria	KAYK	88.5	RL				

City	Station	Freq	Format	City	Station	Freq	Format
Blanding	KBDX	92.1	SIL	Payson	KTCE	92.3	CH
Bountiful	KJMY	99.5	AC	Pleasant Grve	KPGR	88.1	CHR
Brian Head	KREC	98.1	AC/S	Price	KOAL	750	TLK+
Brigham City	KXOL	1660	OLG	Price	KSLL	1080	CW
Brigham City	KEGH	100.7	CW	Price	KARB	98.3	CW
Brigham City	KRAR	106.9	CH	Price	KWSA	100.9	AC
Cedar City	KSUB	590	TLK+	Provo	KOVO	960	SPTS
Cedar City	KNNZ	940	N/TLK	Provo	KSRR	1400	AC/S
Cedar City	KSUU	91.1	RK/H	Provo	KEYY	1450	RL
Cedar City	KXFF	92.5	OLG	Provo	KBYU	89.1	CLA
Cedar City	KXBN	94.9	CHR	Provo	KHTB	94.9	RK/C
Centerville	KSGO	1600	SP/OM	Provo	KXRK	96.3	RK/H
Centerville	KXRV	105.7	CH	Randolph	KDUT	102.3	SP/MEX
Coalville	KJQN	103.1	AH	Richfield	KSVC	980	TLK+
Delta	KNAK	540	N/TLK	Richfield	KCYQ	93.7	CW
Delta	KMGR	97.5	AC/S	Richfield	KLGL	97.5	AC/S
Ephraim	KAGJ	89.5	RK/H	Roosevelt	KNEU	1250	CW
Heber City	KTMP	1340	CW	Roosevelt	KXRQ	94.3	CH
Kanab	KPLD	101.1	RK/H	Roosevelt	KIFX	98.5	AC
Levan	KCFM	96.7	SP/MEX	Roy	KANN	1120	RL/CC
Logan	KVNU	610	TLK+	Roy	KUDD	107.9	CH
Logan	KLGN	1390	AS	Salt Lake City	KUER	90.1	JAZ
Logan	KUSR	89.5	CLA	Salt Lake City	KNRS	570	TLK+
Logan	KZCL	90.5	N/A	Salt Lake City	KKAT	860	CW
Logan	KUSU	91.5	CLA	Salt Lake City	KWDZ	910	YTH
Logan	KBLQ	92.9	AC	Salt Lake City	KSL	1160	N/TLK
Logan	KVFX	94.5	CHR	Salt Lake City	KZNS	1280	SPTS
Manti	KMTI	650	CW	Salt Lake City	KFNZ	1320	SPTS
Manti	KNJQ	105.1	AH	Salt Lake City	KCPW	88.3	N/TLK
Midvale	KQMB	102.7	AC	Salt Lake City	KRCL	90.9	VAR
Moab	KZMU	89.7	VAR	Salt Lake City	KUFR	91.7	RL
Moab	KCYN	97.1	CW	Salt Lake City	KUBL	93.3	CW
Murray	KJQS	1230	SPTS	Salt Lake City	KODJ	94.1	OLG
Naples	KCUA	92.5	RK/C	Salt Lake City	KZHT	97.1	CH
Nephi	KUDE	103.9	CH	Salt Lake City	KBEE	98.7	AC
North Ogden	KNKL	88.7	RL/CC	Salt Lake City	KSFI	100.3	AC/S
No Salt Lake	KALL	700	TLK+	Salt Lake City	KRSP	103.5	CH
Oakley	KEGA	101.5	CW	Salt Lake City	KSOP	104.3	CW
Ogden	KSVN	730	SP/MEX	Sandy	KTKK	630	TLK+
Ogden	KLO	1430	TLK+	Sandy	KBJA	1640	SP/MEX
Ogden	KYFO	1490	RL	Smithfield	KGNT	103.9	OLG
Ogden	KWCR	88.1	AD/A	South Jordan	KUUU	92.5	CH
Ogden	KYFO	95.5	RL	So Salt Lake	KDYL	1060	AS
Ogden	KBZN	97.9	JAZ	So Salt Lake	KSOP	1370	CW
Ogden	KBER	101.1	RK	Spanish Fork	KHQN	1480	N/A
Ogden	KPQP	101.9	CH	Spanish Fork	KOSY	106.5	AC/S
Orem	KOHS	91.7	AD/A	St. George	KDXU	890	TLK+
Orem	KENZ	107.5	AD/A	St. George	KZNU	1450	TLK+
Park City	KPCW	91.9	AD/A	St. George	KAER	89.5	N/A
Parowan	KENT	1400	EZL	St. George	KZBS	90.7	VAR

City	Station	Freq	Format		City	Station	Freq	Format
St. George	KSNN	93.5	AC					
St. George	KZHK	95.9	CH					
St. George	KONY	99.9	CW					
Taylorsville	KUTR	820	TLK+					
Tooele	KCPW	1010	N/TLK					
Tremonton	KNFL	1470	N/A					
Tremonton	KBNZ	104.9	CW					
Vernal	KVEL	920	TLK+					
Vernal	KLCY	105.9	CW					
Washington	KUNF	1210	OLG					
West Jordan	KLLB	1510	RL/G					
West Valley City	KMRI	1550	SP/OM					

Radio on the Road   Vermont

City	Station	Freq	Format		City	Station	Freq	Format
Addison	WXAL	93.7	AC/H		Rutland	WFTF	90.5	RL
Barre	WSNO	1450	TLK+		Rutland	WJEN	94.5	CW
Barre	WCMD	89.9	RL/CC		Rutland	WZRT	97.1	CHR
Barre	WORK	107.1	CHR		Rutland	WJJR	98.1	AC
Bellows Falls	WZLF	107.1	CW		Saint Albans	WWSR	1420	TLK+
Bennington	WBTN	1370	TLK+		Saint Albans	WLFE	102.3	CW
Bennington	WBTN	94.3	CLA		St Johnsbury	WSTJ	1340	AS
Berlin	WWFY	100.9	CW		St. Johnsbury	WVPA	88.5	CLA
Bolton	WGLY	91.5	RL/CC		St Johnsbury	WCKJ	90.5	RL/CC
Brandon	WEXP	101.5	RK/C		St Johnsbury	WKXH	105.5	CW
Brattleboro	WTSA	1450	SPTS		So Burlington	WXXX	95.5	CHR
Brattleboro	WKVT	1490	TLK+		Springfield	WNBX	1480	OLG
Brattleboro	WKVT	92.7	RK/C		Springfield	WTSM	93.5	N/TLK
Brattleboro	WTSA	96.7	AC/H		Stowe	WCVT	101.7	CLA
Burlington	WVMT	620	TLK+		Sunderland	WJAN	95.1	CW
Burlington	WJOY	1230	AS		Vergennes	WIZN	106.7	RK
Burlington	WVAA	1390	CW		Warren	WDEV	96.1	N/TLK
Burlington	WRUV	90.1	AD/A		Waterbury	WDEV	550	N/TLK
Burlington	WEZF	92.9	AC/S		Waterbury	WLKC	103.3	AC/H
Burlington	WOKO	98.9	CW		Wells River	WTWN	1100	RL
Burlington	WVPS	107.9	CLA		White River Jct	WNHV	910	SPTS
Castleton	WIUV	91.3	AD/A		White River Jct	WXLF	95.3	CW
Colchester	WWPV	88.7	VAR		Wilmington	WMTT	100.7	RK/C
Danville	WDOT	95.7	AD/A		Windsor	WVPR	89.5	CLA
Derby Center	WMOO	92.1	AC/H		Woodstock	WGLV	91.7	RL/CC
Hartford	WWOD	104.3	OLG		Woodstock	WMXR	93.9	RK/C
Johnson	WJSC	90.7	AD/A					
Killington	WEBK	105.3	AD/A					
Lyndon	WGMT	97.7	AC					
Lyndonville	WWLR	91.5	AD/A					
Manchester	WEQX	102.7	RK/H					
Marlboro	WRSY	101.5	AD/A					
Middlebury	WFAD	1490	OLG					
Middlebury	WRMC	91.1	VAR					
Montpelier	WSKI	1240	OLG					
Montpelier	WNCS	104.7	AD/A					
Morrisville	WLVB	93.9	CW					
Newport	WIKE	1490	CW					
Northfield	WNUB	88.3	AD/A					
Norwich	WNCH	88.1	CLA					
Plainfield	WGDR	91.1	VAR					
Poultney	WVNR	1340	AC					
Putney	WCMK	91.9	RL/CC					
Randolph	WWWT	1320	TLK+					
Randolph	WCVR	102.1	RK/C					
Randolph Ctr	WVTC	90.7	AD/A					
Royalton	WRJT	103.1	AD/A					
Rupert	WMNV	104.1	RL					
Rutland	WSYB	1380	TLK+					
Rutland	WRVT	88.7	CLA					

City	Station	Freq	Format	City	Station	Freq	Format
Abingdon	WABN	1230	OLG	Charlottesville	WMVE	90.1	N/A
Abingdon	WFHG	92.7	N/TLK	Charlottesville	WTJU	91.1	VAR
Accomac	WVES	99.3	CW	Charlottesville	WNRN	91.9	RK/H
Alberta	WSMY	103.1	URB	Charlottesville	WUVA	92.7	URB
Alexandria	WKDL	730	SP/MEX	Charlottesville	WQMZ	95.1	AC
Altavista	WKDE	1000	N/TLK	Charlottesville	WWWV	97.5	RK/C
Altavista	WKDE	105.5	CW	Charlottesville	WCJZ	107.5	JAZ
Amherst	WAMV	1420	RL/G	Chase City	WMEK	980	RL
Amherst	WYYD	107.9	CW	Chase City	WMEK	980	RL
Appomattox	WOWZ	1280	BIZ	Chase City	WFXQ	99.9	AS
Appomattox	WSNZ	102.7	AD/A	Chatham	WKBY	1080	RL/G
Appomattox	WTTX	107.1	RL/G	Cheriton	WWIP	89.1	RL/CC
Arlington	WABS	780	RL	Chesapeak	WPYA	93.7	AH
Arlington	WZHF	1390	N/A	Chesapeake	WCPK	1600	RL/G
Arlington	WAVA	105.1	RL	Chesapeake	WFOS	88.7	VAR
Ashland	WHAN	1430	TLK+	Chester	WGGM	820	RL/CC
Ashland	WYFJ	100.1	RL	Chester	WDYL	101.1	RK/C
Bassett	WCBX	900	SPTS	Chincoteague	WCTG	96.5	CH
Bayside	WBVA	1450	SIL	Christiansburg	WBZV	100.7	CH
Bedford	WBLT	1350	AD/A	Churchville	WNLR	1150	RL/CC
Bedford	WBWR	106.9	RK/C	Churchville	WBOP	106.3	RK/C
Berryville	WAPP	105.5	CW	Claremont	WPMH	670	RL
Berryville	WWRE	105.5	RK/C	Clarksville	WLUS	98.3	CW
Big Stone Gap	WLSD	1220	RL/G	Clifton Forge	WXCF	1230	CW
Big Stone Gap	WAXM	93.5	CW	Clifton Forge	WXCF	103.9	AC
Blacksburg	WFNR	710	TLK+	Clinchco	WDIC	1430	CW
Blacksburg	WKEX	1430	SPTS	Clinchco	WDIC	92.1	OLG
Blacksburg	WUVT	90.7	VAR	Coeburn	WVSG	99.7	AC/S
Blacksburg	WBRW	105.3	RK/C	Collinsville	WFIC	1530	RL/G
Blackstone	WKLV	1440	SPTS	Colonial Beach	WGRQ	95.9	OLG
Blackstone	WBBC	93.5	CW	Colonial Hts	WDZY	1290	YTH
Bluefield	WBDY	1190	SPTS	Colonial Hts	WKHK	95.3	CW
Bluefield	WHKX	106.3	CW	Covington	WKEY	1340	OLG
Bowling Green	WWUZ	96.9	RK/C	Covington	WIQO	100.9	CW
Bridgewater	WBHB	105.1	OLG	Crewe	WSVS	800	CW
Bristol	WZAP	690	RL	Crewe	WKJS	104.7	RL/G
Bristol	WFHG	980	N/TLK	Crozet	WSUH	102.3	OLG
Broadway	WBTX	1470	RL/G	Crozet	WSUH	102.3	OLG
Broadway	WJDV	96.1	AC/S	Crozet	WMRY	103.5	CLA
Brookneal	WODI	1230	RL/G	Culpeper	WCVA	1490	AS
Buena Vista	WWZW	96.7	AC	Culpeper	WPER	89.9	RL/CC
Buffalo Gap	WZXI	96.1	TLK+	Culpeper	WARN	91.5	RL
Cape Charles	WAZP	90.7	RL/CC	Culpeper	WJMA	103.1	CW
Cape Charles	WROX	96.1	RK/C/NU	Danville	WVOV	970	RL/G
Cedar Bluff	WYRV	770	RL/CC	Danville	WDVA	1250	RL/G
Charles City	WAUQ	89.7	RL	Danville	WBTM	1330	OLG
Charlottesville	WINA	1070	TLK+	Danville	WILA	1580	RL/G
Charlottesville	WCHV	1260	N/TLK	Danville	WOKD	91.1	RL/CC
Charlottesville	WKAV	1400	SPTS	Danville	WAKG	103.3	CW
Charlottesville	WTW	88.5	CLA	Deltaville	WSRV	92.3	OLG
Charlottesville	WTU	89.3	CLA	Dillwyn	WBNN	105.3	CW

City	Station	Freq	Format	City	Station	Freq	Format
Dublin	WPIN	810	SPTS	Harrisonburg	WKCY	1300	TLK+
Dublin	WPIN	91.5	RL/CC	Harrisonburg	WHBG	1360	SPTS
Duffield	WDUF	1120	RL/G	Harrisonburg	WEZI	1700	VAR
Dumfries	WPWC	1480	SP/MEX	Harrisonburg	WXJM	88.7	VAR
Earlysville	WKTR	840	SPTS	Harrisonburg	WMBA	90.7	CLA
Edinburg	WOTC	88.3	RL	Harrisonburg	WEMC	91.7	AD/A
Elkton	WACL	98.5	RK/C	Harrisonburg	WQPO	100.7	CHR
Emory	WEHC	90.7	AD/A	Harrisonburg	WKCY	104.3	CW
Emporia	WEVA	860	AC	Heathsville	WCNV	89.1	N/A
Emporia	WJYA	89.3	RL	Highland Spgs	WCLM	1450	CW
Emporia	WYTT	99.5	URB	Highland Spgs	WHCE	91.1	CH
Ettrick	WJZV	93.1	CW	Hillsville	WHHV	1400	RL/G
Exmore	WROX	96.1	RK/C	Hopewell	WHAP	1340	AS
Fairfax	WDCT	1310	RL	Hot Springs	WCHG	107.1	CW
Fairlawn	WKNV	890	RL/G	Jonesville	WJNV	99.1	CW
Falls Church	WFAX	1220	RL	Kenbridge	WPEX	90.9	RL/G
Falmouth	WGRX	104.5	CW	Kilmarnock	WKWI	101.7	AC
Farmville	WFLO	870	CW	Lawrenceville	WLES	580	OLG
Farmville	WPAK	1490	RL/G	Lawrenceville	WHFD	105.5	RL/G
Farmville	WMLU	91.3	VAR	Lebanon	WLRV	1380	CW
Farmville	WVHL	92.9	CW	Lebanon	WXLZ	107.3	CW
Farmville	WFLO	95.7	AC	Leesburg	WAGE	1200	TLK+
Farmville	WXJK	101.3	RK/C	Lexington	WREL	1450	TLK+
Ferrum	WFFC	89.9	N/TLK	Lexington	WMRL	89.9	CLA
Fieldale	WODY	1160	SPTS	Lexington	WLUR	91.5	VAR
Floyd	WGFC	1030	RL/G	Louisa	WOJL	105.5	OLG
Fort Lee	WKLR	96.5	RK/C	Luray	WRAA	1330	CW
Franklin	WLQM	1250	RL/G	Luray	WYFT	103.9	RL
Franklin	WLQM	101.7	CW	Luray	WMXH	105.7	AS
Fredericksburg	WFVA	1230	TLK+	Lynchburg	WLVA	590	AS
Fredericksburg	WYSK	1350	SP/MEX	Lynchburg	WLLL	930	RL/G
Fredericksburg	WJYJ	90.5	RL	Lynchburg	WBRG	1050	TLK+
Fredericksburg	WFLS	93.3	CW	Lynchburg	WVGM	1320	SPTS
Fredericksburg	WBQB	101.5	AC/H	Lynchburg	WKPA	1390	RL/G
Front Royal	WFTR	1450	CW	Lynchburg	WRVL	88.3	RL/CC
Front Royal	WZRV	95.3	OLG	Lynchburg	WWMC	90.9	RL/CC
Front Royal	WFQX	99.3	RK/C	Lynchburg	WZZU	97.9	OLG
Galax	WWWJ	1360	RL/G	Lynchburg	WVBE	100.1	URB
Galax	WOKG	90.3	RL/CC	Lynchburg	WJJX	101.7	CH
Galax	WBRF	98.1	CW	Lynchburg	WLNI	105.9	TLK+
Gate City	WGAT	1050	RL/G	Manassas	WKDV	1460	SP/MEX
Gloucester	WTOX	1420	TLK+	Manassas	WJFK	106.7	TLK+
Gloucester	WXGM	99.1	AS	Marion	WMEV	1010	RL/G
Goochland	WZEZ	100.5	AS	Marion	WOLD	1330	RL
Gretna	WMNA	730	CW	Marion	WVTR	91.9	CLA
Gretna	WMNA	106.3	TLK+	Marion	WMEV	93.9	CW
Grundy	WNRG	940	RL/G	Marion	WOLD	102.5	AC/S
Grundy	WMJD	97.7	CW	Marion	WZVA	103.5	RK/C
Hampden	WWHS	92.1	VAR	Martinsville	WHEE	1370	TLK+
Hampton	WLRT	1490	TLK+	Martinsville	WMVA	1450	AC
Hampton	WHOV	88.1	JAZ	Martinsville	WPIM	90.5	RL/CC

Radio on the Road   Virginia

City	Station	Freq	Format		City	Station	Freq	Format
					Richlands	WGTH	105.5	RL/G
Midlothian	WCUL	98.9	SIL		Richmond	WRNL	910	SPTS
Moneta	WCQY	880	AS		Richmond	WXGI	950	SPTS
Monterey	WVLS	89.7	CW		Richmond	WLEE	990	AS
Mount Jackson	WSVG	790	CW		Richmond	WRVA	1140	N/TLK
Mount Jackson	WSIG	96.9	CW		Richmond	WVNZ	1320	SP/OM
Narrows	WNRV	990	RL/CC		Richmond	WBTK	1380	RL
Narrows	WZFM	101.3	OLG		Richmond	WREJ	1540	RL/G
Nassawadox	WJCN	90.1	RL		Richmond	WFTH	1590	RL/G
New Market	WLTK	103.3	RL/CC		Richmond	WRIH	88.1	RL
Newport News	WTJZ	1270	RL/G		Richmond	WCVE	88.9	CLA
Newport News	WCMS	1310	SPTS		Richmond	WDCE	90.1	VAR
Newport News	WGH	97.3	CW		Richmond	WRVQ	94.5	CHR
Norfolk	WHRV	89.5	VAR		Richmond	WTVR	98.1	AD/A
Norfolk	WHRO	90.3	CLA		Richmond	WRXL	102.1	RK/C
Norfolk	WNSB	91.1	URB		Richmond	WMXB	103.7	AC/S
Norfolk	WKUS	105.3	URB		Richmond	WKJS	105.7	URB
Norfolk/Virg Bch	WNIS	790	N/TLK		Richmond	WBTJ	106.5	URB
Norfolk/Virg Bch	WTAR	850	TLK+		Roanoke	WVBE	610	URB
Norfolk/Virg Bch	WVXX	1050	SP/MEX		Roanoke	WWWR	910	RL/G
Norfolk/Virg Bch	WCKO	1110	RL/G		Roanoke	WFIR	960	TLK+
Norfolk/Virg Bch	WJOI	1230	AS		Roanoke	WGMN	1240	SPTS
Norfolk/Virg Bch	WVAB	1550	SILNT		Roanoke	WRIS	1410	RL
Norfolk/Virg Bch	WVKL	95.7	URB		Roanoke	WVTF	89.1	CLA
Norfolk/Virg Bch	WNOR	98.7	RK/C		Roanoke	WRXT	90.3	RL/CC
Norfolk/Virg Bch	WYFI	99.7	RL		Roanoke	WXLK	92.3	CHR
Norfolk/Virg Bch	WXMM	100.5	RK/C		Roanoke	WSLC	94.9	CW
Norfolk/Virg Bch	WOWI	102.9	URB		Roanoke	WSLQ	99.1	AC
Norfolk/Virg Bch	WNVZ	104.5	CH		Roanoke	WZBL	104.9	CW
Norton	WNVA	1350	RL/G		Rocky Mount	WYTI	1570	CW
Norton	WNVA	106.3	CW		Ruckersville	WHTE	101.9	CHR
Onley	WESR	1330	CW		Rural Retreat	WCRR	660	RL/G
Onley	WESR	103.3	AC		Rural Retreat	WXBX	95.3	OLG
Orange	WVCV	1340	AS		Salem	WTOY	1480	URB
Orange	WJMA	96.7	CW		Salem	WPAR	91.3	RL/CC
Pennington Gap	WSWV	1570	RL/G		Salem	WSNV	93.5	AC
Pennington Gap	WSWV	105.5	CW		Saltville	WXMY	1600	CW
Petersburg	WROU	1240	RL/G		Smithfield	WKGM	940	RL
Petersburg	WVST	91.3	JAZ		South Boston	WSBV	1560	RL/G
Petersburg	WKJM	99.3	URB		South Boston	WHLF	95.3	AC/H
Petersburg	WARV	100.3	OLG		South Boston	WQOK	97.5	URB
Poquoson	WZNR	106.1	CH		South Hill	WSHV	1370	RL/G
Portsmouth	WRJR	1010	SP/RL		South Hill	WKSK	101.9	CW
Portsmouth	WGPL	1350	RL/G		Spotsylvania	WYSK	99.3	RK/H
Portsmouth	WPCE	1400	RL/G		St. Paul	WXLZ	1140	CW
Portsmouth	WHKT	1650	YTH		Stanleytown	WZBB	99.9	CW
Pound	WDXC	102.3	CW		Staunton	WKDW	900	CW
Powhatan	WBBT	107.3	OLG		Staunton	WTON	1240	SPTS
Pulaski	WBLB	1340	RL/G		Staunton	WSVO	93.1	AC
Pulaski	WPSK	107.1	CW		Staunton	WTON	94.3	CH
Radford	WRAD	1460	SPTS		Staunton	WCYK	99.7	CW
Radford	WVRU	89.9	VAR		Stephens City	WKSI	98.3	CH
					Strasburg	WWRT	104.9	RK/C

Virginia

City	Station	Freq	Format	City	Station	Freq	Format
Stuart	WHEO	1270	CW				
Suffolk	WFOG	92.9	OLG				
Suffolk	WAFX	106.9	RK/C				
Sweet Briar	WNRS	89.9	RK/H				
Tappahannock	WRAR	1000	RL/G				
Tappahannock	WRAR	105.5	AC				
Tazewell	WTZE	1470	TLK+				
Tazewell	WKQY	100.1	RK/C				
Vinton	WKBA	1550	RL/G				
Vinton	WZZI	101.5	OLG				
Vinton	WJJS	106.1	CH				
Virginia Beach	WVAB	1550	SILNT				
Virginia Beach	WJLZ	88.5	RL/CC				
Virginia Beach	WPTE	94.9	AC				
Virginia Beach	WWHV	102.1	URB				
Warrenton	WPRZ	1250	RL/CC				
Warrenton	WKCW	1420	SP/OM				
Warrenton	WBPS	94.3	SP/OM				
Warrenton	WTOP	107.7	N/TLK+				
Warsaw	WNNT	100.9	CW				
Waynesboro	WKCI	970	TLK+				
Waynesboro	WPVA	90.1	RL				
West Point	WTYD	107.9	OLG				
White Stone	WNDJ	104.9	AS				
Williamsburg	WMBG	740	AS				
Williamsburg	WCWM	90.7	AD/A				
Winchester	WNTW	610	SPTS				
Winchester	WINC	1400	N/TLK				
Winchester	WTRM	91.3	RL/G				
Winchester	WINC	92.5	AC/H				
Winchester	WUSQ	102.5	CW				
Windsor	WJCD	107.7	AC/S				
Wise	WISE	90.5	CLA				
Woodbridge	WJZW	105.9	JAZ				
Woodstock	WAMM	1230	SIL				
Woodstock	WAZR	93.7	CH				
Wytheville	WYVE	1280	CW				
Yorktown	WYCS	91.5	RL				
Yorktown	WXEZ	94.1	RL/G				

Washington

City	Station	Freq	Format	City	Station	Freq	Format
Aberdeen	KXRO	1320	TLK+	Deer Park	KAZZ	107.1	JAZ
Aberdeen	KBKW	1450	SP/TK	Dishman	KEYF	1050	AS
Aberdeen	KDUX	104.7	RK/C	Dishman	KSPO	106.5	RL
Airway Heights	KXLX	700	SIL	E Wenatchee	KTRJ	88.1	N/A
Anacortes	KLKI	1340	AS	E Wenatchee	KYSN	97.7	CW
Asotin	KCLK	1430	SPTS	Eatonville	KFNK	104.9	RK/H
Auburn	KDDS	1210	SP/MEX	Edmonds	KCIS	630	RL
Auburn	KGRG	89.9	AD/A	Edmonds	KCMS	105.3	RL/CC
Basin City	KOLW	97.5	OLG	Ellensburg	KNWR	90.7	CLA
Bellevue	KBCS	91.3	JAZ	Ellensburg	KXLE	1240	TLK+
Bellevue	KXPA	1540	SP/MEX	Ellensburg	KCWU	88.1	RK/C
Bellevue	KASB	89.3	AD/A	Ellensburg	KCSH	88.9	RL
Bellevue	KLSY	92.5	AC/H	Ellensburg	KXLE	95.3	CW
Bellingham	KZAZ	91.7	CLA	Ellensburg	KQBE	103.1	AC
Bellingham	KGMI	790	TLK+	Elma	KAYO	99.3	CW
Bellingham	KIXT	930	AS	Elma	KSWW	102.1	AC
Bellingham	KPUG	1170	SPTS	Enumclaw	KENU	1330	CH
Bellingham	KUGS	89.3	AD/A	Ephrata	KULE	730	TLK+
Bellingham	KISM	92.9	RK/C	Ephrata	KTBI	810	RL
Bellingham	KAFE	104.3	AC/S	Ephrata	KULE	92.3	CW
Benton City	KZTB	96.7	SP/MEX	Ephrata	KTAC	93.9	RL
Blaine	KARI	550	RL	Everett	KWYZ	1230	N/A
Blaine	KVRI	1600	N/A	Everett	KRKO	1380	SPTS
Bremerton	KBRO	1490	CH	Everett	KSER	90.7	VAR
Bremerton	KRWM	106.9	AC/S	Ferndale	KRPI	1550	N/A
Burien	KGNW	820	RL	Forks	KVAC	1490	N/A
Camas	KNRK	94.7	RK/C	Forks	KLLM	103.9	AC/S
Cashmere	KZPH	106.7	RK/C	Gig Harbor	KGHP	89.9	AD/A
Castle Rock	KRQT	107.1	RK/C	Goldendale	KLCK	1400	OLG
Centralia	KELA	1470	TLK+	Goldendale	KYYT	102.3	CW
Centralia	KCED	91.3	AC/H	Grand Coulee	KEYG	1490	CW
Centralia	KNBQ	102.9	CW	Grand Coulee	KEYG	98.5	CH
Chehalis	KITI	1420	OLG	Grandview	KARY	100.9	OLG
Chehalis	KACS	90.5	RL/CC	Hoquiam	KWOK	1490	SPTS
Chelan	KOZI	1230	AC	Hoquiam	KXXX	95.3	CW
Chelan	KOZI	93.5	AC	Ilwaco	KVAS	103.9	CW
Cheney	KEWU	89.5	JAZ	Kelso	KLOG	1490	CH
Cheney	KEYF	101.1	OLG	Kelso	KTJC	91.1	RL
Clarkston	KCLK	94.1	OLG	Kelso	KLYK	94.5	AC/H
Clarkston	KVAB	102.9	RK/C	Kennewick	KONA	610	TLK+
Clarkstonh	KNWV	90.5	CLA	Kennewick	KTCR	1340	TLK+
Cle Elum	KXAA	93.7	CH	Kennewick	KTCV	88.1	RK/H
Colfax	KMAX	840	TLK+	Kennewick	KBLD	91.7	RL
Colfax	KCLX	1450	CW	Kennewick	KONA	105.3	AC/S
Colfax	KRAO	102.5	RK/C	Kirkland	KARR	1460	RL
College Place	KGTS	91.3	RL/CC	Lacey	KBRD	680	AS
Colville	KCVL	1240	CW	Lacey	KLDY	1280	CLA
Colville	KCRK	92.1	AC	Lakewood	KLAY	1180	TLK+
Davenport	KKRS	97.3	RL	Lakewood	KNTB	1480	RK/C
Dayton	KZHR	92.5	SP/MEX	Leavenworth	KOHO	97.7	AD/A

City	Station	Freq	Format	City	Station	Freq	Format
Long Beach	KAQX	94.3	AC/H	Port Angeles	KVIX	89.3	JAZ
Longview	KBAM	1270	CW	Port Angeles	KNWP	90.1	CLA
Longview	KEDO	1400	OLG	Prosser	KZXR	1310	TLK+
Longview	KJVH	89.5	RL	Prosser	KMNA	101.7	SP/MEX
Longview	KWYQ	90.3	RL/CC	Pullman	KUUX	650	N/A
Longview	KUKN	105.5	CW	Pullman	KQQQ	1150	N/TLK
Lynden	KWPZ	106.5	RL/CC	Pullman	KWSU	1250	N/TLK
Mabton	KLES	98.7	SP/OM	Pullman	KRLF	88.5	RL/CC
McCleary	KGY	96.9	CW	Pullman	KZUU	90.7	VAR
Medical Lake	KTSL	101.9	RL/CC	Pullman	KFFR	97.7	RL/CC
Mercer Island	KIXI	880	AS	Pullman	KZZL	99.5	CW
Mercer Island	KMIH	104.5	CHR	Pullman	KHTR	104.3	RK/H
Moses Lake	KBSN	1470	TLK+	Puyallup	KSUH	1450	N/A
Moses Lake	KMLW	88.3	RL/CC	Quincy	KWNC	1370	CW
Moses Lake	KLWS	91.5	N/TLK	Quincy	KZML	95.9	SP/MEX
Moses Lake	KDRM	99.3	AC	Quincy	KWWW	96.7	AC/H
Moses Lake	KWIQ	100.3	CW	Raymond	KFMY	97.7	CH
Moses Lake No	KWIQ	1020	SPTS	Renton	KRIZ	1420	JAZ
Mount Vernon	KAPS	660	CW	Renton	KYIZ	1620	URB
Mount Vernon	KBRC	1430	OLG	Richland	KALE	960	SPTS
Mount Vernon	KSVR	90.1	N/TLK	Richland	KFAE	89.1	CLA
Mt. Vernon	KAPS	660	CW	Richland	KIOK	94.9	CW
Mt. Vernon	KSVR	91.7	SP/OM	Richland	KORD	102.7	CW
Naches	KZTA	96.9	SP/MEX	Richland	KEGX	106.5	RK/C
Naches	KQSN	99.3	CW	Rock Island	KXAA	99.5	AC/S
Newport	KUBS	91.5	VAR	Roy	KWFJ	89.7	RL
Newport	KMJY	104.5	CH	Royal City	KRCW	96.3	SP/MEX
Nile	KLRO	88.1	RL/CC	Seattle	KVI	570	TLK+
Oak Harbor	KWDB	1110	AC	Seattle	KIRO	710	N/TLK
Ocean Park	KLOP	88.1	N/A	Seattle	KNWX	770	TLK+
Olympia	KGTK	920	TLK+	Seattle	KJR	950	SPTS
Olympia	KGY	1240	AC/S	Seattle	KOMO	1000	N/TLK
Olympia	KAOS	89.3	VAR	Seattle	KBLE	1050	RL
Olympia	KWGV	90.1	N/A	Seattle	KPTK	1090	TLK+
Olympia	KXXO	96.1	AC	Seattle	KKNW	1150	NWS+
Omak	KOMW	680	AS	Seattle	KKDZ	1250	YTH
Omak	KOMW	680	N/TLK	Seattle	KKOL	1300	TLK+
Omak	KNCW	92.7	CW	Seattle	KLFE	1590	RL/CC
Omak	KZBE	104.3	AC/H	Seattle	KAZJ	1680	N/A
Opportunity	KXLI	630	SPTS	Seattle	KNHC	89.5	CH/D
Opportunity	KIXZ	96.1	CW	Seattle	KEXP	90.3	AD/A
Othello	KRSC	1400	SP/MEX	Seattle	KUBE	93.3	URB
Othello	KZLN	97.5	AC	Seattle	KMPS	94.1	CW
Pasco	KFLD	870	TLK+	Seattle	KUOW	94.9	N/TLK
Pasco	KOLU	90.1	RL	Seattle	KJR	95.7	CH
Pasco	KGSG	93.7	RL/G	Seattle	KJAQ	96.5	AH
Pasco	KEYW	98.3	AC/H	Seattle	KING	98.1	CLA
Pasco	KGDN	101.3	RL	Seattle	KWJZ	98.9	JAZ
Port Angeles	KIKN	1290	TLK+	Seattle	KISW	99.9	RK/C
Port Angeles	KONP	1450	N/TLK	Seattle	KQBZ	100.7	TLK+

Radio on the Road   Washington

City	Station	Freq	Format	City	Station	Freq	Format
Seattle	KPLZ	101.5	AC/H	Vancouver	KRVO	105.9	CH
Seattle	KZOK	102.5	RK/C	Walla Walla	KGDC	1320	N/TLK
Seattle	KNDD	107.7	RK/H	Walla Walla	KUJ	1420	TLK+
Selah	KJOX	980	TLK+	Walla Walla	KTEL	1490	OLG
Shelton	KMAS	1030	AC/S	Walla Walla	KWWS	89.7	N/TLK
Shelton	KRXY	94.5	AC/H	Walla Walla	KWCW	90.5	VAR
Silverdale	KITZ	1400	TLK+	Walla Walla	KRKL	93.3	RL/CC
South Bend	KJET	105.7	AC/H	Walla Walla	KNLT	95.7	OLG
Spokane	KQNT	590	N/TLK	Walla Walla	KXRX	97.1	RK
Spokane	KJRB	790	TLK+	Walla Walla	KUJ	99.1	CHR
Spokane	KXLY	920	N/TLK	Walla Walla	KHSS	100.9	RL
Spokane	KTRW	970	RL	Wapato	KSOH	89.5	RL
Spokane	KSBN	1230	BIZ	Wenatchee	KPQ	560	N/TLK
Spokane	KUDY	1280	TLK+	Wenatchee	KKRT	900	SPTS
Spokane	KMBI	1330	RL	Wenatchee	KWWX	1340	SP/MEX
Spokane	KGA	1510	N/TLK	Wenatchee	KPLW	89.9	RL/CC
Spokane	KAGU	88.7	CLA	Wenatchee	KPQ	102.1	CH
Spokane	KWRS	90.3	VAR	Wenatchee	KKRV	104.9	CW
Spokane	KPBX	91.1	N/TLK	White Salmon	KBNO	89.3	SP/RL
Spokane	KSFC	91.9	N/TLK	Wilson Creek	KWLN	103.3	SP/MEX
Spokane	KZZU	92.9	CH	Winlock	KITI	95.1	AC/H
Spokane	KDRK	93.7	CW	Yakima	KYAK	930	RL
Spokane	KEZE	96.9	OLG	Yakima	KIT	1280	N/TLK
Spokane	KISC	98.1	AC	Yakima	KJOX	1390	RL
Spokane	KKZX	98.9	RK/C	Yakima	KUTI	1460	CW
Spokane	KXLY	99.9	AC	Yakima	KYVT	88.5	AD/A
Spokane	KYWL	103.9	AH	Yakima	KNWY	90.3	CLA
Spokane	KEEH	104.7	RL/CC	Yakima	KYPL	91.1	RL/CC
Spokane	KZBD	105.7	RK/C	Yakima	KDNA	91.9	SP/MEX
Spokane	KMBI	107.9	RL/CC	Yakima	KATS	94.5	RK
Sumner	KZIZ	1560	RL/G	Yakima	KHHK	99.7	CHR
Sunnyside	KZTS	1210	SP/MEX	Yakima	KXDD	104.1	CW
Sunnyside	KAYB	88.1	RL	Yakima	KRSE	105.7	AC/S
Sunnyside	KPLU	88.5	JAZ	Yakima	KFFM	107.3	CHR
Tacoma	KHHO	850	SPTS				
Tacoma	KKMO	1360	SP/MEX				
Tacoma	KUPS	90.1	AD/A				
Tacoma	KVTI	90.9	CH/D				
Tacoma	KXOT	91.7	AD/A				
Tacoma	KBSG	97.3	OLG				
Tacoma	KMTT	103.7	AD/A				
Tacoma	KBKS	106.1	CHR				
Toppenish	KYNR	1490	VAR				
Toppenish	KDBL	92.9	CW				
Tumwater	KVSN	1340	RL/CC				
Twisp	KVLR	106.3	CW				
Union Gap	KYXE	1020	SP/MEX				
Vancouver	KKSN	910	OLG				
Vancouver	KBMS	1480	URB				
Vancouver	KKAD	1550	AS				

West Virginia

City	Station	Freq	Format	City	Station	Freq	Format
Barrackville	WBVQ	93.1	OLG	Fairmont	WMMN	920	SPTS
Beckley	WJLS	560	RL/G	Fairmont	WTCS	1490	N/TLK
Beckley	WWNR	620	TLK+	Fairmont	WRLF	94.3	RK/C
Beckley	WIWS	1070	CH	Fairmont	WKKW	97.9	CW
Beckley	WVPB	91.7	CLA	Fisher	WELD	690	CW
Beckley	WJLS	99.5	CW	Fisher	WQWV	103.7	AC/H
Beckley	WCIR	103.7	CHR	Fort Gay	WFGH	90.7	VAR
Berkeley Sprgs	WCST	1010	CW	Frost	WVMR	1370	CW
Berkeley Sprgs	WDHC	92	CW	Gary	WHQX	107.7	CW
Bethany	WVBC	88.1	AD/A	Grafton	WTBZ	1260	AC
Bethlehem	WukI	105.5	OLG	Grafton	WDKL	95.9	RL/CC
Bluefield	WKEZ	1240	OLG	Green Valley	WAMN	1050	RL/G
Bluefield	WHIS	1440	TLK+	Hinton	WMTD	1380	CH
Bluefield	WPIB	91.1	RL/CC	Hinton	WMTD	102.3	RK/C
Bluefield	WHAJ	104.5	AC/H	Huntington	WVHU	800	TLK+
Bridgeport	WDCI	104.1	AC/S	Huntington	WRVC	930	N/TLK
Buchannon	WVPW	88.9	CLA	Huntington	WHRD	1470	RL/G
Buckhannon	WBUC	1460	AS	Huntington	WMUL	88.1	VA
Buckhannon	WWWC	92.1	AD/A	Huntington	WWWV	89.9	CLA
Buckhannon	WBTQ	93.5	OLG	Huntington	WKEE	100.5	CHR
Buckhannon	WBUC	101.3	CW	Huntington	WTCR	103.3	CW
Charles Town	WXVA	1550	AS	Huntington	WEMM	107.9	RL/G
Charleston	WCHS	580	TLK+	Hurricane	WOKU	1080	SPTS
Charleston	WCAW	680	RL/G	Kenova	WTCR	1420	RL/CC
Charleston	WVTS	950	TLK+	Keyser	WKLP	1390	AS
Charleston	WBES	1240	SPTS	Keyser	WQZK	94.1	RK/C
Charleston	WSWW	1490	SPTS	Keyser	WCBC	107.1	OLG
Charleston	WVPN	88.5	CLA	Kingwood	WFSP	1560	RL/G
Charleston	WXAF	90.9	RL/G	Kingwood	WKMM	96.7	CW
Charleston	WKWS	96.1	CW	Kingwood	WFSP	107.7	AC
Charleston	WQBE	97.5	CW	Lewisburg	WKCJ	103.1	CW
Charleston	WVAF	99.9	AC/H	Lindside	WHFI	106.7	VAR
Charleston	WVAF	99.9	AC/H	Logan	WVOW	1290	AC
Charleston	WVSR	102.7	CHR	Logan	WVOW	101.9	AC
Clarksburg	WPDX	750	AS	Lost Creek	WOTR	96.3	RL
Clarksburg	WXKX	1340	SPTS	Mannington	WTUS	102.7	CW
Clarksburg	WOBG	1400	AS	Marmet	WKWW	93.3	RL/CC
Clarksburg	WKJL	88.1	RL/G	Martinsburg	WRNR	740	TLK+
Clarksburg	WZWA	90.1	RL/G	Martinsburg	WEPM	1340	SPTS
Clarksburg	WGIE	92.7	CW	Martinsburg	WVEP	88.9	CLA
Clarksburg	WPDX	104.9	CW	Martinsburg	WLTF	97.5	AC/S
Clarksburg	WFBY	106.5	AC	Matewan	WHJC	1360	CW
Danville	WZAC	92.5	CW	Matewan	WVKM	106.7	CW
Dunbar	WZJO	94.5	RK/C	Miami	WKAZ	107.3	OLG
Elizabeth	WRZZ	106.1	RK/C	Middlebourne	WRSG	91.5	VAR
Elkins	WDNE	1240	AS	Milton	WZZW	1600	RL/CC
Elkins	WCDE	90.3	VAR	Milton	WAMX	106.3	RK/C
Elkins	WBHZ	91.9	RL	Montgomery	WMON	1340	RL
Elkins	WELK	94.7	AC	Morgantown	WCLG	1300	OLG
Elkins	WDNE	98.9	CW	Morgantown	WAJR	1440	N/TLK

City	Station	Freq	Format	City	Station	Freq	Format
Morgantown	WVPM	90.9	CLA	South Charleston	WPDD	89.3	N/A
Morgantown	WWVU	91.7	AD/A	South Charleston	WMXE	100.9	AC/H
Morgantown	WCLG	100.1	RK/C	Spencer	WVRC	1400	RL/G
Morgantown	WVAQ	101.9	CHR	Spencer	WVRC	104.7	CW
Moundsville	WVLY	1370	TLK+	St. Albans	WJYP	1300	RL
Moundsville	WRKP	96.5	RL/CC	St. Albans	WKLC	105.1	RK
Mount Hope	WTNJ	105.9	CW	St. Marys	WJAW	630	SPTS
Mullens	WPMW	92.7	RK/C	St. Marys	WRRR	93.9	AC
New Mart'sville	WETZ	1330	AS	Summersville	WMLJ	90.5	RL
New Mart'sville	WXCR	92.3	RK/C	Summersville	WCWV	92.9	AC/H
New Mart'sville	WYMJ	99.5	OLG	Sutton	WSGB	1490	AC/H
New Mart'sville	WETZ	103.9	CW	Sutton	WDBS	97.1	CW
Oak Hill	WOAY	860	RL/CC	Vienna	WDMX	100.1	OLG
Oak Hill	WAXS	94.1	CH	Webster Springs	WAFD	100.3	RL/G
Parkersburg	WADC	1050	AS	Weirton	WEIR	1430	AS
Parkersburg	WVNT	1230	TLK+	Welch	WELC	1150	AC/S
Parkersburg	WHNK	1450	CW	Welch	WELA	1340	N/A
Parkersburg	WVPG	90.3	CLA	Welch	WELC	102.9	AC/S
Parkersburg	WXIL	95.1	CH	West Liberty	WGLZ	91.5	VA
Parkersburg	WGGE	99.1	CW	Weston	WHAW	980	CW
Parkersburg	WHBR	103.1	RK	Weston	WFBY	102.3	RK/C
Petersburg	WAUA	89.5	CLA	Westover	WZST	100.9	CW
Petersburg	WELD	101.7	CW	Wheeling	WWVA	1170	RL
Phillippi	WQAB	91.3	AC	Wheeling	WBBD	1400	AS
Pineville	WWYO	970	VAR	Wheeling	WKKK	1600	SPTS
Pocatalico	WRVZ	98.7	JAZ	Wheeling	WVNP	89.9	CLA
Point Pleasant	WBGS	1030	RL/G	Wheeling	WPHP	91.9	CH
Point Pleasant	WPCN	88.1	RL/G	Wheeling	WKWK	97.3	AC
Point Pleasant	WBYG	99.5	CW	Wheeling	WOVK	98.7	CW
Princeton	WAEY	1490	RL/G	Wheeling	WEGW	107.5	RK/C
Princeton	WPWV	90.1	RL	White Sulphur Springs	WSLW	1310	AS
Princeton	WSTG	95.9	CH	Williamson	WBTH	1400	OLG
Princeton	WKOY	100.9	RK/C	Williamson	WVVV	96.5	AC
Rainelle	WRRL	1130	RL/G				
Rainelle	WRLB	95.3	RL				
Ravenswood	WMOV	1360	TLK+				
Ravenswood	WLWF	93.1	CW				
Richwood	WVAR	600	CW				
Ripley	WLKV	90.7	RL/CC				
Ripley	WCEF	98.3	CW				
Romney	WDZN	100.1	YTH				
Romney	WVSB	104.1	CW				
Ronceverte	WRON	1400	TLK+				
Ronceverte	WRON	97.7	OLG				
Rupert	WYKM	1250	CW				
Salem	WITB	91.1	CHR				
Salem	WAJR	103.3	N/TLK				
Salem	WOBG	105.7	RK/C				
Shepherdstown	WSHC	89.7	AD/A				
So Charleston	WSCW	1410	RL/G				

City	Station	Freq	Format	City	Station	Freq	Format
Adams	WDKM	106.1	CH	Dodgeville	WDMP	99.3	CW
Algoma	WBDK	96.7	AS	Durand	WQOQ	1430	RK/C
Algoma	WRLU	104.1	CW	Durand	WRDN	95.9	RK/C
Allouez	WJLW	106.7	RK/C	Eagle River	WERL	950	AS
Altoona	WISM	98.1	AC	Eagle River	WRJO	94.5	OLG
Amery	WXCE	1260	N/TLK	Eau Claire	WAYY	790	N/TLK
Antigo	WATK	900	CW	Eau Claire	WDVM	1050	RL
Antigo	WRLO	105.3	RK/C	Eau Claire	WBIZ	1400	SPTS
Antigo	WACD	106.1	AC	Eau Claire	WUEC	89.7	CLA
Appleton	WRJQ	1570	SPTS	Eau Claire	WVCF	90.5	RL
Appleton	WLFM	91.1	N/TLK	Eau Claire	WHEM	91.3	RL/CC
Appleton	WEMI	91.9	RL/CC	Eau Claire	WIAL	94.1	AC/H
Appleton	WAPL	105.7	RK	Eau Claire	WBIZ	100.7	CHR
Ashland	WATW	1400	AS	Eau Claire	WAXX	104.5	CW
Ashland	WBSZ	93.3	CW	Elk Mound	WECL	92.9	RK/C
Ashland	WJJH	96.7	RK/C	Elm Grove	WGLB	1560	RL/G
Auburndale	WLBL	930	N/TLK	Evansville	WKPO	105.9	CH
Balsam Lake	WLMX	104.9	AC/H	Fond Du Lac	KFIZ	1450	N/TLK
Baraboo	WRPQ	740	AC/S	Fond Du Lac	WVFL	89.9	N/A
Baraboo	WOLX	94.9	OLG	Fond Du Lac	WLWR	91.7	N/A
Barron	WAQE	97.7	AC/H	Fond Du Lac	KFON	107.1	AC
Beaver Dam	WBEV	1430	TLK+	Forestville	WRKU	102.1	AC
Beaver Dam	WXRO	95.3	CW	Fort Atkinson	WFAW	940	TLK+
Beloit	WGEZ	1490	OLG	Fort Atkinson	WSJY	107.3	AC/S
Beloit	WBCR	90.3	VAR	Goodman	WMVM	90.7	RL
Berlin	WISS	1090	CW	Green Bay	WTAQ	1360	TLK+
Berlin	WBJZ	104.7	JAZ	Green Bay	WDUZ	1400	SPTS
Birnamwood	WYNW	92.9	RL	Green Bay	WNFL	1440	SPTS
Black River Fls	WWIS	1260	OLG	Green Bay	WHID	88.1	N/TLK
Black River Fls	WWIS	99.7	AC	Green Bay	WPNE	89.3	CLA
Bloomer	WQRB	95.1	CW	Green Bay	WORQ	90.1	RL/CC
Brillion	WDUZ	107.5	SPTS	Green Bay	WEMY	91.5	RL/CC
Brookfield	WFMR	106.9	CLA	Green Bay	WQLH	98.5	AC/H
Brule	WHSA	89.9	CLA	Green Bay	WIXX	101.1	CHR
Burlington	WBSD	89.1	AD/A	Greenfield	WMCS	1290	TLK+
Chetek	WATQ	106.7	CW	Hallie	WOGO	680	N/TLK
Chilton	WMBE	1530	SPTS	Hallie	WWIB	103.7	RL/CC
Chippewa Falls	WEAQ	1150	SPTS	Hartford	WTKM	1540	CW
Chippewa Falls	WCFW	105.7	AC	Hartford	WTKM	104.9	CW
Cleveland	WLKN	98.1	AC	Hayward	WHSM	910	AS
Clintonville	WFCL	1380	AS	Hayward	WRLS	92.3	AC
Clintonville	WJMQ	92.3	CW	Hayward	WHSM	101.1	AC
Columbus	WTLX	100.5	TLK+	Highland	WHHI	91.3	N/TLK
Cornell	WDRK	99.9	RK/C	Holmen	WKBH	1570	RL
De Forest	WHIT	93.1	CH	Hudson	WDGY	630	SP/MEX
De Pere	WKSZ	95.9	AC/H	Hudson	WMIN	740	SP/OM
Delafield	WHAD	90.7	N/TLK	Hurley	WHRY	1450	OLG
Denmark	WPCK	104.9	CW	Iron River	WNXR	107.3	OLG
Dickeyville	WVRE	101.1	CW	Jackson	WZER	540	RL
Dodgeville	WDMP	810	CW	Janesville	WCLO	1230	N/TLK

Wisconsin Radio on the Road

City	Station	Freq	Format	City	Station	Freq	Format
Janesville	WJVL	99.9	CW	Mayville	WMDC	98.7	OLG
Kaukauna	WJOK	1050	RL	Medford	WIGM	1490	SPTS
Kaukauna	WOGB	103.1	OLG	Medford	WKEB	99.3	AC
Kenosha	WLIP	1050	AS	Menom'ee Falls	WJMR	98.3	URB
Kenosha	WGTD	91.1	CLA	Menomonie	WMEQ	880	TLK+
Kenosha	WIIL	95.1	RK	Menomonie	WHWC	88.3	N/TLK
Kewaunee	WAUN	92.7	RK/C	Menomonie	WVSS	90.7	CLA
Kiel	WSTM	91.3	RL/CC	Menomonie	WMEQ	92.1	RK/C
Kimberly	WHBY	1150	N/TLK	Merrill	WJMT	730	CW
La Crosse	WKTY	580	SPTS	Merrill	WMZK	104.1	RK
La Crosse	WIZM	1410	TLK+	Middleton	WWQM	106.3	CW
La Crosse	WLFN	1490	AS	Milladore	WGNV	88.5	RL/CC
La Crosse	WSLU	88.9	CLA	Milwaukee	WTMJ	620	N/TLK
La Crosse	WHLA	90.3	N/TLK	Milwaukee	WNOV	860	URB
La Crosse	WIZM	93.3	CHR	Milwaukee	WOKY	920	AS
La Crosse	WRQT	95.7	RK	Milwaukee	WISN	1130	TLK+
La Crosse	WLXR	104.9	AC	Milwaukee	WSSP	1250	SPTS
La Crosse	WQCC	106.3	CW	Milwaukee	WJYI	1340	RL/CC
Ladysmith	WLDY	1340	AS	Milwaukee	WMWK	88.1	RL
Ladysmith	WJBL	93.1	OLG	Milwaukee	WYMS	88.9	JAZ
Lake Geneva	WZRK	1550	RL	Milwaukee	WUWM	89.7	N/TLK
Lake Geneva	WLKG	96.1	AC	Milwaukee	WMSE	91.7	AD/A
Lancaster	WGLR	1280	SPTS	Milwaukee	WJZI	93.3	JAZ
Lancaster	WJTY	88.1	RL	Milwaukee	WKTI	94.5	AC/H
Lancaster	WGLR	97.7	CW	Milwaukee	WRIT	95.7	OLG
Lomira	WFDL	97.7	AC	Milwaukee	WKLH	96.5	CH
Madison	WTSO	1070	SPTS	Milwaukee	WQBW	97.3	OLG
Madison	WIBA	1310	N/TLK	Milwaukee	WMYX	99.1	AC
Madison	WLMV	1480	SP/OM	Milwaukee	WLUM	102.1	RK
Madison	WTUX	1550	AS	Milwaukee	WLZR	102.9	RK
Madison	WTDY	1670	TLK+	Milwaukee	WVCY	107.7	RL
Madison	WERN	88.7	CLA	Minocqua	WLKD	1570	SPTS
Madison	WORT	89.9	VAR	Minocqua	WMQA	95.9	AC/S
Madison	WSUM	91.7	VAR	Mishicot	WZOR	94.7	RK/H
Madison	WMGN	98.1	AC	Monroe	WEKZ	1260	CW
Madison	WIBA	101.5	RK/C	Monroe	WEKZ	93.7	AC/S
Madison	WNWC	102.5	RL/CC	Mosinee	WOFM	94.7	OLG
Madison	WZEE	104.1	CHR	Mukwanago	WFZH	105.3	RL/CC
Madison	WHA	970	N/TLK	Neenah	WNAM	1280	AS
Manitowoc	WOMT	1240	AC/S	Neenah	WROE	94.3	AC/S
Manitowoc	WLTU	92.1	OLG	Neenah	WNCY	100.3	CW
Manitowoc	WQTC	102.3	CH	Neilsville	WCCN	1370	AS
Marathon	WKQH	104.9	RK/C	Neilsville	WPKG	92.7	AC/H
Marinette	WMAM	570	SPTS	Neilsville	WCCN	107.5	RK/C
Marinette	WLST	95.1	AC/H	Nekoosa	WMMA	93.9	RL
Marshall	WJWD	90.3	RL/CC	Nekoosa	WUSP	105.5	SIL
Marshfield	WDLB	1450	NWS+	New London	WOZZ	93.5	RK/C
Marshfield	WLJY	106.5	AC/S	New Richmond	WIXK	1590	CW
Mauston	WRJC	1270	AS	New Richmond	WIXK	107.1	AS
Mauston	WRJC	92.1	AC	Oconto	WOCO	1260	CW

198

City	Station	Freq	Format	City	Station	Freq	Format
Oconto	WOCO	107.1	AS	Seymour	WECB	104.3	AC
Omro	WPKR	99.5	CW	Shawano	WTCH	960	CW
Oshkosh	WVCY	690	RL	Shawano	WOWN	99.3	OLG
Oshkosh	WOSH	1490	N/TLK	Sheboygan	WCLB	950	SPTS
Oshkosh	WRST	90.3	N/TLK	Sheboygan	WHBL	1330	TLK+
Oshkosh	WWWX	96.9	RK	Sheboygan	WSHS	91.7	VAR
Park Falls	WNBI	980	SPTS	Sheboygan	WBFM	93.7	CW
Park Falls	WHBM	90.3	N/TLK	Sheboygan Fls	WHBZ	106.5	RK/C
Park Falls	WHBM	90.3	N/TLK	Shell Lake	WCSW	940	TLK+
Park Falls	WCQM	98.3	CW	Shell Lake	WGMO	95.3	RK/C
Peshtigo	WSFQ	96.3	OLG	Siren	WXCX	105.7	CH
Platteville	WPVL	1590	SPTS	Sister Bay	WHND	89.7	CLA
Platteville	WSUP	90.5	RK/C	Sister Bay	WHDI	91.9	N/TLK
Platteville	WPVL	107.1	OLG	Sparta	WKLJ	1290	TLK+
Plymouth	WJUB	1420	AS	Sparta	WCOW	97.1	CW
Plymouth	WXER	104.5	AC/H	Spencer	WOSQ	92.3	CW
Pt Washington	WGLB	100.1	OLG	Spooner	WPLT	106.3	CW
Portage	WPDR	1350	TLK+	Stevens Point	WSPT	1010	TLK+
Portage	WBKY	95.9	CW	Stevens Point	WWSP	89.9	VAR
Portage	WDDC	100.1	CW	Stevens Point	WSPT	97.9	AC/H
Poynette	WHFA	1240	RL	Sturgeon Bay	WDOR	910	AC
Prairie du Chien	WPRE	980	OLG	Sturgeon Bay	WRGX	88.5	RL/CC
Prairie du Chien	WQPC	94.3	CW	Sturgeon Bay	WPFF	90.5	RL/CC
Racine	WRJN	1400	N/TLK	Sturgeon Bay	WDOR	93.9	AC
Racine	WBJX	1460	SP/MEX	Sturgeon Bay	WSRG	97.7	AC
Racine	WEZY	92.1	AC/S	Sturgeon Bay	WLYD	99.7	CH
Racine	WKKV	100.7	URB	Sturtevant	WEXT	104.7	CW
Reedsburg	WRDB	1400	AC	Sun Prairie	WNWC	1190	RL
Reedsburg	WBDL	102.9	AC	Sun Prairie	WXXM	92.1	TLK+
Reedsburg	WNFM	104.9	CW	Superior	WDSM	710	TLK+
Reserve	WOJB	88.9	N/TLK	Superior	WGEE	970	SPTS
Rhinelander	WOBT	1240	SPTS	Superior	KUWS	91.3	VAR
Rhinelander	WXPR	91.7	N/TLK	Superior	KRBR	102.5	RK/H
Rhinelander	WHDG	97.5	CW	Suring	WRVM	102.7	RL
Rhinelander	WRHN	100.1	AC/H	Sussex	WKSH	1640	YTH
Rice Lake	WAQE	1090	CW	Three Lakes	WLSL	93.7	CH
Rice Lake	WJMC	1240	N/TLK	Tomah	WBOG	1460	OLG
Rice Lake	WJMC	96.3	CW	Tomah	WTMB	94.5	CW
Rice Lake	WKFX	99.1	CH	Tomah	WXYM	96.1	AC
Richland Ctr	WRCO	1450	AS	Tomah	WVCX	98.9	RL
Richland Ctr	WRCO	100.9	CW	Tomahawk	WJJQ	810	SPTS
Ripon	WRPN	1600	TLK+	Tomahawk	WJJQ	92.5	AC/S
Ripon	WRPN	90.1	VAR	Trempealeau	WFBZ	105.5	SPTS
Ripon	WTCX	96.1	CH	Two Rivers	WCUB	980	CW
River Falls	WEVR	1550	AC/S	Two Rivers	WTRW	1590	OLG
River Falls	WRFW	88.7	VAR	Verona	WMMM	105.5	AD/A
River Falls	WEVR	106.3	AC/S	Viroqua	WVRQ	1360	OLG
Rudolph	WIZD	99.9	OLG	Viroqua	WVRQ	102.3	CW
Sauk City	WMAD	96.3	RK/H	Washburn	WEGZ	105.9	RL
Schofield	WRIG	1390	AS	Watertown	WTTN	1580	OLG

City	Station	Freq	Format		City	Station	Freq	Format
Watertown	WJJO	94.1	RK/C					
Waukesha	WAUK	1510	SPTS					
Waukesha	WCCX	104.5	VAR					
Waukesha	WMIL	106.1	CW					
Waunakee	WCHY	105.1	AH					
Waupaca	WDUX	800	CW					
Waupaca	WDUX	92.7	AC/H					
Waupun	WFDL	1170	N/TLK					
Wausau	WSAU	550	N/TLK					
Wausau	WXZO	1230	SPTS					
Wausau	WCLQ	89.5	RL/CC					
Wausau	WHRM	90.9	CLA					
Wausau	WXPW	91.9	VAR					
Wausau	WLBW	91.9	N/TLK					
Wausau	WIFC	95.5	CHR					
Wausau	WDEZ	101.9	CW					
Wausau	WLRK	107.9	CH					
Wautoma	WAUH	102.3	CH					
Wauwatosa	WXSS	103.7	CHR					
West Bend	WBKV	1470	CW					
West Bend	WBWI	92.5	CW					
West Salem	WKBH	100.1	RK/C					
Whitehall	WHTL	102.3	OLG					
Whitewater	WSUW	91.7	JAZ					
Whitewater	WSLD	104.5	CW					
Whitewater	WKCH	106.5	OLG					
Whiting	WYTE	96.7	CW					
Winneconne	WVBO	103.9	OLG					
Wisconsin Dells	WDLS	900	TRAV					
Wisconsin Dells	WNNO	106.9	AC					
Wisconsin Rapids	WFHR	1320	TLK+					
Wisconsin Rapids	WGLX	103.3	RK/C					
Wittenberg	WVRN	88.9	N/A					

City	Station	Freq	Format	City	Station	Freq	Format
Afton	KRSV	1210	CW	Gillette	KLWD	91.9	RL
Afton	KUWA	91.3	CLA	Gillette	KAML	96.9	RK/C
Afton	KRSV	98.7	CW	Gillette	KGWY	100.7	CW
Albin	KKAW	107.3	CW	Gillette	KXXL	103.9	RK/C
Buffalo	KBBS	1450	OLG	Glendo	KYOD	100.1	AC/H
Buffalo	KBUW	90.5	CLA	Glenrock	KGRK	98.3	N/A
Buffalo	KLGT	92.9	CW	Green River	KUGR	1490	AC
Burns	KIG N	101.9	CH	Green River	KFRZ	92.1	CW
Casper	KTWO	1030	TLK+	Green River	KZWB	97.9	N/A
Casper	KVOC	1230	AS	Greybull	KZMQ	1140	CW
Casper	KKTL	1400	SPTS	Greybull	KZMQ	100.3	CW
Casper	KLWC	89.1	JAZ	Guernsey	KANT	104.1	N/A
Casper	KCSP	90.3	RL/CC	Jackson	KSGT	1340	CW
Casper	KUWC	91.3	CLA	Jackson	KINL	88.3	N/A
Casper	KMLD	94.5	OLG	Jackson	KURT	89.1	N/A
Casper	KWYY	95.5	CW	Jackson	KUWJ	90.3	CLA
Casper	KMGW	96.7	AC	Jackson	KJAX	93.3	CW
Casper	KHOC	102.5	AC/H	Jackson	KMTN	96.9	AD/A
Casper	KQLT	103.7	CW	Kemmerer	KMER	950	OLG
Casper	KTRS	104.7	CHR	Kemmerer	KAOX	107.3	AC
Casper	KASS	106.9	RK/C	Lander	KOVE	1330	CW
Cheyenne	KFBC	1240	N/TLK	Lander	KDLY	97.5	RK/C
Cheyenne	KJUA	1380	SP/OM	Laramie	KHAT	1210	OLG
Cheyenne	KRAE	1480	CW	Laramie	KOWB	1290	TLK+
Cheyenne	KWYH	88.1	RL	Laramie	KAIW	88.9	N/A
Cheyenne	KQLF	97.9	AC	Laramie	KUWR	91.9	CLA
Cheyenne	KKPL	99.9	RK/H	Laramie	KCGY	95.1	CW
Cheyenne	KOLZ	100.7	CW	Laramie	KIMX	96.7	CH
Cheyenne	KRRR	104.9	OLG	Laramie	KHIH	98.7	N/A
Cheyenne	KLEN	106.3	AC/S	Laramie	KARS	102.9	OLG
Chugwater	KLWV	90.9	RL/CC	Laramie	KRQU	104.5	RK/C
Chugwater	KCUG	99.5	N/A	Lost Cabin	KWYW	99.1	CW
Cody	KODI	1400	TLK+	Lovell	KROW	107.1	N/A
Cody	KTAG	97.9	AC	Midwest	KRVK	107.9	CH
Diamondville	KDWY	105.3	CW	Mills	KHAD	105.5	N/A
Douglas	KKTY	1470	OLG	Newcastle	KASL	1240	CW
Douglas	KDUW	91.7	CLA	Newcastle	KUWN	90.5	CLA
Douglas	KBOG	92.5	N/A	Newcastle	KRKI	99.5	OLG
Douglas	KKTY	99.3	CW	Orchard Valley	KGAB	650	TLK+
Douglas	KTED	100.9	N/A	Orchard Valley	KWYC	90.3	N/A
Douglas	KTED	100.9	CW	Pine Bluffs	KREO	105.3	CW
Ethete	KWRR	89.5	N/A	Pinedale	KUWX	90.9	CLA
Evanston	KEVA	1240	CW	Pinedale	KPIN	101.1	CW
Evanston	KBGMG	106.1	SP/TO	Powell	KPOW	1260	TLK+
Evansville	KUYO	830	RL/G	Powell	KUWP	90.1	CLA
Fox Farm	KRND	1630	SP/MEX	Powell	KLZY	92.5	N/A
Gillette	KIML	1270	TLK+	Powell	KCGL	104.1	RK/C
Gillette	KYPR	88.9	CLA	Rawlins	KRAL	1240	AC/H
Gillette	KAXG	89.7	RL	Rawlins	KIQZ	92.7	AC/H
Gillette	KUWG	90.9	AD/A	Riverton	KVOW	1450	OLG

Wyoming

Wyoming

Radio on the Road

City	Station	Freq	Format	City	Station	Freq	Format
Riverton	KCWC	88.1	JAZ				
Riverton	KTRZ	93.1	RK/C				
Riverton	KTAK	93.9	CW				
Rock River	KVAN	95.9	N/A				
Rock Springs	KRKK	1360	OLG				
Rock Springs	KUWZ	90.5	CLA				
Rock Springs	KYCS	95.1	AC/H				
Rock Springs	KQSW	96.5	CW				
Rock Springs	KSIT	104.5	RK/C				
Saratoga	KTGA	99.3	N/A				
Sheridan	KROE	930	TLK+				
Sheridan	KWYO	1410	AS				
Sheridan	KPRQ	88.1	VAR				
Sheridan	KOHR	88.9	RL				
Sheridan	KWCF	89.3	N/A				
Sheridan	KSUW	91.3	CLA				
Sheridan	KYTI	93.7	CW				
Sheridan	KZWY	94.9	RK/C				
Story	KZZS	98.3	AC/H				
Sundance	KUWD	91.5	CLA				
Sundance	KYDT	103.1	CW				
Thermopolis	KTHE	1240	AC				
Thermopolis	KUWT	91.3	CLA				
Thermopolis	KDNO	101.7	CW				
Torrington	KGOS	1490	CW				
Torrington	KERM	98.3	CW				
West Laramie	KRWT	89.9	N/A				
Wheatland	KYCN	1340	CW				
Wheatland	KZEW	101.7	AC				
Worland	KWOR	1340	OLG				
Worland	KKLX	96.1	AC/H				

Talk Radio

Rush Limbaugh

Location	Sta	Freq	Airtime	Location	Sta	Freq	Airtime
Alaska				**California**			
Anchorage	KENI	650	9a-12n	Alturas	KKFJ	570	9a-12n
Fairbanks	KFBX	970	8a-11a	Armed Forces	AFRTS	- - -	12n-1p
Juneau	KJNO	630	8a-11a	Bakersfield	KZTK	970	9a-12p
Kenai-Soldotna	KPEN	101.7	8a-11a	Barstow	KSZL	1230	9a-12n
Ketchikan	KTKN	930	8a-11a	Buckley	KNZR	1560	11a-2p
Alabama				Chico	KPAY	1290	9a-12n
Athens	WVNN	770	11a-2p	El Centro	KAMP	1430	9a-12n
Birmingham	WERC	960	12n-3p	Eureka	KINS	980	9a-12n
Brewton	WEBJ	1240	11a-2p	Fresno	KMJ	580	9a-12n
Cullman	WKUL	92.1	1p-4p	Ft. Bragg	KDAC	1230	9a-12n
Dothan	WWNT	1450	11a-2p	Grass Valley	KNCO	830	9a-12n
Ft. Payne	WFPA	1400	11a-2p	Lake Isabella	KQAB	1140	9a-12n
Gadsden	WAAX	570	12n-3p	Lancaster	KIIS	1220	12n-3p
Greenville	WGYV	1380	11a-2p	Lompoc	KTME	1410	9a-12n
Grove Hill	WFOW	106.1	11a-2p	Los Angeles	KFI	640	9a-12n
Guntersville	WGSV	1270	11a-2p	Marysville	KMYC	1410	9a-12n
Guntersville	WTW	95.9	11a-2p	Merced	KYOS	1480	9a-12n
Mobile	WNTM	710	11a-2p	Modesto	KFIV	1360	9a-12n
Monroeville	WVMA	930	6a-8a	Monterey Area	KSCO	1080	9a-12n
Montgomery	WLWI	1440	11a-2p	Mt. Shasta	KMJC	620	9a-12n
Opelika	WANI	1400	11a-2p	Palm Springs	KPSI	920	9a-12n
Selma	WHBB	1490	11a-2p	Paso Robles	KPRL	1230	9a-12n
Troy	WTBF	970	11a-2p	Quincy	KPCO	1370	9a-12n
Tuscaloosa	WTBC	1230	11a-2p	Quincy	KSPY	100.3	9a-12n
Arkansas				Redding	KQMS	1400	9a-12n
Cabot	KKRN	102.5	12n-3p	Sacramento	KFBK	1530	9a-12n & 10p-1a
Clinton	KGFL	1110	11a-2p	San Diego	KOGO	600	9a-12n
Dardanelle	KCAB	980	11a-2p	San Francisco	KSFO	560	9a-12n
El Dorado	KELD	1400	11a-2p	San Luis Obispo	KVEC	920	1p-4p
Fayetteville	KFAY	1030	11a-2p	Santa Barbara	KTMS	990	9a-12n
Fort Smith	KWHN	1320	11a-2p	Santa Maria	KSMA	1240	9a-12n
Hope	KXAR	1490	11a-2p	Sonora	KVML	1450	9a-12n
Hot Springs	KZNG	1340	11a-2p	S Lake Tahoe	KOWL	1490	9a-12n
Humnoke	KARN	101.7	12n-3p	Ukiah	KUKI	1400	9a-12n
Jonesboro	KBTM	1230	11a-2p	Victorville	KIXW	960	9a-12n
Little Rock	KARN	920	12n-3p	Yreka	KSYC	1490	9a-12n
Little Rock	KCAB	980	11a-2p	**Colorado**			
Searcy	KWCK	1300	11a-2p	Basalt	KNFO	106.1	10a-1p
Cabot	KKRN	102.5	12n-3p	Colorado Spr	KVOR	740	10a-1p
Clinton	KGFL	1110	11a-2p	Cortez	KVFC	740	10a-1p
Arizona				Denver	KOA	850	12n-3p
Bullhead City	KFLG	1000	10a-1p	Fort Collins	KCOL	600	10a-1p
Flagstaff	KAFF	930	10a-1p	Fort Morgan	KSIR	1010	10a-12n/1p-2p
Kingman	KAAA	1230	10a-1p	Grand Jct	KNZZ	1100	10a-1p
Parker	KLPZ	1380	10a-1p	Greeley	KFKA	1310	10a-12n/1p-2p
Phoenix	KFYI	550	10a-1p	Pueblo	KCSJ	590	10a-1p
Prescott	KYCA	1490	10a-1p	Salida	KVRH	1340	10a-1p
Safford	KATO	1230	10a-1p	Steamboat Spr	KRMR	107.3 100.5	10a-1p
Show Low	KVSL	1450	10a-1p				
Sierra Vista	KTAN	1420	10a-1p				
Tucson	KNST	790	10a-1p				
Yuma	KBLU	560	10a-1p				

Rush Limbaugh

Location	Sta	Freq	Airtime	Location	Sta	Freq	Airtime
Connecticut				**Hawaii**			
Groton	WSUB	980	12n -3p	Hilo	KPUA	670	10a-1p
Hartford	WTIC	1080	12n-3p	Honolulu	KHVH	830	10a-1p
New Haven	WELI	960	12n-3p	Wailuku	KAOI	1110	9a-12n
D of C				**Idaho**			
Washington	WMAL	630	12n-3p	Boise	KIDO	580	10a-1p
Delaware				Bonners Ferry	KBFI	1450	9a-12n
Dover	WDOV	1410	12n-3p	Idaho Falls	KID	590	10a-1p
Rehoboth	WGMD	92.7	12n-3p	Pocatello	KWIK	1240	10a-1p
Wilmington	WILM	1450	12n-3p	Rupert	KBAR	970	10a-1p
Florida				Sandpoint	KSPT	1400	9a-12n
Arcadia	WFLN	1480	12n-3p	Twin Falls	KLIX	1310	10a-1p
Daytona Beach	WNDB	1150	12n-3p	**Illinois**			
Fort Myers	WINK	1240	12n-3p	Carbondale	WCIL	1020	11a-2p
Fort Pierce	WJNX	1330	12p-3p	Champaign	WDWS	1400	11a-2p
Fort Pierce	WJNX	1330	12n-3p	Chicago	WLS	890	11a-2p
Ft Walton Beach	WFTW	1260	11a-2p	Danville	WDAN	1490	11a-2p
Gainesville	WSKY	97.3	12n-3p	Decatur	WSOY	1340	11a-2p
Jacksonville	WOKV	690	12n-3p	Effingham	WCRA	1090	11a-2p
Lakeland	WLKF	1430	12n-3p	Fairfield	WFIW	1390	1p-4p
Marathon	WFFG	1300	12n-3p	Galesburg	WGIL	1400	12n-3p
Melbourne	WMMB	1240	12n - 3p	Herrin	WJPF	1340	11a-2p
Miami	WIOD	610	12n-3p	Kewanee	WKEI	1450	2p-5p
Naples	WNOG	1270	12n-3p	Mt. Vernon	WMIX	940	11a-2p
Ocala	WOCA	1370	12n-3p	Peoria	WMBD	1470	11a-2p
Orlando	WFLA	540	12n-3p	Quincy	WTAD	930	11a-2p
Panama City	WLTG	1430	11a-2p	Rockford	WROK	1440	11a-2p
Punta Gorda	WCCF	1580	12n-3p	Springfield	WTAX	1240	12n-3p
Sarasota	WSPB	1450	12n-3p	**Indiana**			
Sebring	WWTK	730	12n-3p	Anderson	WHBU	1240	12n-3p
St. Augustine	WFOY	1240	12n-3p	Bedford	WBIW	1340	12n-3p
Tallahassee	WNLS	1270	6a-9a	Bloomington	WGCL	1370	1p-4p
Tallahassee	WNLS	1270	12n-3p	Centerville	WHON	930	12n-3p
Tampa	WFLA	970	12n-3p	Columbus	WCSI	1010	12n-3p
Vero Beach	WAXE	1370	12p-3p	Evansville	WGBF	1280	11a-2p
W. Palm Beach	WJNO	1290	12n-3p	Ft. Wayne	WOWO	1190	1p-4p
Georgia				Indianapolis	WIBC	1070	1p-4p
Albany	WALG	1590	12n-3p	Jasper	WITZ	990	12n-3p
Athens	WGAU	1340	12n-3p	Kokomo	WIOU	1350	12n-3p
Atlanta	WGST	640	12n-3p	Logansport	WSAL	1230	1p-4p
Augusta	WGAC	580	12n-3p	Marion	WGOM	860	12n-3p
Brunswick	WGIG	1440	12n-3p	South Bend	WSBT	960	12n-3p
Columbus	WDAK	540	12n-3p	Tell City	WTCJ	1230	12n-3p
Dalton	WBLJ	1230	12n-3p	Vincennes	WAOV	1450	9a-10a
Douglas	WDMG	860	12n-3p	**Iowa**			
Gainesville	WGGA	1240	12n-3p	Cedar Rapids	WMT	600	1p-4p
Greensboro	WDDK	103.9	12n-3p	Davenport	WOC	1420	11a-2p
Helen	WHEL	105.1	12n-3p	Des Moines	WHO	1040	1p-4p
LaGrange	WGSE	720	12n-3p	Dubuque	WDBQ	1490	11a-2p
Lakeland	WVGA	1001	11a-2p	Estherville	KILR	1070	11a-2p
Macon	WMAC	940	12n-3p	Fort Dodge	KVFD	1400	11a-2p
Rome	WLAQ	1410	12n-3p	FT. Madison	KBKB	1360	11a-2p
Savannah	WTKS	1290	12n-3p	Mason City	KGLO	1300	1p-4p
Statesboro	WWNS	1240	12n-3p	Sheldon	KIWA	1550	1p-4p
Tifton	WTIF	107.5	1p-4p	Sheldon	KIWA	105.3	1p-4p
Tifton	WTIF	1340	1p-4p	Sioux City	KSCJ	1360	11a-2p
				Waterloo	KXEL	1540	11a-2p

Rush Limbaugh

Location	Sta	Freq	Airtime	Location	Sta	Freq	Airtime
Kansas				**Michigan**			
Chanute	KKOY	1460	11a-2p	Muskegon	WKBZ	1520	12n-3p
Garden City	KBUF	1030	11a-12n 1p-3p	Adrian	WABJ	1490	12n-3p
Hutchinson	KWBW	1450	11a-12n, 1p-3p	Alpena	WATZ	1450	12n-3p
Kansas City	KMBZ	980	11a-2p	Battle Creek	WBCK	930	12n-3p
Liberal	KSCB	1270	11a - 2p	Benton Harbor	WSJM	1400	12n-3p
Manhattan	KMAN	1350	11a-2p	Big Rapids	WBRN	1460	12n-3p
Salina	KSAL	1150	1p-4p	Cadillac	WATT	1240	12n-3p
Topeka	KMAJ	1440	11a-2p	Detroit	WJR	760	12p-3p
Wichita	KNSS	1240	11a-2p	Escanaba	WCHT	600	12n-3p
Kentucky				Grand Rapids	WOOD	1300	12n-3p
Bowling Green	WKCT	930	11A-2P	Grand Rapids	WMUS	1090	12n-3p
Corbin	WKDP	1330	12n-3p	Greenville	WPLB	1380	12n-3p
Hopkinsville	WHOP	1230	11a-2p	Hancock	WMPL	920	12n-3p
Lexington	WLAP	630	12n -3p	Iron Mountain	WMIQ	1450	11a-2p
Louisville	WHAS	840	12n-3p	Ironwood	WJMS	590	11a-2p
Madisonville	WTTL	1310	11a-2p	Ishpeming	WIAN	1240	12n-3p
Mayfield	WNGO	1320	11a-2p	Jackson	WKHM	970	12n-3p
Owensboro	WOMI	1490	11a-2p	Kalamazoo	WKMI	1360	12n -3p
Paducah	WKYX	570	11a-2p	Lansing	WJIM	1240	12n-3p
Prestonsburg	WDOC	1310	12n-3p	Ludington	WKLA	1450	12n-3p
Somerset	WSFC	1240	1p-4p	Manistee	WMTE	1340	12n-3p
Louisiana				Marquette	WDMJ	1320	12n-3p
Alexandria	KSYL	970	11a-2p	Mt. Pleasant	WMMI	830	6a-9a
Baton Rouge	WJBO	1150	11a-2p	Mt. Pleasant	WCEN	1150	12n-3p
Global/short wav	WRNO	15.420	11a-2p	Muskegon	WMUS	1090	6a-8a
Lafayette	KPEL	105.1	11a-2p	Saginaw	WSGW	790	12n-3p
Lake Charles	KAOK	1400	11a-2p	Sandusky	WMIC	660	2p-5p
Lake Charles	KAOK	97.9	11a-2p	S. Ste Marie	WKNW	1400	12n-3p
Monroe	KMLB	1440	11a-2p	Traverse City	WTCM	580	1p-4p
Natchitoches	KNOC	1450	11a-2p	**Minnesota**			
New Orleans	WWL	870	11a-2p	Alexandria	KXRA	1490	2p-5p
Shreveport	KEEL	710	11a-2p	Austin / Rochester	KNFX	970	11a-2p
Massachussetts				Bemidji	KKBJ	1360	11a-2p
Boston	WRKO	680	12n-3p	Brainerd	WWWI	1270	11a-2p
Pittsfield	WBEC	1430	12n-3p	Duluth	WEBC	560	11a-2p
Springfield	WHYN	560	12n-3p	Fergus Falls	KBRF	1250	1p-4p
West Yarmouth	WXTK	94.9	12n-3p	Minneapolis	KSTP	1500	12p-3p
Worcester	WTAG	580	12n - 3p	Park Rapids	KPRM	870	11a-2p
Maryland				St. Cloud	KNSI	1450	11a-2p
Baltimore	WBAL	1090	12n-3p	Staples/Wad'na	KSKK	94.7	11a-2p
Frederick	WFMD	930	12n - 3p	Willmar	KDJS	1540	11a-2p
Frostburg	WFRB	560	12n-3p	Winona	KWNO	1230	11a-12n 1p-3p
Hagerstown	WHAG	1410	12n-3p	Worthington	KWOA	730	1p-4p
Lexington Park	WMDM	1690	12n-3p	**Missouri**			
Salisbury	WICO	1320	12n-3p	C. Girardeau	KZIM	960	11a-2p
Maine				Columbia	KFRU	1400	11a-2p
Boothbay Harbor	WCME	96.7	12n-3p	Farmington	KREI	800	11a-2p
Howland-Bangor	WVOM	103.9	12n-3p	Hannibal	WTAD	930	11a-2p
Lewiston	WTME	1240	12n-3p	Jefferson City	KWOS	950	11a-2p
Monticello	WREM	710	12n-3p	Joplin	KQYX	1560	11a-2p
Portland	WGAN	560	12n-3p	Kansas City	KMBZ	980	11a-2p
Presque Isle	WEGP	1390	12n-3p				
Waterville	WHQO	107.9	12n-3p				

Rush Limbaugh

Location	Sta	Freq	Airtime	Location	Sta	Freq	Airtime
Missouri				**North Carolina**			
Moberly	KWIX	1230	11a-2p	Statesville	WSIC	1400	12n-3p
Osage Beach	KRMS	1150	11a-2p	Washington	WDLX	930	12n-3p
Poplar Bluff	KWOC	930	11a-2p	Whiteville	WTXY	1540	12n-3p
Sedalia	KSIS	1050	11a-2p	Wilmington	WAAV	980	12n-3p
Sikeston	KSIM	1400	11a-2p	Winston-Salem	WSJS	600	12n-3p
Springfield	KWTO	560	11a-2p	**North Dakota**			
St. James	KTTR	99.7	11a-2p	Bismarck	KXMR	710	11a-2p
St. Joseph	KFEQ	680	2p-5p	Jamestown	KQDJ	1400	11a-2p
St. Louis	KMOX	1120	11a-2p	Minot	KRRZ	1390	11a-2p
Thayer	KALM	1290	11a-2p	**Nebraska**			
Tupelo	WKMQ	1060	6a-8a	Columbus	KTTT	1510	11a-2p
Tupelo	WKMQ	1060	11a-2p	Cozad	KAMI	1580	11a-2p
Waynesville	KJPW	1390	11a-2p	Kearney	KGFW	1340	11a-2p
West Plains	KWPM	1450	11a-2p	Lincoln	KLIN	1400	11a-2p
Willow Springs	KUKU	1330	11a-2p	McCook	KSWN	93.9	11a-2p
Mississippi				Norfolk	WJAG	780	1p-4p
Columbia	WCJU	1450	11a-2p	North Platte	KODY	1240	11a-2p
Greenwood	WABG	960	11a-2p	Omaha	KFAB	1110	11a-2p
Jackson	WJNT	1180	11a-2p	Omaha	KFAB	1110	11a-2p
McComb	WAPF	980	11a-2p	Scottsbluff	KOLT	1320	10a-1p
Meridian	WMOX	1010	11a-2p	**N. Hampshire**			
Natchez	WNAT	1450	11a-2p	Exeter	WGIP	1540	12n-3p
Philadelphia	WHOC	1490	11a-2p	Hanover	WTSL	1400	12n-3p
Vicksburg	WQBC	1420	11a-2p	Keene	WKBK	1220	12n-3p
Montana				Laconia	WEMJ	1490	12n-3p
Billings	KBUL	970	10a-1p	Lebanon	WTSL	1400	12n - 3p
Bozeman	KMMS	1450	10a-1p	Manchester	WGIR	610	12n-3p
Butte	KXTL	1370	10a-1p	Rochester	WGIN	930	12n-3p
Hamilton	KLYQ	1240	10a-1p	**New Jersey**			
Helena	KBLL	1240	10a-1:30p	Pleasantville	WOND	1400	12p-3p
Libby	KTNY	101.7	10a-1p	**New Mexico**			
Missoula	KGVO	1290	10a-1p	Alamogordo	KRSY	1230	10a-1p
Whitefish	KJJR	880	10a-12n 1p-2p	Albuquerque	KKOB	770	10a-1p
North Carolina				Artesia	KSVP	990	10a-1p
Albemarle	WSPC	1010	12n-3p	Bayard	KNFT	950	10a-1p
Asheville	WTZY	880	12n-3p	Farmington	KENN	1390	10a-1p
Atlantic	WTKF	107.3	12n-3p	Hobbs	KYKK	1110	10a-1p
Blowing Rock	WXIT	1200	12n-3p	Las Cruces	KOBE	1450	10a-1p
Brevard	WSQL	1240	12n-3p	Roswell	KBIM	910	10a-1p
Burlington	WSML	1200	12n-3p	Ruidoso Dwns	KRUI	1490	10a-1p
Charlotte	WBT	1110	12n-3p	**Nevada**			
Chester	WBT	99.3	12n-3p	Elko	KTSN	1340	9a-12n
Elizabeth City	WGAI	560	12n-3p	Ely	KELY	1230	9a-12n
Fayetteville	WFNC	640	12n-3p	Las Vegas	KXNT	840	9a-12n
Goldsboro	WGBR	1150	12n-3p	Reno	KKOH	780	9a-12n
Hendersonville	WHKP	1450	12n-3p	Elko	KTSN	1340	9a-12n
Hickory	WHKY	1290	12n-3p	**New York**			
Jacksonville	WJNC	1240	12n-3p	Albany	WGY	810	12n-3p
Raleigh-Durham	WPTF	680	12n-3p	Binghamton	WNBF	1290	12n-3p
Rutherfordton	WCAB	590	12n-3p	Buffalo	WBEN	930	12n-3p
				Elmira	WWLZ	820	12n-3p
				Hornell	WLEA	1480	12n - 3p
				Massena	WYBG	1050	12n-3p
				Middletown	WALL	1340	12n-3p
				New York	WABC	770	12n-3p

Rush Limbaugh

Location	Sta	Freq	Airtime	Location	Sta	Freq	Airtime
New York				**Oregon**			
Ogdensburg	WSLB	1400	12n-3p	Roseburg	KQEN	1240	9a-12n
Olean	WMNS	1360	12n - 3p	Salem	KYKN	1430	9a-12n
Poughkeepsie	WEOK	1390	12n-3p	The Dalles	KACI	1300	9a-12n
Poughkeepsie	WGHQ	920	12p-3p	**Pennsylvania**			
Rochester	WHAM	1180	2p-5p	Allentown	WAEB	790	12n-3p
Syracuse	WSYR	570	1p-4p	Altoona	WRTA	1240	12n-3p
Utica	WIBX	950	12n-3p	Bellefonte	WBLF	970	6a-8a
Watertown	WTNY	790	12n-3p	Bellefonte	WBLF	970	12n-3p
Jamestown	WKSN	1340	12p-3p	Bellefonte	WBLF	970	12n-3p
Ohio				Erie	WLKK	1400	12n-3p
Akron	WHLO	640AM	12-3p	Harrisburg	WHP	580	12n-3p
Ashtabula	WFUN	970	12n-3p	Johnstown	WNTJ	1490	12n-3p
Chillicothe	WBEX	1490	12n-3p	Lewistown	WIEZ	670	12n-3p
Cincinnati	WKRC	550	12n-3p	Meadville	WMGW	1490	12n-3p
Cleveland	WTAM	1100	12n-3p	Milton	WMLP	1380	12n -3p
Columbus	WTVN	610	1p-4p	Mt. Pocono	WILP	1300	12n-3p
Dayton	WHIO	1290	12n-3p	Oil City	WOYL	1340	12n-3p
Defiance	WONW	1280	12n-3p	Philadelphia	WPHT	1210	12n-3p
Findlay	WFIN	1330	12n-3p	Pittsburgh	WPGB	104.7	12p-3p
Ironton	WIRO	1230	12n-3p	Pottsville	WPPA	1360	12n-3p
Ironton	WIRO	1230	6a-8a	Reading	WEEU	850	12n-3p
Lima	WIMA	1150	1p-4p	Scranton	WGBI	910	12n-3p
Mansfield	WMAN	1400	12n-3p	State College	WMAJ	1450	12n-3p
Marion	WMRN	1490	1p-4p	Wilkes-Barre	WILK	980	12n-3p
Steubenville	WSTV	1340	12n-3p	Williamsport	WRKK	1190	12n - 3p
Toledo	WSPD	1370	12n-3p	Williamsport	WRAK	1400	12n - 3p
Wheeling	WOMP	1290	12n-3p	York	WSBA	910	12n-3p
Wooster	WKVX	960	12n-3p	**Rhode Island**			
Oklahoma				Providence	WPRO	630	12n-3p
Ardmore	KVSO	1240	11a-2p	**So. Carolina**			
Bartlesville	KWON	1400	11a-2p	Anderson	WAIM	1230	12n-3p
Blackwell	KOKB	1580	11a-2p	Bamberg	WWBD	92.1	12n-3p
Enid	KGWA	960	11a-2p	Charleston	WSCC	94.3	12n-3p
Oklahoma City	KTOK	1000	1p-4p	Columbia	WVOC	560	12n-3p
Perry	KOKP	1020	11a-2p	Florence	WJMX	970	12n-3p
Stillwater	KSPI	780	1p-4p	Greenville	WYRD	1330	12n-3p
Tulsa	KRMG	740	1p-4p	Greenwood	WCRS	1450	12n-3p
Woodward	KWFX	93.5	11a-2p	Kingstree	WDKD	1310	12n - 3p
Oregon				Myrtle Beach	WRNN	94.5	12n-3p
Astoria	KAST	1370	9a-12n	Spartanburg	WORD	910	12n-3p
Baker City	KBKR	1490	9a-12n	Sumter	WDXY	1240	12n-3p
Bend	KBND	1110	9a-12n	**South Dakota**			
Coquille	KWRO	630	9a-12n	Pierre	KGFX	1060	1p-4p
Enterprise	KWVR	1340	9a-12n	Rapid City	KOTA	1380	10a-1p
Eugene	KPNW	1120	9a-12n	Sioux Falls	KELO	1320	11a-2p
Grants Pass	KAJO	1270	1p-4p	Watertown	KSDR	1480	11a-2p
Klamath Falls	KAGO	1150	9a-12n	**Tennessee**			
La Grande	KLBM	1450	9a-12n	Athens	WYXI	1390	12n-3p
Medford	KCMX	580	9a-12n	Chattanooga	WGOW	1150	12n-3p
Medford	KMED	1440	9a-12n	Columbia	WKRM	1340	11a-2p
Milton-Freewater	KTEL	1490	9a-12n	Cookeville	WPTN	780	11a-2p
Ontario	KSRV	1380	10a-1p	Crossville	WAEW	1330	1p-4p
Pendleton	KTIX	1240	9a-12n	Jackson	WTJS	1390	11a-2p
Pendleton	KUMA	1290	1p-4p				
Portland	KEX	620	9a-12n				
Portland	KEX	1190	9a-12p				

Rush Limbaugh

Location	Sta	Freq	Airtime	Location	Sta	Freq	Airtime
Tennessee				**Virginia**			
Johnson City	WJCW	910	12n-3p	Lexington	WREL	1450	12n-3p
Knoxville	WNOX	100.3	12n-3p	Norfolk	WNIS	790	12n-3p
Martin	WCMT	1410	11a-2p	Onley	WESR	1330	12n-3p
McMinnville	WAKI	1230	11a-2p	Richmond	WRVA	1140	12n-3p
Memphis	WREC	600	11a-2p	Roanoke	WFIR	960	12n-3p
Nashville	WLAC	1510	12n-3p	Waynesboro	WINF	970	12n-3p
Texas				White Stone	WNDJ	104.9	12n - 3p
Abilene	KWKC	1340	11a-2p	Winchester	WINC	1400	12n-3p
Abilene	KXYL	104.7	11a-2p	**Vermont**			
Abilene	KXYL	104.7	6a-8a	Barre	WSNO	1450	12n-3p
Amarillo	KGNC	710	11a-2p	Burlington	WKDR	960	12n-3p
Austin	KLBJ	590	12n-3p	Plattsburgh	WEAV	960	12n-3p
Beaumont	KLVI	560	11a-2p	Rutland	WSYB	1380	12n-3p
Big Spring	KBST	1490	11a-2p	**Washington**			
Brenham	KWHI	1280	1p-4p	Bellingham	KGMI	790	9a-12n
College Station	WTAW	1620	11a-2p	Centralia	KELA	1470	9a-12n
Corpus Christi	KRYS	1360	11a-2p	Ellensburg	KXLE	1240	9a-12n
Dallas	WBAP	820	12n-3p	Moses Lake	KBSN	1470	2p-5p
Edinburg	KURV	710	11a-2p	Omak	KOMW	680	9a-12n
El Paso	KTSM	690	10a -1p	Pasco	KFLD	870	9a - 12n
Fredericksburg	KNAF	910	1p-4p	Port Angeles	KONP	1450	9a-12n
Houston	KPRC	950	11a-2p	Pullman	KQQQ	1150	9a-12p
Laredo	KDOS	1490	11a-2p	Seattle	KTTH	770	9a-12n
Longview	KEES	1430	11a-2p	Wenatchee	KPQ	560	9a-12n
Lubbock	KFYO	790	11a-2p	Yakima	KIT	1280	9a-12n
Midland	KCRS	550	6a-8a	**Virginia**			
Midland	KCRS	550	11a-2p	Beckley	WWNR	620	12n-3p
Mt. Pleasant	KIMP	960	11a-2p	Bluefield	WHIS	1440	12n-3p
Nacogdoches	KSFA	860	11a-2p	Charleston	WCHS	580	12n-3p
Plainview	KKYN	1090	2p-5p	Clarksburg	WHAR	1340	12n-3p
San Angelo	KKSA	1260	11a-2p	Huntington	WVHU	800	12n-3p
San Antonio	WOAI	1200	11a-2p	Lewisburg	WRON	1400	12n-3p
San Antonio	KAMG	1340	11a-2p	Logan	WVOW	1290	12n-3p
Temple	KTEM	1400	11a-2p	Morgantown	WAJR	1440	12n-3p
Texarkana	KTFS	940	11a-2p	Parkersburg	WLTP	1450	12n-3p
Tyler	KTBB	600	1p-4p	Wheeling	WWVA	1170	12n-3p
Waco	KWTX	1230	11a-2p	**Wyoming**			
Wichita Falls	KWFS	1290	11a-2p	Casper	KTWO	1030	10a-1p
Utah				Cheyenne	KGAB	650	10a-1p
Cedar City	KSUB	590	10a-1p	Cody	KODI	1400	10a-1p
Logan	KVNU	610	10a-1p	Gillette	KIML	1270	10a-1p
Price	KOAL	750	10a-1p	Laramie	KOWB	1290	10a-12n 1p-2p
Richfield	KSVC	980	1p-4p	Rock Springs	KRKK	1360	6AM-8AM
Salt Lake City	KNRS	570	10a-1p	Sheridan	KWYO	1410	10a-1p
St. George	KDXU	890	10a-1p				
Vernal	KVEL	920	10a-1p				
Virginia							
Blacksburg	WFNR	710	12n-3p				
Charlottesville	WCHV	1260	12n-3p				
Danville	WBTM	1330	12n-3p				
Frederickburg	WFVA	1230	12n-3p				
Harrisonburg	WKCY	1300	12n-3p				
Harrisonburg	WKCY	1300	12n-3p				

Bill O'Reilly - The Radio Factor   M-F

Alabama	Stat	Freq	Time	Connecticut	Stat	Freq	Time
Anniston	WHMA	1390	11a-1p	Hartford	WDRC	1360	3p- 5p
Birmingham	WAPI	1070	11a-1p	Hartford/New Haven	WMMW	1470	N - 2p
Cullman	WFMH	1340	1p-3p	Hartford/Torrington	WSNG	610	N - 2p
Dothan	WRJM	93.7	11a-1p	Hartford/Waterbury	WWCO	1240	N - 2p
Eufaula	WULA	1240	1p-3p	New London	WSUB	980	N - 2p
Florence	WBCF	1240	4p-6p	Delaware			
Mobile	WABB	1480	12p-2p	Hagerstown	WARK	1490	2p - 4p
Mobile/Evergreen	WPGG	93.3	11a-1p	D of C			
Mobile/Pensacola	WCOA	1370	9p-11p	DC/Cumberland	WCBC	1270	1p - 3p
Mobile/Pensacola	WNWF	1120	11a-1p	Washington DC	WJFK	106.7	1p - 3p
Montgomery	WLWI	1440	5p-7p	Florida			
Selma	WMRK	1340	11a-1p	Auburndale	WTWB	1570	12p-2p
Alaska				De Funiak Sprgs	WGTX	1280	12p-2p
Anchorage	KFQD	750	2p-4p	Ft. Myers	WPTK	1200	4p-6p
Fairbanks	KFAR	660	8a-10a	Gainesville	WBSY	99.5	6p-8p
Arizona				Jacksonville	WYMM	1530	12p-2p
Flagstaff	KVNA	600	9a-11a	Jupiter	WJBW	1000	9p-11p
Phoenix	KFNN	1510	9a-11a	Marathon	WFFG	1330	6p-8p
Phoenix	KAAA	1230	3p-5p	Mariana	WTYS	1340	12p-2p
Kingman	KZZZ	1490	3p-5p	Melbourne	WMEL	920	12p-2p
Lake Havasu	KNTR	980	1P-3P	Miami/Ft Lauder.	WFTL	850	9p-11p
Prescott	KYCA	1490	3P-5P	Panama City	WYOO	101.1	12p-2p
Tucson	KJLL	1330	5p-7p	Pinecastle	WAMT	1190	1p-3p
Yuma	KJOK	1400	11a-1p	Stuart	WSTU	1450	12p-2p
Arkansas				Tallahassee	WTAL	1450	2p-4p
Bald Knob	KAPZ	710	5p-7p	Tampa	WLSS	93o	1p-3p
Dardanelle	KCAB	980	11a-1p	Tampa	WWBA	1040	1p-3p
Eldorado	KLBQ	98.7	11a-1p	Titusville	WIXC	1060	12p-2p
Fayetteville	KFAY	1030	3p-5p	Venice	WENG	1530	12p-2p
Fort Smith	KLOX	94.5	1p-3p	Vero Beach	WTTB	1490	12p-2p
Heber Springs	KAWW	1370	5p-7p	W. Palm Beach	WDGA	1420	9p-11p
Hot Springs	KZNG	1340	5p-7p	Georgia			
Little Rock	KARN	920	10p12a	Albany	WALG	1590	3p-5p
Little Rock	KARN	102.5	10p12a	Atlanta	WALR	1340	12p-2p
Searcy	KWCK	1300	3p-5p	Atlanta	WCNN	680	12p-2p
California				Atlanta	WFOM	1230	12p-2p
Bakersfield	KNZR	1560	9a-11a	Augusta	WRDW	1630	12p-2p
Eureka	KWSW	790	9a-11a	Columbus	WRCG	1420	1p-3p
Los Angeles	KABC	790	9a-11a	Cuthbert	WCUG	850	12p-2p
LA/Ventura	KVTA	1520	9a-11a	Dalton	WBLJ	1030	10p12a
Palm Springs	KPSI	920	1p-3p	Dry beach	WMMR	1670	12p-2p
Palmdale	KUTY	1470	12p-2p	Macon	WNNG	1350	12p-2p
Redding	KQMS	1400	6p-8p	Savannah	WBMQ	630	12p-2p
Sacramento	KSTE	650	6p-8p	Trenton	WKWN	1420	12p-2p
San Diego	KFMB	760	10a12p	Waycross	WWGA	1230	6p-8p
San Francisco	KNEW	910	7p-9p	Hawaii			
Santa Barbara	KXTI	99.7	9a-11a	Honolulu	KHNR	97.5	10p12a
Santa Maria	KUHL	1440	9a-11a	Idaho			
Santa Maria	KTME	1410	9a-11a	Boise	KBOI	670	7p - 9p
So. Lake Tahoe	KTHO	590	7p-9p	Pocatello	KSLJ	690	10a - N
Susanville	KSUE	1240	12p-2p	Pocatello	KSSL	1260	10a - N
Watsonville	KOMY	1340	9a-11a	Illinois			
Colorado				Bloomington	WJBC	1230	10p12a
Colorado Sprgs	KVOR	740	9p-11p	Chicago	WCKG	105.9	10p12a
Canon City	KRLN	1400	10a12p	Danville	WDAN	1490	11a-1p
Denver	KHOW	630	12p-2p	Decatur	WSOY	1340	5p-7p
Brush	KSIR	1010	2p-4p	Effingham	WCRA	1090	5p-7p
Loveland	KSXT	1570	9a-11a	Geneseo	WGEN	1500	3p-5p
Greeley	KFKA	1310	1p-3p				

Bill O'Reilly - The Radio Factor M-F

Illinois	Sta	Freq	Time	Louisiana	Sta	Freq	Time
Bloomington	WJBC	1230	10p12a	Alexandria	KSYL	970	4p - 6p
Chicago	WCKG	105.9	10p12a	Baton Rouge	WIBR	1300	11a-1p
Danville	WDAN	1490	11a-1p	Lafayette	KPEL	105.1	5p - 7p
Decatur	WSOY	1340	5p-7p	Lake Charles	KAOK	1400	5p - 7p
Effingham	WCRA	1090	5p-7p	Monroe	KNOE	540	11a 1p
Geneseo	WGEN	1500	3p-5p	New Or./McComb	WHNY	1250	11a 1p
Joliet	WJOL	1340	1p-3p	New Orleans	WSMB	1350	2p - 4p
Kewanee	WKEI	1450	3p-5p	Shreveport	KRMD	1340	11a 1p
Metropolis	WMOK	920	11a-1p	Shreveport/Texarkan	KKTK	1400	11a 1p
Mt. Carmel	WVMC	1360	11a-1p	**Massachusetts**			
Quincy	WGAB	930	4p-6p	Boston	WTKK	96.9	1p - 3p
Rockford	WROK	1440	5p-7p	Boston/Gardner	WGAW	1340	12p 2p
Springfield	WGAX	1240	3p-5p	Springfield	WHMP	1400	3p - 5p
Sterling	WSDR	1240	11a-1p	Springfield	WHNP	1600	3p - 5p
Waukegan	WKRS	1220	11a-1p	Springfield	WHMQ	1240	3p - 5p
Indiana				**Michigan**			
Evansville	WGAB	1180	12p 2p	Charlevoix/Petoskey	WMKT	1270	12p-2
Ft. Wayne/Portland	WPGW	1440	1p- 3p	Detroit	WKRK	97.1	11p- 1a
Harrisburg/Lebanon	WLBR	1270	1p - 3p	Escanaba	WCHT	600	3p - 5p
Indianapolis	WIBC	1070	11p-1a	Flint	WLSP	1530	12p 2p
Marion/Indianapolis	WGOM	860	3p - 5p	Flint	WSGW	790	3p - 5p
South Bend	WSBT	960	2p- 4p	Grd Rapids Area	WKMI	1360	12p 2p
Terre Haute	WTAY	1570	1p - 3p	Greenville	WSCG	1380	4p - 6p
Vincennes	WAOV	1450	2p - 4p	Iron Mountain	WMIQ	1450	3p - 5p
Iowa				Ironwood	WJMS	590	11a- 1p
Davenport Area	KCPS	1150	5p-7p	Lansing	WJIM	1240	10p M
Davenport Area	KJOC	1170	11a-1p	Marquette	WDMJ	1320	3p - 5p
Des Moines	KRNT	1350	10p 12	Marquette	WIAN	1240	3p - 5p
Des Moines/Boone	KWBG	1590	1p-3p	Menominee	WAGN	1340	1p - 3p
Dubuque	WDBQ	1490	1p-3p	Mt. Pleasant	WMMI	830	3p - 5p
Keokuk	KOKX	1310	1p-3p	South Bend/St. Joe	WHFB	1060	12p 2p
Ottumwa	KLEE	1480	11a-1p	Traverse City	WTCM	580	5p - 7p
Shenandoah	KMA	960	2p-4p	Traverse City Area	WKLA	1450	3p - 5p
Sioux City	KSCJ	1360	12p-2p	Traverse City Area	WMTE	1340	3p - 5p
Spencer	KICD	1240	3p-5p	**Minnesota**			
Kansas				Austin	KAUS	1480	3p - 5p
Dodge City	KGNO	1370	11a-1p	Duluth	KDAL	610	9p 11p
Goodland	KLOE	730	10a 12	Fargo/Fosston	KKCQ	1480	11a- 1p
Iola	KALN	1370	11a 1p	Golden Valley	KYCR	1570	11a-1p
Kansas City	KMBZ	980	2p - 4p	Minneapolis/Pillager	KBKK	95.9	11a-1p
Lawrence	KLWN	1320	12p 2p	Minneapolis/St. Cld	WJON	1240	11a-1p
Manhattan	KMAN	1350	2p - 4p	Minneapolis/Stillwat'r	WMGT	1220	11a-1p
Topeka	KMAJ	1440	5p - 7p	Pillager	WWWI	95.9	11a-1p
Tulsa/Coffeyville	KGGF	690	1p - 3p	Rochester	KROC	1340	11a-1p
Wichita	KNSS	1240	2p - 4p	Waseca	KOWZ	1170	11a-1p
Wichita/Hutchinson	KWBW	1450	3p - 5p	**Mississippi**			
Kentucky				Lucedale	WRBE	1440	11a- 1p
Bowling Green	WKCT	930	10a - N	Lucedale	WRBE	106.9	4p - 6p
Cadiz	WKDZ	1110	1p - 3p	Meridian	WALT	910	12p- 2p
Drakesboro	WMSK	95.3	11a 1p	New Orleans	WHNY	1250	4p - 6p
Lexington	WLXO	96.1	N - 2p	Starkville	WKOR	980	5p - 7p
Lexington/London	WFTG	1400	N - 2p	Tupelo	WKMQ	1060	5p - 7p
Louisville	WTSZ	1600	2p -4p	**Missouri**			
Mayfield	WNGO	1320	5p - 7p	Cabool	KOZX	98.1	9p 11p
Murray	WNBS	1340	1p - 3p	Cape Girardeau	KAPE	1550	11a 1p
Owensboro	WOMI	1490	5p - 7p	Columbia	KSSZ	93.9	12p- 2p
Paducah	WKYX	570	4p - 6p	Columbus/Tupelo	WJWF	1400	5p - 7p

City	Call	Freq	Time
Missouri			
Festus	KJFF	1400	11a 1p
Jefferson City	KLIK	1240	11a- 1p
Lebanon	KLWT	1230	5p - 7p
Poplar Bluff	KWOC	930	2p - 4p
Springfield	KSGF	1260	11a 1p
Springfield	KTTR	1490	11a- 1p
St. Joseph's	KFEQ	680	4p - 6p
St. Louis	KFTK	97.1	11a 1p
St. Louis	KFMO	1240	11a 1p
Montana			
Bozeman	KBOZ	1090	10a12p
Great Falls	KQDI	1450	10a12p
Helena	KCAP	1340	7p - 9p
Kalispell	KOFI	1180	10a12p
Missoula	KYLT	1340	10a 2p
Nebraska			
Lincoln	KLIN	1400	4p - 6p
Lincoln/Grand Isle	KRGI	1430	12p-2p
North Platte	KODY	1240	2p - 4p
Omaha	KKAR	1290	5p - 7p
Scottsbluff	KOLT	1320	2p - 4p
New Hampshire			
Boston/Concord	WKXL	1450	3p - 5p
Boston/Dover	WTSN	1270	2p - 4p
Boston/Keene	WKBK	1290	6p - 8p
Boston/Manchester	WKBR	1250	12p-2p
New Mexico			
Albuquerque	KTBL	1050	10p 12
Albuquerque	KRSY	1230	2p - 4p
Albuquerque/Bayard	KNFT	950	2p - 4p
Farmington	KENN	1390	2p - 4p
Las Cruces	KOBE	1450	4p - 6p
Los Alamos	KRSN	1490	12p-2p
New York			
Binghamton	WYOS	1360	6p - 8p
Buffalo	WLVL	1340	1p - 3p
Elmira	WWLZ	820	6p - 8p
Glens Falls	WWSC	1450	12 - 2p
Ithaca/Syracuse	WKRT	920	12 - 2p
New York	WOR	710	2p - 4p
Newburgh	WGNY	1220	12 - 2p
Ithaca/Syracuse	WKRT	920	12 - 2p
New York	WOR	710	2p - 4p
Newburgh	WGNY	1220	12 - 2p
Rochester/Newark	WACK	1420	4p - 6p
Springfield	WSPQ	1330	4p - 6p
Syracuse	WFBL	1390	12 - 2p
Utica	WIBX	950	3p - 5p
Watertown	WTNY	790	4p - 6p
North Carolina			
Charlotte/Gastonia	WZRH	960	12 - 2p
Charlotte/Lenoir	WJRI	1340	12 - 2p
Charlotte/Salisbury	WSTP	1490	1p - 3p
Greensboro	WMFR	1230	12 - 2p
Greensboro	WSML	1200	10p12a
Greenville	WNCT	1070	12 - 2p
North Carolina			
Raleigh/Fayetteville	WFNC	640	5p - 7p
Raleigh/Fayetteville	WFNC	102.3	5p - 7p
Salisbury	WQMR	101.1	12 - 2p
Wilmington	WAAV	980	3p - 5p
Winston-Salem	WSJS	600	10p 2a
North Dakota			
Bismarck	KLXX	1270	11a 1p
Fargo/Fosston	KKCQ	1480	11a -1p
Jamestown	KQDJ	1400	11a -1p
West Fargo	KQWB	1660	11a -1p
Ohio			
Canton/Cleveland	WCER	900	4p - 6p
Cincinnati	WBOB	1160	9p 11p
Cleveland/Conneaut	WWOW	1360	1p- 3p
Columbus	WCLT	1430	3p - 5p
Dayton/Middletown	WPFB	910	12 - 2p
Lancaster	WLOH	1320	12 - 2p
Portsmouth	WPAY	1400	12 - 2p
Steubenville	WSTV	1340	1p - 3p
Toledo	WTOD	1560	12 - 2p
Toledo	WFOB	1430	4p - 6p
Youngstown	WPIC	790	12 - 2p
Oklahoma			
Cushing	KUSH	1600	11a-1p
Duncan	KPNS	1350	11a-1p
Magnum	KHIM	97.7	11a 1p
Oklahoma City	KOKC	1520	3p - 5p
Tulsa	KFAQ	1170	12 - 2p
Tulsa/Coffeyville	KGGF	690	1p - 3p
Oregon			
Bend	KBND	1110	3p- 5p
Coquille	KWRO	630	3p- 5p
Eugene	KUGN	590	3p - 5p
Medford	KCMX	880	9a 11a
Portland	KXL	750	9a 11a
Portland/Astoria	KAST	1370	1p - 3p
Portland/Lebanon	KGAL	1580	9a 11a
The Dalles	KACI	1300	3p - 5p
Pennsylvania			
Altoona	WFBG	1290	N - 2p
Bethlehem	WGPA	1100	2p - 4p
Buffalo/Bradford	WESB	1490	1p - 3p
Erie	WJET	1400	6p - 8p
Franklin	WFRA	1450	3p - 5p
Harrisburg/Lebanon	WLBR	1270	1p - 3p
Johnstown/St Coll	WRSC	1390	N - 2p
Johnstown/St Coll	WBLF	970	N - 2p
Lansford	WLSH	1410	N - 2p
Meadville	WMGW	1490	3p - 5p
Oil City	WOYL	1340	3p - 5p
Philadelphia	WPHT	1210	10p M
Pittsburgh	KDKA	1020	N - 2p
Punxsutawny	WECZ	1540	N - 2p
Titusville	WTIV	1230	3p - 5p

Pennsylvania				Virginia			
Wilkes Barre	WILK	980	6p - 8p	Abington	WFHG	980	N- 2p
Wilkes Barre/Hazlet	WOGY	1300	6p - 8p	Abington	WFHG	92.7	N- 2p
Wilkes Barre/Milton	WMLP	1380	6p - 8p	Buffalo Gap	WZXI	95.5	1p - 3p
Wilkes Barre/Scrant	WGBI	910	6p - 8p	Charlottesville	WCHV	1260	3p - 5p
Wilkes Barre/Will's	WWPA	1340	N - 2p	Glen Allen	WTOX	1480	N- 2p
Rhode Island				Norfolk	WNIS	790	6p - 8p
Providence	WPRO	630	10p M	Roanoke	WBRG	1050	N- 2p
Providence	WBSM	1420	N - 2p	Roanoke	WRIS	1410	N- 2p
South Carolina				Roanoke/Lexington	WREL	1450	4p - 6p
Char/Georgetown	WGTN	1400	N - 2p	Roanoke/Lynchburg	WFNR	100.7	6p - 8p
Charleston	WTMA	1250	1p - 3p	Roanoke/Martinsville	WHEE	1370	N- 2p
Columbia	WISW	1320	N - 2p	Washington			
Florence/Myrtle Bch	WQJM	1450	N - 2p	Oak Harbor	KWDB	1110	9a 11a
Fountain Inn	WFIS	1600	N - 2p	Port Angeles	KONP	1450	6p 8p
Greenville	WORD	910	6p - 8p	Seattle	KTTH	770	3p - 4p
Greenville	WYRD	1330	6p - 8p	Seattle	KTTH	770	7p-8p
Sumter	WDXY	1240	3p - 5p	Spokane	KGA	1510	9a-11a
South Dakota				Spokane	KGA	1510	9p-11p
Rapid City	KTOQ	1340	10a - N	Spokane/Colfax	KMAX	840	9a-11a
Sioux Falls	KSOO	1140	10a - N	Spokane/Ephrata	KULE	730	11a-1p
Sioux Falls/Huron	KIJV	1340	11a 1p	West Virginia			
Yankton	WNAX	570	4p - 6p	Charleston	WSWW	1490	4p - 6p
Tennessee				Charleston	WCHS	580	10p M
Chattanooga	WGOW	1150	9p-11p	Clarksbr/Buckhanon	WBUC	1460	N- 2pm
Harrogate	WRWB	740	N - 2p	Fairmont	WMMN	920	N- 2pm
Knoxville	WNOX	990	9p 11p	Huntington	WRVC	930	N- 2pm
Knoxville	WNOX	99.1	9p 11p	Logan	WVOW	1290	N- 2pm
Nashville	WWTN	99.7	10p M	Parkersburg	WVNT	1230	4p - 6p
Memphis	WXRZ	94.3	4P-6P	Ravenswood	WMOV	1360	N - 2p
Texas				Wheeling	WVLY	1370	N - 2p
Abilene	KSLI	1280	11a - 1	Wisconsin			
Amarillo	KGNC	710	5p - 7p	Chippewa Falls	WOGO	680	11a 1p
Austin	KJCE	1370	N - 2p	Eau Claire	WAYY	790	1p- 3p
Beaumont	KRCM	1380	11a - 1	Fort Atkinson	WFAW	940	1p - 3p
Beaumont	KOLE	1340	11a - 1	Green Bay	WNGB	1400	11a-1p
Corpus Christi	KEYS	1440	11a - 1	Green Bay/Oshk	WOSH	1490	9 - 11p
Dallas	KLIF	570	N - 2p	Green Bay/Ripon	WRPN	1600	1p - 3p
El Paso	KROD	600	N - 2p	Kewaunee	WAUN	92.7	11a 1p
Houston	KSEV	700	2p - 4p	Madison	WTDY	1670	11a-1p
Houston/Coll Sta	KWBC	1550	1p - 3p	Marshfield, WI	WDLB	1450	1p - 3p
Longview/Tyler	KFRO	1370	11a - 1	Menominee/Marinett	WAGN	1340	1p - 3p
Lubbock	KJTV	950	11a - 1	Menomonie	WMEQ	880	5p - 7p
Lufkin	KRBA	1340	5p - 7p	Milwaukee/Racine	WRJN	1400	11a 1p
Midland/Odessa	KWEL	1070	11a - 1	Rice Lake	WJMC	1240	1p - 3p
San Angelo	KGKL	960	11a - 1	Sparta	WKLJ	1290	4p - 6p
San Antonio	KTSA	550	10p-M	Wausau/Merrill	WJMT	730	1p - 3p
Temple	KTEM	1400	5p - 7p	Wyoming			
Utah				Cheyenne	KRAE	1480	10a - N
Salt Lake City	KSL	1160	9p - 11	Powell	KPOW	1260	10a - N
St. George	KZNU	1450	10a - N	Riverton	KVOW	1450	10a - N
Vermont							
Bennington	WBTN	1370	2p - 4p				
Boston/Brattleboro	WKVT	1490	6p - 8p				
Burlington	WVMT	620	1p - 3p				

The Clark Howard Show

ALABAMA

Anniston	WDNG	1450	5-6p
Geneva	WRJM	93.7	Sat. noon-3p
Huntsville/Athens	WVNN	770	8-9a; Sun. 3-5p
Mobile	WABB	1480	1-2p
Montgomery	WLWI	1440	6-8p
Troy	WTBF	970	3-6p
Tuscumbia	WVNA	1590	6-8a

ALASKA

Anchorage	KFQD	750	10a-noon

ARIZONA

Bullhead City	KZZZ	1490	8-10p
Kingman	KAAA	1230	8 - 10p
Phoenix	KTAR	620	10p-mid

CALIFORNIA

Arroyo Grande	KXTK	1280	10a-1p
Eureka	KGOE	1480	6-8p
Grass Valley	KNCO	830	2-4p
Lake Tahoe	KOWL	1490	2-4p
Modesto/Stockton	KTRB	860	11a-12p/ 1-2p
Porterville -	KTIP	1450	1-4p
San Diego	KFMB	760	Sat 10p-12a/Sun 12-4a
Santa Cruz	KSCO	1080	3-4p
Sonora	KVML	1450	5-8p; Sat. 3-6p

COLORADO

Grand Junction	KNZZ	1100	11p-1a

CONNECTICUT

Bridgeport -	WICC	600	1-4p
Hartford	WTIC	1080	7-10p
Stamford/Norwalk	WSTC/WNLK	1400/1350	9-10a M-W, 2-4p M-F

FLORIDA

Gainesville	WSKY	97.3	Sun., 6-9a, 3-6p
Jacksonville	WOKV	690	6-9p
Milton/Pensacola	WEBY	1330	1-3p
Orlando	WDBO	580	1-3p
Panama City	WY00	101.1	1-2p
Naples	WGUF	98.9 FM	tba

The Clark Howard Show

Sarasota	WIBQ	1220	5-7p
Tallahassee	WHBT	1410	1-3p
GEORGIA			
Albany	WALG	1590	Sat. 12-3p
Athens	WGAU	1340	6-8p
Atlanta	WSB	750	1-4p
Augusta	WGAC	580	6-8p
Carrollton	WLBB	1330	2-4p
Clayton	WGHC	1370	1-3p
Ellijay	WPGY	1560	2-4p
Macon	WMAC	940	6-7p
Rome	WRGA	1470	4-6p
Savannah	WBMQ	630	2-3p
Toccoa	WNEG	630	1-3p
Warrenton	WRFN	93.1	9-10a; 6-8p
ILLINOIS			
Crystal Lake	WAIT	850	4-5p
Fairfield	WFIW	1390	4-7p
Joliet	WJOL	1340	2-4p
Quincy -	WGEM	1440	11a-noon; 1-2p
Springfield	WTAX	1240	7-9p
Streator	WSPL	1250	3-6p; Sat. 11-noon
INDIANA			
Bloomington	WGCL	1370	9-11p
Elkhart New!	WGBF	1340	7-10p
Fort Wayne	WOWO	1190	Sat. 6-7p; Sun. 1-3,8-9p
Indianapolis	WIBC	1070	Sun 1-4p
Princeton	WRAY	1250	4-6p/Sat. 1-4p
IOWA			
Davenport - New!	KJOC	1170	3-6p
Ottumwa - New!	KBIZ	1240	9a-noon
Marshalltown	KFJB	1230	2-3p/6-7p
Spencer	KICD	1240	2-4p
KENTUCKY			
Lexington	WVLK	590	12-2p
LOUISIANA			
Jennings	KJEF	1290	12-3p

The Clark Howard Show

Lake Charles	KLCL	1470	12-3p
Monroe	KNOE	540	11a-noon
New Orleans	WSMB	1350	9-11a
Shreveport	KRMD	1340	Sat. 11a-2p
MAINE			
Bangor	WABI	910	1-3p, 6-7p
MARYLAND			
Salisbury	WICO	1320	5-7p
MASSACHUSETTS			
Springfield	WHMP	1400	Sat. 2-4p
MICHIGAN			
Detroit - New!	WKRK	97.1	Sat. 4-7p
Flint/Saginaw Bay	WSGW	790	6-9p
Kalamazoo	WKZO	590	Sat. 1-6p
Kalamazoo -	WKLZ	1470	5-7p
Traverse City	WTCM	580	6-8p
MINNESOTA			
Rochester	KROC	1340	Sat. 1-3p
Stillwater	WMGT	1220	1-4p
St. Paul	WFMP	107.1	9p-12a
MISSISSIPPI			
Biloxi	WBUV	104.9FM	2-5p
Columbus	WJWF	1400	1-3p
Greenville	WGVM	1260	1-3p
McComb	WHNY	1250	2-4p
Natchez	WNAT	1450	3-5p
Starkville	WSSO	1230	1-3p
MISSOURI			
Ash Grove	KZRQ	104.1	10p-1a
Cabool -	KOZX	98.1	noon-2p, Sat. 9-noon
Columbia	KFRU	1400	7-9p, Sun. 12-3p
Jefferson City	KLIK	1240	6-9p; Sat. 3-6p
Kansas City	KMBZ	980	Sun. 7-9p
Springfield	KSGF	1260	1-4p
MONTANA			
Laurel	KBSR	1490	8-10p
NEW HAMPSHIRE			

217

The Clark Howard Show

Keene	WKBK	1290	8-10p
NEW JERSEY			
New Brunswick	WCTC	1450	2-4p
NEW YORK			
Warwick	WTBQ	1400	2-4p
Auburn	WAUB	1590	3-4p/6-7p
Binghamton	WEBO	1330	7-9a
Canadaigua	WCGR	1550	3-4p, 6-7p
Geneva	WGVA	1240	3-4p, 6-7p
Ithaca	WHCU	870	10p-12a
Oswego	WEBO	1330	1-3p
Seneca Falls	WSFW	1590	3-4p, 6-7p
Utica	WIBX	950	Sat. 12-3p
Dundee	WFLR	1570	2-4p
NEVADA			
Las Vegas	KNUU	970	6-8p
NORTH CAROLINA			
Charlotte	WBT	1110	Sun. 7-9p
Charlotte	WJRI	1340	5-7p
Greensboro	WZTK	101.1	1-3p
Raleigh/Durham	WPTF	680	9a-12p, Sat. noon-3p
Wilmington	WLTT	106.3/103.9	1-3p, Sat. noon-3p
NORTH DAKOTA			
Lisbon -	KQLX	570	1-4p
OHIO			
Bellaire	WOMP	1290	5-7p
Dayton/Springfield	WHIO	1290	6-9p, Sun. 6-10p
Elyria	WEOL	930	1-4p
Heath/Newark	WHTH	790	2-3p, 6-7p
Steubenville	WSTV	1340	5-7p
OKLAHOMA			
Oklahoma City	KOMA	1520	1-3p
Tulsa	KRMG	740	7-10p, Sun 3-5a
OREGON			
Astoria	KAST	1370	Sat. 9a-noon
Bend	KBND	1110	3-5p

The Clark Howard Show

Eugene	KUGN	590	6-9p
Phoenix	KCMX	880	10p-1a
Portland	KPAM	860	9a-noon
PENNSYLVANIA			
Landsdale	WNPV	1440	3-5p
Sharon	WPIC	790	2-5p
	WPTT	1360	6-10p
SOUTH CAROLINA			
Charleston	WTMA	1250	Sat. 2-4p, Sun 1-2p
Fountain Inn	WFIS	1600	1-4p
Honea Path	WRIX	103.1	1-4p
Myrtle Beach	WQJM	1450	2-3p
Socastee	WRNN	99.5	5-7p
Union -	WBCU	1460	1-3p
Walhalla	WWOF	1000	1-4p
TENNESSEE			
McMinnville	WAKI	1230	11a-noon
Murfreesboro	WGNS	1450	noon-3p
TEXAS			
Austin	KLBJ	590	7-10p
Pasadena	KIKK	650	Sun. 1-4p
Plainview	KVOP	1400	3-6p
Waco-Marlin	KBBW	1010	8-10p
UTAH			
Richfield	KSVC	980	4-5p
VERMONT			
Burlington	WVMT	620	6-7p; Sat. 1-3p
VIRGINIA			
Bristol/Abingdon	WFHG	620/980	6-7p/5-7p
Charlottesville	WINA	1070	Sat. & Sun. 3-6p
Lynchburg	WLNI	105.9	1-3p; Sun. 3-5p
Norfolk	WTAR	850	7-10p, Sun. 2-5p
WASHINGTON			
Aberdeen	KXRO	1320	12:30- 3p
Bellingham	KGMI	790	7-10p
Pullman	KQQQ	1150	11a-1p
Spokane	KXLY	920	Sat. 11a-noon

The Clark Howard Show

WEST VIRGINIA

Beckley	WWNR	620	3-5p
Charleston	WVTS	950	4-6p/Sat. 12-3p

WISCONSIN

Eau Claire -	WAYY	790	5-7p
Fond du Lac	KFIZ	1450	1-3p, Sun. 9a-12p
Green Bay-Appleton	WHBY	1150	4-6p
Janesville	WCLO	1230	2-4p/Sat. 11a-2p
Milwaukee	WTMJ	620	2-3p
Stevens Point	WSPT	1010	12:30-3p; Sat. 12-2p

Airtime is M-F unless noted
otherwise

AM Stations contain no
decimals

FM Stations always contain
decimals

Visit Clark's website:
www.clarkhoward.com

Talk Radio   Dr Laura

Alaska	Sta	Freq	Airtime
Anchorage	KENI	650	1-4p
Fairbanks	KFBX	970	1-3p
Juneau	KJNO	630	11a-2p
Soldotna	KSRM	920	1-4p
Alabama			
Bridgeport	WYMR	1480	3-6p
Gadsden	WAAX	570	9a-10:45a
Mobile	WIJD	1270	2-5p
Arkansas			
Fayetteville	KUOA	1290	11a-2p
Jonesboro	KBTM	1230	5-8p
Arizona			
Flagstaff	KVNA	600	1-3p
Kingman	KAAA	1230	1-4p
Sierra Vista	KTAN	1420	12-3p
South Tucson	KJLL	1330	2-5p
Yuma	KBLU	560	8-11p
California			
Eureka	KGOE	1480	12p-1p
Eureka	KGOE	1480	2p-3p
Eureka	KGOE	1480	2a-3a
Los Angeles	KFI	640	12-3p
Marysville	KMYC	1410	12-3p
Merced	KYOS	1480	12n-3p
Needles	KTOX	1340	7-10p
Palm Springs	KNWT	1270	12p-3p
Paso Robles	KPRL	1230	8-11p
Sacramento	KSTE	650	8-10p
San Diego	KOGO	600	6-9p
San Francisco	KSFO	560	12-3p
Santa Barbara	KTMS	990	3p-6p
Santa Maria	KSMA	1240	12-3p
Colorado			
Denver	KHOW	630	7-10p
Hayden	KRMR	107.3	1-4p
Kimball	KIMB	1260	1-4p
Delaware			
Denver	KHOW	630	7-10p
Hayden	KRMR	107.3	1-4p
Florida			
Cocoa	WMMV	1350	10p-1a
Ft. Pierce	WZTA	1370	10a-12p
Graceville	WTOT	101.7	2-5p
Jacksonville	WIOJ	1010	3-6pm
Melbourne	WMMB	1240	10p-1a
Panama City	WGTX	1280	2-5p
Punta Gorda	WCCF	1580	3-5p
Sarasota -	WIBQ	1220	3-6p
Tampa	WHNZ	1250	7-10p
W. Palm Bch	WBZT	1230	9p-12a
Atlanta	WLBB	1330	3-6p
Cairo	WGRA	790	3-6p
Columbus	WDAK	540	6-9p
Unadilla	WQSA	99.9	9a-12p

Location	Sta	Freq	Airtime
Georgia			
Atlanta	WLBB	1330	3-6p
Cairo	WGRA	790	3-6p
Columbus	WDAK	540	6-9p
Unadilla	WQSA	99.9	9a-12p
Hawaii			
Honolulu	KHBZ	990	10a-1p
Iowa			
Ames	KASI	1430	2-5p
Davenport	WOC	1420	9a-12p
Des Moines	WHO	1040	6-9p
Ft. Madison	KBKB	1360	2-5p
Norfolk	WJAG	780	9-11a
Ottumwa	KBIZ	1240	6-9p
Sioux City	KMNS	620	6-9p
Idaho			
Boise	KIDO	630	1-3p
Bonners Ferry	KBFI	1450	2-5p
Idaho Falls	KID	590	6-9p
Pocatello	KWIK	1240	6-9p
Sandpoint	KSPT	1400	2-5p
Twin Falls	KLIX	1310	7-10p
Illinois			
Galesburg	WGIL	1400	3-6p
Kankakee	WKAN	1320	2-5p
Mount Vernon	WMIX	940	2-5pm
Quincy	WTAD	930	2-4p
Springfield	WTAX	1240	9a-12p
Indiana			
Bedford	WBIW	1340	3-6p
Bloomington	WGCL	1370	9a-12p
Columbus	WCSI	1010	3-6p
Kansas			
Goodland	KLOE	730	1-4p
Kansas City	KCMO	710	11a-2p
Moberly	KWIX	1230	2-5p
Osage Beach	KRMS	1150	2-5p
Springfield	KWTO	560	4-6p
Waynesville	KJPW	1390	2-5p
Louisiana			
Baton Rouge	WYNK	1380	2-5p
Shreveport	KIMP	960	2-5p
MA			
Boston	WTTT	1150	9p-12a
Springfield	WHYN	560	9a-11:30a
Worcester	WTAG	580	7-10p
Maine			
Bangor	WABI	910	3-6p
Boothbay Hbr	WCME	96.7	9a-12p
Howland	WVOM	103.9	9a-12p

Michigan	Sta	Freq	Airtime
Alpena	WATZ	1450	3-6p
Battle Creek	WBCK	930	10a-12p
Charlevoix	WMKT	1270	9a-12p
Detroit	WJR	760	10p-1a
Grand Rapids	WTKG	1230	8-10p
Ishperning	WIAN	1240	3-6p
Ludington	WKLA	1450	9a-12p
Manistee	WMTE	1340	9a-12p
Manistique	WPIQ	99.9	9a-12n
Marquette	WDMJ	1320	3-6p
Muskegon/Grd Rapids	WKBZ	1090	6-9p
St. Joseph	WSJM	1400	9-11a, 6-7p
Minnesota			
Baxter	WWWI	1270	9-11a, 5-6p
Bemidji	KKBJ	1360	8-11a
Duluth	KQDS	1490	11-2p
Minneapolis	WFMP	107.1	12p-3p
Red Wing	KCUE	1250	2-5p
Wascca	KOWZ	1170	9a-12p
Mississippi			
Jackson	WJNT	1180	6-9p
Montana			
Billings	KBUL	970	1-4p
Helena	KBLL	1240	1:30p-4:30p
Missoula	KGVO	1290	3-5p
Missoula	KLYQ	1240	7-9p
North Carolina			
Durham	WDNC	620	9a-12p
Elizabeth City	WGAI	560	3-6p
Greensboro	WMFR	1230	9a-12p
Hickory	WHKY	1290	9a-12p
Monroe	WXNC	1060	9a-12p
Raleigh	WDNZ	570	9a-12p
Salisbury	WSTP	1490	3p-6p
Statesville	WSIC	1400	7-10p
Nebraska			
Kearney	KGFW	1340	2-5p
Lincoln	KMMJ	750	2-5p
McCook	KSWN	93.9	2-5p
Omaha	KKAR	1290	11a-2p
New Hampshire			
Manchester	WKXL	1450	3-6p
Nevada			
Las Vegas	KXNT	840	12-3p
Reno	KHIT	1450	12-3p
New York			
Auburn	WAUB	1590	9-11a
Binghamton	WYOS	1360	3-6 p
Buffalo	WBEN	930	7-10p
Canandaigua	WCGR	1550	9-11a
Geneva	WGVA	1240	9-11a
Massena	WYBG	1050	9a-12p

Location	Sta	Freq	Airtime
New York			
New York	WWDJ	970	2-4p
Ogdensburg	WSLB	1400	9a-12noon
Seneca Falls	WSFW	1110	9-11a
Syracuse	WFLR	1570	9-11a
Watertown	WATN	1240	9am-Noon
Ohio			
Bellefontaine	WBLL	1390	3-6p
Canton	WCER	900	4-7p
Chillicothe	WBEX	1490	3-6p
Cincinnati	WKRC	550	9p-12a
Findlay	WFIN	1330	3-5p
Mansfield	WMAN	1400	3-6p
Oklahoma			
Bartlesville	KWON	1400	7p-10p
Oklahoma City	KEBC	1340	10a-1p
Tulsa	KRMG	740	7p-10p
Oregon			
Baker	KBKR	1490	12n-2p
Bend	KBND	1110	8-10p
Corvallis	KLOO	1340	7-10p
La Grande	KLBM	1450	12n-2p
Medford-Ashland	KMED	1440	3-4p; 7-9p
Portland	KEX	1190	M-F: 1-4p
Pennsylvania			
Altoona	WFBG	1290	4-7p
Johnstown-Altoona	WHUN	1150	4-7p
Pittsburgh	WPIT	730	12p-2p
South Carolina			
Myrtle Beach	WQJM	1450	3-6p
Spartanburg	WSPA	910	9a-12p
Sumter	WDXY	1240	9a-12p
South Dakota			
Sioux City	KSOO	1140	9p-12a
Tennessee			
Cleveland	WBAC	1340	3-6p
Dayton	WDNT	1280	3-6p
Kingsport	WPWT	870	12p-2p
Memphis	KWAM	990	8a-11a
Memphis	KWAM		9p-12a
Spring City	WXQK	970	3-6p
Texas			
Abilene	KWKC	1340	2-5p
Amarillo	KIXZ	940	11a-2p
Austin	KLBJ	590	10p-12a
Beaumont	KLVI	560	9-11a
Big Spring	KBST	1490	2-5p
Dallas	KLIF	570	2p-4p

Location	Sta	Freq	Airtime					
Texas								
Gladewater	KEES	1430	8-11a					
Lubbock	KJTV	950	2-5p					
McAllen - Brownsville	KVNS	1700	1-4p					
Paris	KPLT	1490	2-5p					
Richardson	KKLF	1700	12p-3p					
San Angelo	KKSA	1260	2-5p					
Utah								
Cedar City	KNNZ	940	1-4p					
Green River	KUGR	1490	1-4p					
Logan	KVNU	610	1-4p					
Price	KOAL	750	1-4p					
Salt Lake City	KNRS	570	1-4p					
St. George	KDXU	890	1-4p					
Virginia								
Cumberland	WCBC	1270	6-8p					
Frederick	WFMD	930	9a-12p					
Martinsburg	WRNR	740	3-6p					
Norfolk	WTAR	850	2-5p					
Richmond	WRVA	1140	10p-1a					
Salisbury	WGMD	92.7	9a-12p					
Vermont								
Burlington	WRSA	1420	12p-3p					
Washington								
Centralia	KELA	1470	2-5p					
Pasco	KFLD	870	1-4p					
Pullman	KQQQ	1150	12-2p					
Spokane	KQNT	590	12-3p					
Yakima	KIT	1280	9p-12a					
Wisconsin								
Eau Claire	WMEQ	880	8-11a					
Oshkosh	WOSH	1490	2-5p					
Superior	KQDS	1490	11a-2p					
Wisconsin Rapids	WFHR	1320	3-5p					
Wyoming								
Casper	KKTL	1400	3-6p					

Location	Sta	Freq	Airtime	Location	Sta	Freq	Airtime
AK				**MI**			
Anchorage	KBYR	700	8p11p	Detroit	WDTK	1400	12p-2p
AL				Detroit	WDTK	1400	1p6p
Florence	WBCF	1240	TBA	Petoskey	WJML	1110	1p3p
AZ				**MN**			
Phoenix	KKNT	960	10 a-1p	Little Falls	KLTF	960	1p2p
Tucson	KVOI	690	10 a-1p	Minneapolis	WWTC	1280	11a2p
CA				**MO**			
Los Angeles	KRLA	870	9a12p	Hannibal	KHMO	1030	12p2p
Monterey Area	KNRY	1240	9a12p	Kansas City	KSIS	1050	2p4p
Palm Springs	KGAM	1450	9a12p	**MS**			
Riverside	KTIE	590	9a12p	Vicksburg	WQBC	1420	11p2a
Sacramento	KTKZ	1380	6p9p	**NE**			
San Diego	KCBQ	1170	9a12p	Lincoln	KLIN	1400	3p5p
San Diego	KNTS	1220	9a12p,9p12a	**NY**			
Visalia	KTIP	1450	9a12p	Albany	WROW	590	12p3p
CO				Nassau-Suff'lk	WLIE	540	12p3p
Colorado Spngs	KZNT	1460	10a1p	**OH**			
Denver	KNUS	710	10 a-1p	Cincinnati	WBOB	1160	12p3p
FL				Cleveland	WHK	1420	12a3a
Ft Myers-N'ples	WCNZ	1660	12p3p	Columbus	WHTH	790	12p-2p
Jacksonville	WJGR	1320	12p3p	**OK**			
Miami	WKAT	1360	12p3p	Tulsa	KCFO	970	12a2a
Panama City	WLTG	1430	11a-12p	**OR**			
Pensacola	WEBY	1330	12p3p	Portland	KPDQ	800	1a4a
GA				Salem	KYKN	1430	9p12a
Atlanta	WGKA	920	12p-2p	**PA**			
St. Simons Is.	WCGA	1100	12p-2p	Butler	WISR	680	2p3p
HI				Philadelphia	WNTP	990	12p3p
Honolulu	KHNR	97.5	8a9a	Pittsburgh	WPIT	730	1p3p
Honolulu	KHNR	97.5	7p9p	**SC**			
IA				Greenville	WSPA	910	12p3p
Mason City	KILR	1070	11p2a	**TN**			
ID				Knoxville	WETR	760	12p-2p
Idaho Falls	KZNR	690	11a1p	**TX**			
IL				Dallas	KSKY	660	11a1p
Carbondale	WINI	1420	11p2a	Houston	KNTH	1070	11a1p
Chicago	WIND	560	11a2p	Houston	KNTH	1070	8p-9p
Chicago	WJJG	1530	1p2p	San Antonio	KLUP	930	11a2p
IN				**VA**			
Evansville	WRAY	1250	11a2p	Blacksburg	WPIN	810	12p-2p
KS				Richmond	WTOX	1480	2p3p
Hutchinson	KWBW	1450	12a-1a	**WA**			
KY				Colfax	KMAX	840	10p12a
Bowling Green	KPCR	1530	12p2p	Seattle	KKOL	1300	9a12p
Lexington	WWFT	1250	9p12a	Yakima	KJOX	980	9a12p
Louisville	WGTK	970	12p3p	Yakima	KGDC	1320	9a12p
MA				**WI**			
Boston	WTTT	1150	12-3p	Green Bay	WRPN	1600	11a1p
MD				Milwaukee	WRRD	540	12p3p
Baltimore	WITH	1230	12p3p				
ME							
Bangor	WWNZ	1400					

AK	Station	Freq	Airtime
Fairbanks	KFBX	970	Sa 8a-10
Fairbanks	KFBX	970	Su 9a12
AL			
Athens	WVNN	770	Sa 5p6p,
Athens	WVNN	770	Su 6a7a,
Athens	WVNN	770	Su 5p6p
Birmingham	WERC	960	Sa 9p12
Dothan	WWNT	1450	M-Fr 6p7
Guntersville	WGSV	1270	M-F 5p6p
Sylacauga	WFEB	1340	M-F 2p3p
Tuscumbia	WVNA	1590	Sa 5a8a;
Tuscumbia	WVNA	1590	Su 6a12a
AR			
Jonesboro	KBTM	1230	M-F 5p6p
AZ			
Bullhead City	KZZZ	1490	M-F 6p7p
Kingman	KAAA	1230	M-F 6p7p
Prescott	KYCA	1490	M-F 6p7p
Tucson	KNST	790	Su 4p7p
Yuma	KBLU	560	M-F 3p4p
CA			
Eureka	KGOE	1480	M-F 1p2p
Fresno	KMJ	580	Sa 4p7p
Grass Valley	KNCO	830	M-F 1p2p
Lompoc	KTME	1410	M-F 9p10
Merced/Atwater	KYOS	1480	M-F 3p4p
Modesto	KFIV	1360	M-F 3p4p
Palm Springs	KPSI	920	Sa 4p6p,
Palm Springs	KPSI	920	Su 4p7p
Sacramento	KFBK	1530	Sa 6p9p
Sacramento	KFBK	1530	Su7p10p
San Francisco	KGO	810	M-F 1p2p
Santa Barbara	KTMS	990	Su1p6p
Santa Maria	KUHL	1440	M-F 9p10
So. Lake Tahoe	KOWL	1490	M-F 7p8p
CO			
Basalt	KNFO	106.1	Sa 8a10
Basalt	KNFO	106.1	Su 8a10
Colorado Springs	KVOR	740	Sa1 p2p
Colorado Springs	KVOR	740	Sa7p8p,
Colorado Springs	KVOR	740	Su11a2p
Cortez	KVFC	740	M-F 7p8
Durango	KDGO	1240	M-F 7p8
Hayden	KRMR	107.3	M-Th 4p5
Hayden	KRMR	107.3	Sa 12p1
CT			
Danbury	WLAD	800	Sa 12p1
Danbury	WLAD	800	Su 9a1p
New Haven	WELI	960	Su 6p10p
D of C			
Washington	WMAL	630	Su 7p9p
Dover	WDOV	1410	Su-Th 6p7

FL	Station	Freq	Airtime
Cocoa/Melbourne	WMMV	1350	Sa 4p7p
Cocoa/Melbourne	WMMV	1350	Su 4p6p
Ft Walton Bch	WFTW	1260	M-F 5p6p
Melbourne	WMMB	1240	Sa 4p7p,
Melbourne	WMMB	1240	Su 4p6p
Panama City	WLTG	1430	M-F 7a8a
Pensacola	WCOA	1370	Sa 1p3p
Pensacola	WCOA	1370	Su12p3p
Pine Island Ctr	WPTK	1200	M-F 4p5p
Punta Gorda	WCCF	1580	M-F 11a12
St Augustine	WFOY	1240	M-F 4p5p
Tampa	WHNZ	1250	Sa7p10p,
Tampa	WHNZ	1250	Su 8p10p
W. Palm Beach	WJNO	1290	Su 5a6a,
W. Palm Beach	WJNO	1290	Su 4p8p
GA			
Albany	WALG	1590	M-F 8a9a
Atlanta	WGST	640	Sa9p1a
Augusta	WGAC	580	Sa 6a7a
Augusta	WGAC	580	Sa 7p9p
Augusta	WGAC	580	Su 12p1p
Augusta	WGAC	580	Su 5p6p
Dalton	WBLJ	1230	Sa 11a2p
Dalton	WBLJ	1230	Su 6a8a
Lakeland	WVGA	105.9	M-F 6p7p
Lakeland	WVGA	105.9	Su 6a9a
Marietta	WLTP	910	M-F 6p7p
GU			
Agana	KGUM	910	M-F 7p8p,
Agana	KGUM	910	Sa 6a8a
HI			
Hilo	KPUA	670	M-F 1p2p
Honolulu	KHBZ	990	Sa 10p11
Honolulu	KHVH	830	S-Su 5a7
Kihei	KAOI	1110	M-F 3p4p
IA			
Ames	KASI	1430	M-F 6p7p
Burlington	KBUR	1490	Sa 5p8p,
Burlington	KBUR	1490	Su 2p4p
Davenport	WOC	1420	Sa12p1p,
Davenport	WOC	1420	Su11a3p
Des Moines	WHO	1040	Su12p3p
Iowa City	KXIC	800	M-F10a11
Ottumwa	KLEE	1480	M-F 12a1
Sioux City	KMNS	620	Sa11a3p
ID			
Boise	KIDO	580	Su 5a8a
Boise	KIDO	580	Su12p1p
Boise	KIDO	580	Su 2p3p
Bonner's Ferry	KBFI	1450	M-F 1p2p
Pocatello	KWIK	1240	Su 6a8a,
Pocatello	KWIK	1240	Su 4p7p

Dr. Dean Edell

IL	**Station**	**Freq**	**Airtime**
Bloomington	WJBC	1230	Su 6p9p
Chicago	WLS	890	Su 5a10
Galesburg	WGIL	1400	F 10a11a
Peoria	WMBD	1470	Sa 6p8p,
Peoria	WMBD	1470	Su 10a1
Princeton	WZOE	1490	M-F 3p4p
Quincy	WTAD	930	Sa 4p6p,
Quincy	WTAD	930	Su 4p6p
Springfield	WTAX	1240	Su 11a3p
Streator	WSPL	1250	M-F 3p4p
IN			
Columbus	WCSI	1010	M-F 6p7p
Fort Wayne	WOWO	1190	Su 3p6p
Indianapolis	WIBC	1070	Su 4p6p
Logansport	WSAL	1230	M-F 5p6p
South Bend	WSBT	960	Sa 2p5p,
South Bend	WSBT	960	Su 9a11
KS			
Goodland	KLOE	730	M-F 5p6p
Hutchinson	KWBW	1450	M-F 6p7p
Manhattan	KMAN	1350	M-F 10a11
Topeka	KMAJ	1440	M-F 4p5p
KY			
Owensboro	WOMI	1490	Sa 12p3
Owensboro	WOMI	1490	Su 12p3
Somerset	WSFC	1240	M-F 4p5p
LA			
Abbeville	KPEL	105.1	Su 6p9p
Alexandria	KSYL	970	Sa 4p7p,
Alexandria	KSYL	970	Su 4p6p
New Orleans	WSMB	1350	M-F 7p8p
MA			
Lowell	WCAP	980	M-F 9a10
Pittsfield	WBEC	1420	Sa 6a8a,
Pittsfield	WBEC	1420	Su 7a8a
Springfield	WHYN	560	Sa 6p9p,
Springfield	WHYN	560	Su 4p6p
MD			
Baltimore	WBAL	1090	Sa 5a7a
Baltimore	WBAL	1090	Su 5a7a
Baltimore	WBAL	1090	Sa 1a2a
Frederick	WFMD	930	Sa 6a7a,
Frederick	WFMD	930	S-Su 6p8
ME			
Portland	WGAN	560	Su 6p10
MI			
Adrian	WABJ	1490	M-F 10a11
Alpena	WATZ	1450	M-F 10a11
Big Rapids	WBRN	1460	M-F 4p5p

Sandpoint	KSPT	1400	M-F 1p2p
MI	**Station**	**Freq**	**Airtime**
Lansing	WJIM	1240	Su 2p4p
Traverse City	WTCM	580	Sa 7p10p
Traverse City	WTCM	580	Su 7p9p
MN			
Nashwauk	WNMT	650	M-F 2p3p
Rochester	KROC	1340	M-F 11a12
MS			
Meridian	WMOX	1010	Sa 5a7a
MT			
Missoula	KGVO	1290	Su11a4p
MT			
Kearney	KGFW	1340	M-F11p12
Kearney	KGFW	1340	Su 3p5p
Kearney	KHLP	1420	M-F 11a12
Lincoln	KLIN	1400	Sa 12a3a
Lincoln	KLIN	1400	Su 10a1p
NH			
Exeter/Portsmouth	WGIP	1540	Sa 3p6p
Hanover	WTSL	1400	Sa 3p6p
Laconia	WEMJ	1490	M-F 6p7p
Manchester	WGIR	610	Sa 3p6p
Rochestr/Portsm.	WGIN	930	Sa 3p6p
NM			
Las Cruces	KOBE	1450	M-F 9a10
Los Alamos/SantaFe	KRSN	1490	M-F 2p3p
NY			
Canandaigua	WCGR	1550	M-F 4p5p
Geneva	WGVA	1240	M-F 4p5p
Jamestown	WKSN	1340	M-F 5p6p
Ogdensburg	WSLB	1400	M-F 6p7p
Rochester	WHAM	1180	M-F 11p12
Schenectady	WGY	810	Sa 10p1a
Seneca Falls	WSFW	1110	M-F 4p5p
Syracuse	WSYR	570	Sa 6p8p,
Syracuse	WSYR	570	Su 6p9p
Utica/Rome	WIBX	950	Su 1p4p
NC			
Asheville	WWNC	570	Su 6a7a
Asheville	WWNC	570	Su 9a10a
Asheville	WWNC	570	Su 4p7p
Charlotte	WBT	1110	Sa 5a6a,
Charlotte	WBT	1110	Su 1p4p
Hickory	WHKY	1290	M-F 6p7p
Raleigh/Durham	WPTF	680	M-F 11p1
Washington	WDLX	930	M-F 6p7p
NV			
Elko	KTSN	1340	M-F 6a7a
North Las Vegas	KXNT	840	Sa 4p7p,
North Las Vegas	KXNT	840	Su 4p6p

Dr. Dean Edell

Detroit	WJR	760	Sa 10p1a,	Reno	KJFK	1230	Sa 5a8a,
Detroit	WJR	760	Su 8p10p	Reno	KJFK	1230	Su 6a8a
Grand Rapids	WTKG	1230	Sa 8p10p				
Grand Rapids	WTKG	1230	Su 6a9a	x	x	x	x

OH	Station	Freq	Airtime	SC	Station	Freq	Airtime
Akron	WHLO	640	Su 12p5	Charleston	WSCC	94.3	Sa 5a7a,
Chillicothe	WBEX	1490	Sa 5a9a,	Charleston	WSCC	94.3	Su 7p10p
Columbus	WTPG	1230	Su 6a9a	Socastee	WRNN	99.5	Sa 8a10
Columbus	WTPG	1230	Su 6a7a	Socastee	WRNN	99.5	*Su 9a1p*
OK				**SD**			
Tulsa	KRMG	740	SaSu 4p6	Rapid City	KOTA	1380	Sa 4p9p
OR				**TN**			
Bend	KBND	1110	Sa 10a1p	Chattanooga	WGOW	1150	Su 7a12p
Bend	KBND	1110	Su 10a12p	Johnsn Cty/Kings	WJCW	910	M-F 9p10
Corvallis	KLOO	1340	SaSu 4p7	Memphis	KWAM	990	M-F 5p6p
Enterprise	KWVR	1340	M-F 8p9p	Nashville	WLAC	1510	Su 2p6p
Eugene/Springfield	KPNW	1120	Su 1p6p	**TX**			
Keizer/Salem	KYKN	1430	Sa 7a10	Abilene	KSLI	1280	M-F 1p2p
Keizer/Salem	KYKN	1430	Su 12p2	Austin	KLBJ	590	Sa 4p6p,
Medford	KMED	1440	M-F 9p10p	Austin	KLBJ	590	Su 3p6p
Newport	KNPT	1310	M-F 4p5p	College Statn/Br	WTAW	1620	SaSu 6p9
North Bend	KBBR	1340	M-F 6p7p	Corpus Christi	KKTX	1360	Su11a4p
Pendleton	KUMA	1290	M-F 3p4p	Edinburg	KURV	710	Sa 12p3
Portland	KEX	1190	Sa 2p4p,	Edinburg	KURV	710	Su 1p3p
Portland	KEX	1190	Su 3p6p	Fort Worth	WBAP	820	Sa 8p12a
PA				Jacksonville	KDVE	103.1	M-F 8p9p
Allentown	WAEB	790	SaSu 6p8	Lubbock	KFYO	790	Sa 9a11
Allentown	WAEB	790	Su 6a7a	Lubbock	KFYO	790	Sa 2p5p
Altoona	WFBG	1290	Sa 9a1p	San Angelo	KKSA	1260	M-F 8a9a
Franklin	WFRA	1450	Sa 6p9p,	San Antonio	WOAI	1200	Su 12p4p
Franklin	WFRA	1450	Su 10a12	Waco	KWTX	1230	Sa 5a8a,
Harrisburg	WHP	580	Su 6a10a	Waco	KWTX	1230	Su 6a8a
Hughsvle/William	WRKK	1200	M-F 6p7p	**UT**			
Huntington	WHUN	1150	Sa 9a1p	Cedar City	KSUB	590	Su 6a9a,
Lewistown	WIEZ	670	M-F 4p5p	Cedar City	KSUB	590	Su 11a2p
Meadville	WMGW	1490	Sa 6p9p,	Logan	KVNU	610	M-F 10p11
Meadville	WMGW	1490	Su10a12p	Salt Lake City	KNRS	570	Su 5a8a
Milton	WMLP	1380	M-F 8a9a	**VA**			
New Castle	WKST	1200	M-F 12a1a	Norfolk	WNIS	790	Sa 12p1p
Oil City	WOYL	1340	Sa 6p9p,	Norfolk	WNIS	790	Sa 7p8p
Oil City	WOYL	1340	Su 10a12	Norfolk	WNIS	790	Su12p1p
Scranton/Wilkes-Barre	WBZU	910	Su 6a9a	Norfolk	WNIS	790	Su7p10p
Titusville	WTIV	1230	Sa 6p9p,	Norfolk	WRVA	1140	Sa 6a8a,
Titusville	WTIV	1230	Su 10a12	Norfolk	WRVA	1140	Su 5p7p
W Hazleton	WKZN	1300	Su 6a9a	Onley-Onancock	WESR	1330	M-F 10a11
Wilkes-Barr/Scranton	WILK	980	Su 6a9a	Roanoke/Lynchbrg	WFIR	960	Sa 1p4p
Williamsport	WRAK	1400	MF 6p7p	Winchester	WINC	1400	M-F 9a10a
PR				**Virgin Islands**			
San Juan	WOSO	1030	M-F 4p5p	Charotte-Amalie	WWVI	1000	M-F 1p2p
RI							
Providence	WPRO	630	Sa 7p8p,				

Dr. Dean Edell

Providence	WPRO	630	Su 8a9a					
Providence	WPRO	630	Su12p1p					
Providence	WPRO	630	Su 7p9p					
SC								
Chester	WBT	99.3	Sa 5a6a,					
Chester	WBT	99.3	*Su 1a4p*					
Columbia	WVOC	560	M-F11p12					
Florence	WJMX	970	Su 8a11a					

Laura Ingraham M-F

Times are M-F unless noted

	Station	Freq	Airtime		Station	Freq	Airtime
Alaska				**DC Metro**			
Anchorage	KBYR	700	5-8a	Silver Spring	WTNT	570	9a-12p
Fairbanks	KFBX	970	6-9p	**DE**			
Alabama				Rehoboth Bch	WGMD	92.7	10p-1a
Anniston	WDNG	1450	Su6-9p	Wilmington	WILM	1450	Su7-10p
Birmingham	WAPI	1070	12p -2p	**Florida**			
Dothan	WTOT	101.7	8-11a	Dade City	WDCF	1350	7-8p
Eufala	WULA	600	7-10p	Daytona Beach	WNDB	1150	Su7-10p
Florence/M.Shoals	WBCF	1240	8-9a	Delray Beach	WDJA	1420	7-9p
Foley	WHEP	1310	6-9p	Ft Myers-Naples	WNOG	1270	9a-12p
Greenville	WGYV	1380	10-11a	Ft. Lauderdale	WFTL	850	9p-12a
Montgomery	WACV	1170	6-9p	Ft.Myers-Naples	WINK	1240	9a-12p
Selma	WHBB	1490	9-10a	Ft. Walton Bch	WNWF	1120	8-11a
Sylacauga	WFEB	1340	8-9a	Gainesville-	WBXY	99.5	9-11a
Arkansas				Gainesville	WSKY	97.3	10p-1a
Hope	KXAR	1490	9-11a	Jacksonville	WJGR	1320	9a-12p
Little Rock	KARN	920	Su7-9p	Jacksonville	WJGR	1320	Sa4-7p
Little Rock	KARN	101.7	Su7-9p	Lakeland	WTWB	1570	10a-12p
Little Rock	KKRN	102.5	Su7-9p	Melbourne	WMMB, WMMV	1240, 1350	6-7p
Arizona				Miami	WKAT	1360	9a-12p
Phoenix	KKNT	960	7-10a	Naples	WNOG	1270	7-10p
Prescott	KYCA	1490	8-9a	Orlando	WAMT	1190	2p-5
Safford	KATO	1230	9-10a	Panama City	WYOO	101.1	Su 6p-9
Sierra Vista	KTAN	1420	6-8a	Pensacola	WCOA	1370	Su 6-9p
Tucson	KVOI	690	6-9a	Pensacola	WVTJ	610	Su 6-9p
California				Sarasota	WENG	1530	Su7p-10
Bakersfield	KNZR	1560	6-9 a	Sarasota	WIBQ	1220	1p-3
Los Angeles	KRLA	870	6-9a	St. Augustine	WFOY	1240	7-10p
Modesto	KFIV	1360	Su12-3p	Tampa-St Pete	WGUL	860	9a-12p
Needles/Mohave	KTOX	1340	9a-12p	**Georgia**			
Palm Springs	KPSI	920	2-4p	Atlanta	WGKA	920	9a-12p
Riverside	KTIE	590	6-9a	Augusta	WRDW	1480	10a-12p
Sacramento	KTKZ	1380	9a-12p	Cairo	WGRA	790	9a-12p
San Diego	KCBQ	1170	6-9a	Hawkinsville	WRPG	103.9	9a-12p
San Francisco	KSFO	560	8-10p	Jasper	WYYZ	1490	5-6a
San Luis Obispo	KPRL	1230	2-4p	Macon	WMWR	156	4-6a
Santa Maria	KTME	1410	9-11a	McRae	WYIS	1410	7-10p
Santa Maria	KUHL	1440	9-11a	**Iowa**			
Sonora	KVML	1450	3-6p	Dubuque	WDBQ	1490	8-11a
Susanville	KSUE	1240	4-5p/6-7	Estherville	KILR	1070	10-11a
Victor Valley	KIXW	960	6-9a	Sioux City	KSCJ	1360	7-9p
Visalia-Tulare	KTIP	1450	7-10p	**Idaho**			
Colorado				Boise	KBOI	670	10a-1p
Colorado Spngs	KZNT	1460	7-10a	Rupert	KBAR	1230	3-6a
Denver-Boulder	KNUS	710	7-10a	**Illinois**			
Ft. Collins	KSXT	1570	7-10a	Aurora	WBIG	1280	M4-7a
Grand Junction	KNZZ	1100	10p-1a	Chicago	WIND	560	8-11a
Pueblo	KCSJ	590	4-6p	Decatur	WSOY	1340	7-9p
Steamboat Sprgs	KRMR	107.3	Su5-8p	Elgin	WRMN	1410	4-7a
Connecticut				Marion-	WCIL	1010	7-9p
Hartford	WDRC	1360	10a-12p	Marion	WJPF	1340	7-9p
Hartford	WMMW	1470	10a-12p	Ottawa	WCMY	1430	6-9p
New Haven	WELI	960	7-10p	Quincy	WGEM	105.1	7-10p
New London	WXLM	102.3	10a-12p	Springfield	WTAX	1240	2-5p
Torrington	WSNG	610	10a-12p				
Waterbury	WWCO	1240	10a-12p				

Laura Ingraham M-F

Times are Monday-Friday unless noted

	Station	Freq	Airtime		Station	Freq	Airtime
Indiana				**Missouri**			
Indianapolis	WXNT	1430	8p-12a	Cape Girardeau	KAPE	1550	8-11a
South Bend	WSBT	960	7-10p	Columbia	KSSZ	93.9	Su7-9p
Kansas				Jefferson City	KLIK	1240	8-11a
Chanute	KKOY	1460	8-10a	Mountain Grove	KELE	1360	8-11a
Garden City	KBUF	1030	Su6-8p	Springfield	KSGF	1260	Su 9a-12
Shawnee Miss.	KMBZ	980	10p-1a	Springfield	KSGF	104.1	Su 9a-12
Wichita	KNSS	1240	Su8-9p	Springfield	KSWM	940	8-11a
				St. Louis	KFTK	97.1	9-11a
Kentucky				West Plains	KWPM	1450	6-9p
Bowling Green	WKCT	930	9-11a	**Mississippi**			
Cadiz	WKDZ	1110	and 6p-9	Biloxi	WTNI	1640	8-11a
Corbin	WKDP	1330	9a-12p	Jackson	WJNT	1180	9-11a
Lexington	WLXO	96.1	10p-1a	Laurel-Hattiesburg	WMXI	98.1	6-9p
Louisville	WGTK	970	9am-12p	Meridian	WMOX	1010	9-1a, 2-3p
Murray	WNBS	1340	9-10a	**Montana**			
Louisiana				Billings	KBSR	1490	11p-2a
Many	KWLA	1400	8-11a	Billings	KHDN	1230	11p-2a
Shreveport	KRMD	1340	6-9p	Great Falls	KQDI	1450	7-9a
Massachusetts				Hamilton	KLYQ	1240	5-6p
Boston	WBET	1460	M 5-8a	Helena	KBLL	1240	Su5-8p
Boston	WTKK	96.9	7-10p	Kalispell	KJJR	880	7-9p
Fitchburg	WEIM	1280	7-10P	Missoula	KGVO	1290	4-5p
Providence	WPEP	1570	9a-12p	**North Carolina**			
Providence	WSAR	1480	8-10p	Aberdeen	WQNX	1350	9a12p
Yarm'th-Cape Cod	WXTK	95.1	7-10p	Albemarle	WSPC	1010	10a-12p
Maryland				Greensboro-	WSJS	600	10a-12p
Baltimore	WITH	1230	9a-12p	-Winston Salem	WSML	1190	10a-12p
Cumberland	WCBC	1270	10a-12p	Greenville	WJNC	1240	7-10p
Maine				Greenville	WNCT	1070	10a-12p
Portland	WLOB	1310	9a-12p	Lenior	WJRI	1340	9-12p
Portland	WLOB	96.3	9a-12p	Raleigh-Durham	WDNC	620	9p-12a
Michigan				Raleigh-Durham	WDNZ	570	9p-12a
Adrian	WABJ	1490	3-5p	**North Dakota**			
Ann Arbor	WAAM	1600	9a-12p	Fargo	WDAY	970	9p-12a
Benton Harbor	WHFB	1060	9a-12p	**Nebraska**			
Big Rapids	WBRN	1460	9a-12p	Lincoln	KLIN	1400	10-11a
Grand Rapids	WMFN	640	8-10p	Omaha	KKAR	1290	8-11am
Grand Rapids	WOOD	1300	10p-12a	**New Hampshire**			
Holland	WHTC	1450	7-10p	Concord	WKXL	1450	Sa5-7p
Jackson	WKHM	970	Su7-10p	Lebanon	WTSL	1400	Su7-10p
Ludington	WKLA	1450	7-10p	Manchester	WGIP	1540	Su7-10p
Ludington	WMTE	1340	7-10p	Manchester	WGIR	610	Su 7-10p
Traverse City-	WJML	1110	7-10p	**New Mexico**			
Petoskey	WWKK	750	9a-12p	Alamagordo	KINN	1270	Su 5-8p
Minnesota				Albuquerque	KTBL	1050	10a-1p
Brainerd	WWWI	95.9	10-11a	Albuquerque	KTBL	920	10a-1p
Brainerd	WWWI	1270	8-11a	Clovis	KICA	980	8-10a
Fosston	KKCQ	1480	8-11a	Farmington	KENN	1390	8a-12p
Minneapolis-	KYCR	1570	8-11a	Roswell	KBIM	910	9-10a
-St Paul	WWTC	1280	8-11a	**Nevada**			
Owatonna	KOWZ	1170	9-11a	Las Vegas	KDOX	1280	Sa1-4p
St. Cloud	KNSI	1450	11a-12p	Las Vegas	KDOX	1280	Su7p

Laura Ingraham M-F

Times are M-F unless noted

New York	Station	Freq	Airtime	Rhode Island	Station	Freq	Airtime
Albany	WCSS	1490	7-10pm				
Albany	WROW	590	11a-12p	Providence	WHJJ	920	1-5a
Binghamton	WNBF	1290	Su 7-10p	Providence	WNRI	1380	1-3a
Elmira-Corning	WCLI	1450	10p-1am	Providence, RI	WADK	1540	10a-12p
Elmira-Corning	WCLI	1450	9a-12p	So Carolina			
Elmira-Corning	WCLI	1450	Sa4-7p	Charleston	WQSC	1340	10a-1p
Elmira-Corning	WENY	1230	9a-12p	Columbia	WISW	840	9-11a
Elmira-Corning	WENY	1230	Sa4-7p	Fountain Inn	WFIS	1600	8-10a
Hornell	WLEA	1000	10a-12p	Greenville-	WAIM	1230	10a-12p
New York	WABC	770	8-10pm	Myrtle Beach	WGTN	1400	Su7-10p
Newark, NY	WACK	1420	10a-12p	South Dakota			
Syracuse	WFBL	1390	9a-12p	Pierre	KSQP	1450	8-11a
Utica-Rome	WIBX	950	9a-12p	Rapid City	KTOQ	1340	8-11p
Watertown	WTNY	790	7-10p	Sioux Falls	KELO	1320	6-9p
Ohio				Tennessee			
Akron	WHLO	640	9p-12a	Chattanooga	WGOW	1150	7-10 p
Akron	WHLO	640	Sa7-10p	Harrogate	WRWB	740	9a-12p
Bellefontaine	WBLL	1390	9-11a	Jackson	WTNE	93.1	Sa3-6p
Cincinnati	WBOB	1160	9a-12p	Knoxville	WATO	1290	10a-12p
Cleveland	WHK	1420	9a-12p	Knoxville	WGAP	1400	10a-12p
Columbus	WLOH	1320	9a-12p	Knoxville	WKVL	850	10a-12p
Middletown	WPFB	910	7-10p	Knoxville	WLOD	1140	10a-12p
Newark	WCLT	1430	10a-12p	Knoxville	WMTY	670	10a-12p
Toledo	WTOD	1560		Nashville	WWTN	99.7	Sa 6-9p
Oklahoma				Springfield	WJQY	1590	8-11a
Oklahoma City	KOKC	1520	8-11a	Texas			
Tulsa	KFAQ	1170	2-5p	Amarillo	KIXZ	940	8-9p
Oregon				Ardmore	KVSO	1240	7-8a,2-3
Astoria	KAST	1370	7-10p	Ardmore	KVSO	1240	4-7a
Bend	KRDM	1240	10a-12p	Austin	KJCE	1370	9a-12p
Coquille	KWRO	630	6-9a	Beaumont-	KOLE	1340	9-11a
Eugene-Sp'gfield	KEED	1600	6-9a	-Pt Arthur	KRCM	1380	9-11a
Eureka	KWSW	790	6-7p	Brownwood	KXYL	96.9	M 4-5a
Isabella	KQAB	1140	7-9a	College Sta	WTAW	1620	6-9p
Medford-Ashland	KCMX	880	4-7p	Corpus Christi	KKTX	1360	6-9p
Pendleton	KUMA	1290	4-7p	Dallas-Ft. Worth	KSKY	660	8-11am
Portland	KPAM	860	9a-12p	Houston-	KSEV	700	9-11a
Quincy	KPCO	1370	6-7p	Lubbock	KJTV	950	7-9p
Quincy	KPCO	1370	SaSu4-7p	Lubbock	KVOP	1090	8-11a
Pennsylvania				McAllen-	KURV	710	7-9p
Beaver Falls	WMBA	1460	M 5-6a	McAllen-	KVJY	840	Sa3-6p
Carlisle	WHYL	960	9 a12p	Odessa-Midland	KWEL	1070	Sa3-6p
Milton	WMLP	1380	Su7-10p	San Angelo	KGKL	960	8-11a
Philadelphia	WNTP	990	9a-12p	San Antonio	KLUP	930	8-11a
Philipsburg	WPHB	1260	7-10p	Texarkana	KTFS	940	7-8a
Pittsburgh	WPTT	1360	7-9a	Tyler-Longview	KFRO	1370	8-11a
Punxsutawney	WECZ	1540	9a-12p	Utah			
State College	WBLF	970	7-10p	Cedar City	KNNZ	940	8-10a
State College	WRSC	1290	7-10p	Delta	KNAK	540	8-10a
Wilkes Barre-	WGBI	910	Su7-10p	Salt Lake City	KLO	1430	7-10a
Wilkes Barre-	WILK	980	Su7-10p	St George	KZNU	1450	7-10a
Wilkes Barre-	WOGY	1300	Su 7-10p	Vernal-Roosevelt	KVEL	920	8-10a
Youngstown	WPIC	790	10a-12p	Vermont			
				Burlington-Platsbgh	WVMT	620	10a-1p

Laura Ingraham M-F

Times are Monday-Friday unless noted

Virginia	Station	Freq	Airtime					
Charlottesville	WCHV	1260	3-6p					
Fredericksburg	WFVA	1230	Sa4-7p					
Fredericksburg	WFVA	1230	7-10p					
Gretna	WMNA	106.3	9a-12p					
Harrisonburg	WKCI	970	8-10p					
Harrisonburg	WKCY	1300	8-10p					
Richmond	WTOX	1480	10a-12p					
Roanoke-L'burg	WFIR	960	8:30-10a					
Roanoke-L'burg	WLNI	105.9	9-11a					
Washington								
Colfax	KMAX	840	6-9a					
Ellensburg	KXLE	1240	SaSu5-7p					
Pasco	KGDC	1320	6-9a					
Seattle	KTTH	770	8-11p					
Spokane	KGA	1510	12-2p					
Tri-Cities	KTCR	1340	6-9a					
Tri-Cities	KTCR	1340	Su4-7p					
Yakima	KUSA	980	4-7p					
Wisconsin								
Eau Claire	WKLJ	1290	9-10a					
Eau Claire	WOGO	680	6-9p					
Shell Lake	WCSW	940	8-11a					
West Virginia								
Martinsburg	WRNR	740	10a-12p					
Roncerverte	WRON	1400	M 5-6a					

Radio on the Road   Satellite Sisters

Location	Sta	Freq	Airtime		Location	Sta	Freq	Airtime
AL					MA			
Bridgeport	WYMR	1480	Sa 1-4p		Cape Cod	WXTK	95.1	9a-12p
Florence	WBCF	1240	8-11a		Springfield	WHMP	1400	Sa12p-2p
AK					Springfield	WHMQ	1240	Sa12p-2p
Anchorage	KBYR	700	5a-8a		Springfield	WHNP	1600	Sa12p-2p
AZ					MI			
Lk Havasu City	KNTR	980	Sa11-1p		Detroit	CIDR	93.9	Su9p-12a
Prescott	KYCA	1490	Su1-4p		Detroit	CKLW	800	Su11a-1p
AR	N/A				Escanaba	WDBC	680	Sa1p-4p
CA					Manistique	WPIQ	99.9	9a-12p
Bakersfield	KNZR	1560	6a8a		MN			
Los Angeles	KABC	790	6a-9a		Duluth	WDSM	710	Sa6p-9p
Merced	KYOS	1480	6A-8A		Minneapolis	WFMP	107.1	Su1p-4p
Palm Springs	KPSI	920	6a-9a		Springfield	KNSG	94.7	Sa8a-11a
Palmdale	KUTY	1470	6a-9a		Willmar	KWLM	1340	Sa12:30-3p
San Diego	KFMB	760	7p-10p		MS			
San Francisco	KVON	1440	Sa10a-1p		Greenwood	WDSK	1410	Sa12p-3p
CO					MO			
Denver	KNUS	710	7a-10a		Columbia	KFRU	1400	Sa12p-3p
CT					Columbia	KFRU	1400	Sa9p-12a
Hartford	WTIC	1080	7p-10p		Jefferson City	KLIK	1240	Sa12p-3p
DE	N/A				Jefferson City	KLIK	1240	Sa9p-12a
D of C					Farmington	KREI	800	Sa12p-3p
Washington	WWRC	1260	9a-12p		Kansas City	KCMO	710	Sa6p-9p
FL					Springfield	KWTO	560	Sa1p-3p
Ft. Meyers	WINK	1240	Sa1p-4p		St. Joseph	KFEQ	680	Sa12p-3p
Jacksonville	WOKV	690	10p-1a		MT			
Melbourne	WMEL	920	8p-10p		Hamilton	KLYQ	1240	Su6p-8p
Naples	WNOG	1270	Sa1-4p		Missoula	KGVO	1290	Su6p-8p
Ocala	WOCA	1370	Su4-6p		NE			
Orlando	WDBO	580	M12a-3a		Hastings	KLIQ	94.5	Su9a-12p
Sarasota	WIBQ	1220			NV			
GA					Reno	KTHO	590	Sa6a-9a
Atlanta	WSBA	750	Sa8p-11p		NH			
HI					Laconia	WEMJ	1490	Sa1p-3p
Honolulu	KHBZ	990	Su11a-2p		Manchester	WXXL	1450	10a-12p
ID					NJ	N/A		
Blackfoot	KSLJ	690	Su2p-4p		NM			
Idaho Falls	KSSL	1260	Su2p-4p		Albuquerque	KTBL	1050	Sa5p-8p
IL					NY			
Champaign	WDWS	1400	Su6p-9p		New York	WLIB	1190	Su10a-1p
Mt. Carmel	WVMC	1360	Sa6p-9p		Olean	WOEN	1360	9a-12p
Princeton	WZOE	1490	8a-11a		Watertown	WTNY	790	Sa7p-9p
Quincy	WGEM	105.1	San-3p		NC			
IN					Charlotte	WRHI	1340	Su6p-9p
Elkhart	WTRC	1340	Su6p-9p		Raleigh	WCHL	1360	9a-12p
IA					Southern Pines	WEEB	990	Sa1p-4p
Davenport	WOC	1420	Sa6p-9p		ND			
Ottumwa	KBIZ	1240	Sa12p-3-p		Grand Forks	KCNN	1590	12p-3p
KS					OH			
Holcomb	KBUF	1030	Sa12p-3p		Dayton	WHIO	1290	Su3p-6p
KY					OK			
Hopkinsville	WHOP	1230	Sa12p-3p		Tulsa	KRMG	740	Sa9p-12a
Sa9p-12a	WSBX	750	8a-11a		OR			
LA					Bend	KBND	1110	Sa7p-10p
Baton Rouge	WYNK	1380	8a-11a		Coos Bay	KBBR	1340	6a-9a

Oregon										
Eugene	KPNW	1120	Sa1p-4p							
Medford	KAGO	1150	Sa4p-7p							
Portland	KPOJ	620	6a-9a							
ME	N/A									
MD										
Salisbury	WQMR	101.1	Sa1p-4p							
PA										
Altoona	WRTA	1240	Sa9a-10a							
State College	WBLF	970	Sa7p10p							
State College	WRSC	1390	Sa7p-10p							
RI	N/A									
SC										
Anderson	WAIM	1230	9a-12p							
Charleston	WGTN	1400	Sa9a-12p							
Greenville	WFIS	1600	Sa7p-10p							
SD										
Rapid City	KOTA	1380	Su5a-8a							
TN										
Knoxville	WETR	760	9a-12p							
TX										
Abilene	KWKC	1340	Sa6p-9p							
Beaumont	KOLE	1340	Sa10p12a							
Beaumont	KRCM	1380	Sa10p-12a							
Clayton	KLMX	1450	Sa1p-4p							
Tyler	KTBB	600	Su9p-12a							
UT										
Cedar City	KNNZ	940	Sa11a-2p							
Logan	KVNU	610	Sa11a-2p							
Salt Lake City	KSL	1160	Su2p-3p							
St. George	KDXU	890	Su6a-9a							
VT	N/A									
VA										
Martinsville	WMVA	1450	9a-12p							
Roanoke	WFIR	960	Su6p-8p							
WA										
Bellingham	KGMI	790	Su7a-10a							
WV										
Wheeling	WVLY	1370	Sa1p-4p							
WI										
Appleton	WHBY	1150	Sa2p-5p							
Ft. Atkinson	WFAW	940	Sa12p-3p							
WY										
Sheridan	KROE	930	7a-10a							

Radio on the Road

Location	Sta.	Freq	Airtime	Location	Sta.	Freq	Airtime
AK				**AZ**			
Anchorage	KSKA	91.1	Sa2pSu10a	Tucson	KUAT	90.5	Sa4pSu10a
Barrow	KBRW	680	Sa2pSu10a	Yuma	KAWC	1320	Th2pSa6p
Bethel	KYUK	640	T10pSa2p	**CA**			
Chevak	KCUK	88.1	Sa2p	Arcata	KHSU	90.5	Sa3p
Dillingham	KDLG	670	Sa2p	Arcata	KNHM	91.5	Sa3pSu12p
Fairbanks	KUAC	89.9	Sa5p	Calexico	KQVO	97.7	Sa5pSu11a
Ft. Yukon	KZPA	900	Sa5p	Chico	KCHO	91.7	Sa6pSu11a
Galena	KIYU	910	Sa5p	Crescent City	KHSR	91.9	Sa3p
Glenallen	KXGA	90.5	Sa5p	Groveland	KXSR	91.7	Sa6p
Haines	KHNS	102.3	Sa2p	Los Angeles	KPCC	89.3	Sa3pSu11a
Homer	KBBI	890	Sa2p	Mendocino	KPMO	1300	Sa3pSu12p
Juneau	KTOO	104.3	Sa4p	MontereyArea	KAZU	90.3	Sa6pSu11a
Kenai	KDLL	91.9	Su10a	Mt Shasta	KMJC	620	Sa3pSu12p
Kenai	KDLL	91.9	Sa2p	Philo	KZYX	90.7	Sa6p
Ketchikan	KRBD	105.9	Sa2p	Quincy	KQNC	88.1	Su3p
Kodiak	KMXT	100.1	Sa2p	Redding	KFPR	88.9	Sa6pSu11a
Kotzebue	KOTZ	720	Sa5p	Redding	KJPR	1330	Sa3pSu12p
McCarthy	KXKM	89.7	Sa5p	Redding	KJPR	1330	Sa3p
McGrath	KSKO	870	Sa5p	Riverside	KVCR	91.9	Su3p
Petersburg	KFSK	100.9	Sa2pSu5p	Riverside	KVCR	91.9	F10aSa3p
Sand Point	KSDP	840	Sa2p	Sacramento	KXJZ	88.9	Su3p
Seward	KSWD	950	Sa2pSu10a	Sacramento	KQEI	89.3	Su11a
Sitka	KCAW	104.7	Sa6p	Sacramento	KQEI	89.3	Sa6p
St. Paul Isl.	KUHB	91.1	Sa4p	Sacramento	KTTO	90.5	Su3p
Talkeetna	KTNA	88.5	Sa5p	Sacramento	KXPR	90.9	Sa6p
Unalakleet	KNSA	930	Sa2p	San Diego	KPBS	89.5	Sa5pSu11a
Unalaska	KIAL	1450	Sa2p	San Francisco	KQED	88.5	Sa6pSu11a
Valdez	KCHU	770	Sa5p	S. Luis Obspo	KCBX	90.1	Sa3p
Wrangell	KSTK	101.7	Sa2p	Santa Barbara	KCLU	88.3	Sa5pSu10a
AL				Santa Barbara	KSBX	89.5	Sa3p
Birmingham	WBHM	90.3	Sa5pSu11a	Stockton	KUOP	91.3	Su3p
Dothan	WRWA	88.7	Sa5pSu11a	Tahoe City	KKTO	90.5	Su3p
Gadsden	WSGN	91.5	Sa5pSu11a	Willits	KZYZ	91.5	Sa6p
Huntsville	WLRH	89.3	Sa5pSu1p	Yreka	KSYC	1490	Sa3pSu12p
Muscle Shls	WQPR	88.7	Sa5pSu2p	Yuba City	KXJS	88.7	Su3p
Selma	WAPR	88.3	Sa5pSu2p	**CO**			
Troy	WTSU	89.9	Sa5pSu2p	Aspen	KCJX	88.9	Sa4pSu1p
Tuscaloosa	WUAL	91.5	Sa5pSu2p	Aspen	KAJX	91.5	Sa4pSu1p
AR				Boulder	KCFC	1490	Sa6pSu10a
El Dorado	KBSA	90.9	Sa5pSu12	Colorado Sprg	KRCC	91.5	Sa6pSu11a
Fayetteville	KUAF	91.3	Sa5Su1p	Crested Butte	KBUT	90.3	Sa4p
Jonesboro	KASU	91.9	Sa5Su2p	Denver-Bldr	KCFR	1340	Sa6pSu10a
Little Rock	KUAR	89.1	11aSa5p	Ft. Collins	KUNC	91.5	Sa4pSu11a
Texarkana	KTXX	91.5	12pSa5p	Grand Junction	KPRN	89.5	Sa6pSu10a
AZ				Grand Junction	KPRU	103.3	Sa6pSu10a
Flagstaff	KNAU	88.7	Sa6p	Ignacio	KUTE	90.1	Sa4pSu12p
Flagstaff	KPUB	91.7	Su10a	Ignacio	KSUT	91.3	Sa4pSu12p
Grand Cnyn	KNAG	90.3	Su6pSa6p	La Junta	KRLJ	89.1	Sa6pSu11a
Page	KNAD	91.7	Su6pSa6p	Montrose	KPRH	88.3	Sa6pSu10a
Phoenix	KJZZ	91.5	Sa5pSu10a	Pueblo	KCFP	91.1	Sa6pSu10a
Phoenix	KJZZ	91.5	Sa5pSu10a	Pueblo	KKPC	1230	Sa6pSu10a
Prescott	KNAQ	89.3	Sa6pSu6p	Vail	KPRE	89.9	Sa6pSu10a
Safford	KUAZ	89.1	Su10a				
Show Low	KNAA	90.7	Su6pSa6p				

Location	Sta.	Freq	Airtime	Location	Sta.	Freq	Airtime
CT				**IA**			
Bridgeport	WSHU	91.1	Sa6p	Mason City	KRNI	1010	T12pSa5p
Hartford	WPKT	90.5	Sa6pSu6p	Mitchelville	KDMR	88.9	Sa5pSu12p
Norwich	WNPR	89.1	Sa6pSu6p	Sioux City	KWIT	90.3	Sa5pSu2p
Stamford	WEDW	88.5	Sa6pSu6p	Spirit Lake	KOJI	90.7	Sa5pSu2p
Westport	WSHU	1260	Sa2PSa6p	Waterloo	KUNI	90.9	Sa5Su12p
D of C				**ID**			
Washington	WETA	90.9	Sa6Su12p	Boise	KBSU	90.3	Sa4Su7pW10
FL				Bonners Ferry	KIBX	92.1	Sa3pSu6p
Ft. Myers	WGCU	90.1	Sa6pSu3p	Cottonwood	KNOW	90.1	Sa5pSu12p
Ft. Pierce	WQCS	88.9	Sa6pSu3p	McCall	KBSM	91.7	W10pSa4pSu7p
Gainesville	WUFT	89.1	Sa6p	Moscow	KRFA	91.7	Sa5pSu12p
Inverness	WJUF	90.1	Sa6p	Pocatello	KISU	91.1	Sa6pSu3p
Jacksonville	WJCT	89.9	Sa6pSu2p	Rexburg	KBYI	100.5	Sa4p
Marco	WMKO	91.7	Sa6pSu3p	Sun Valley	KWRV	91.9	Sa4p
Miami	WLRN	91.3	Sa6pSu12p	Twin Falls	KBSW	91.7	W10pSa4pSu9
Orlando	WMFE	90.7	Sa6pSo10a	**IL**			
Panama City	WFSW	89.1	Sa6pSo10a	Carbondale	WSIU	91.9	Sa5pSu10a
Pensacola	WUWF	88.1	Sa5pSu10a	Chicago	WBEQ	90.7	Sa5p
Tallahassee	WFSU	88.9	Sa6pSu3p	Chicago	WBEZ	91.5	Sa5p
Tampa	WUSF	89.7	Sa6pSu3p	DeKalb	WNIJ	89.5	Sa5pSu10a
GA				Freeport	WNIE	89.1	Sa5pSu10a
Albany	WUNV	91.7	Sa6pSu2p	LaSalle	WNIW	91.3	Sa5pSu10a
Athens	WUGA	91.7	Sa6pSu2p	Macomb	WIUM	91.3	Sa5pSu10a
Atlanta	WABE	90.1	Sa6pSu10a	Mt. Vernon	WVSI	88.9	Sa5pSu10a
Augusta	WACG	90.7	Sa6pSu2p	Olney	WUSI	90.3	Sa5pSu10a
Brunswick	WWIO	88.9	Sa6pSu2p	Peoria	WCBU	89.9	Sa5pSu6p
Carrollton	WUWG	90.7	Sa6pSu2p	Pittsfield	WIPA	89.3	Sa5pSu12p
Columbus	WJSP	88.1	Sa6pSu2p	Quincy	WQUB	90.3	Sa5pSu12p
Columbus	WTJB	91.7	Sa6pSu2p	Rock Island	WVIK	90.3	Sa5pSu5p
Columbus	WTJB	91.7	Sa5p	Springfield	WUIS	91.9	Sa5pSu12p
Dahlonega	WNGU	89.5	Sa6pSu2p	Sterling	WNIQ	91.5	Sa5pSu10a
Demorest	WPPR	88.3	Sa6pSu2p	Urbana	WILL	90.9	Sa5p
Ft. Gaines	WJWV	90.9	Sa6pSu2p	Urbana	WILL	580	Su2p
Macon	WDCO	89.7	Sa6pSu2p	Warsaw	WIUW	89.5	Sa5p
Savannah	WSVH	91.1	Sa6pSu2p	**IN**			
St. Marys	WWIO	1190	Sa6pSu2p	Anderson	WBSB	89.5	Sa6pSu12p
Tifton	WABR	91.1	Sa6pSu2p	Bloomington	WFIU	103.7	Sa6p
Valdosta	WWET	91.7	Sa6pSu2p	Chesterton	WBEW	89.5	Sa5p
Waycross	WXVS	90.1	Sa6pSu2p	Crawfordsville	WVXI	106.3	Sa6pSu12p
HI				Elkhart	WVPE	88.1	Sa6pSu10a
Honolulu	KHPR	88.1	Sa6p	Evansville	WNIN	88.3	Sa5pSu10a
Honolulu	KIPO	89.3	Sa6p	Ft. Wayne	WBOI	91.3	Sa6pSu10a
Honolulu	KANO	91.1	Sa6p	Hagerstown	WBSH	91.1	Sa6pSu12p
Wailuku	KKUA	90.7	Sa6p	Indianapolis	WFYI	90.1	Sa6pSu10a
IA				Marion	WBSW	90.9	Sa6pSu12p
Carroll	KWOI	90.7	Sa5p	Muncie	WBST	92.1	Sa6pSu12p
Council Bluffs	KIOS	91.5	Sa5p	Portland	WBSJ	91.7	Sa6pSu12p
Decorah	KLNI	88.7	Sa5pSu11a	Richmond	WVXR	89.3	Sa6pSu12p
Decorah	KLCD	89.5	Sa5p	Vincennes	WVUB	91.1	Sa6pSu10a
Des Moines-	WOI	90.1	Sa5p	West Lafayette	WBAA	101.3	Sa6p
-Ames	WOI	640	Sa5pSu1p				
Ft. Dodge	KTPR	91.1	Sa5p				

Location	Sta.	Freq	Airtime	Location	Sta.	Freq	Airtime
Iowa City	KSUI	91.7	Sa5p				
Lamoni	KOWI	97.9	Sa5p				
Mason City	KUNY	91.5	T12pSa5p				
KS				**MI**			
Emporia	KANH	89.7	Sa5pSu10a	Alpena	WCML	91.7	Sa6pSu12p
Garden City	KANZ	91.1	Sa5pSu1p	Detroit	WUOM	91.7	Sa6pSu1p
Gt Bend/Hays	KHCT	90.9	Sa5pSu1p	East Jordan	WICV	100.9	Sa6p
Hays	KZAN	91.7	Sa5pSu1p	Flint	WFUM	91.1	Sa6pSu1p
Hill City/Hays	KZNA	90.5	Sa5pSu1p	Grand Rapids	WBLU	88.9	Sa6pSu1p
Hutchinson	KHCC	90.1	Sa5pSu1p	Grand Rapids	WVGR	104.1	Sa6pSu1p
Junction City	KANV	91.3	Sa5pSu10a	Harbor Springs	WCMW	103.9	Sa6pSu12p
Kansas City	KCUR	89.3	Sa5pSu11a	Harrison	WVXH	92.1	Sa6pSu12p
Manhattan	KHCD	89.5	Sa5pSu1p	Houghton	WGGL	91.1	Sa5pSu11a
Pittsburg	KRPS	89.9	Sa5pSu3p	Interlochen	WIAA	88.7	Sa6p
Topeka	KANU	91.5	Sa5pSu10a	Interlochen	WICA	91.5	Su10a
Wichita	KMUW	89.1	Sa6pSu12p	Kalamazoo	WMUK	102.1	Sa6p
KY				Lansing	WKAR	90.5	Sa6pSu10a
Ashland	WWWV	89.9	Sa6Su1p	Manistee	WVXM	97.7	Sa6pSu12p
Bowling Grn	WKYU	88.9	Sa5pSu2p	Marquette	WNMU	90.1	Sa6p
Elizabethtown	WKUE	90.9	Sa5pSu2p	Mount Pleasant	WCMU	89.5	Sa6pSu12p
Henderson	WKPB	89.5	Sa5pSu2p	Muskegon	WBLV	90.3	Sa6p
LeRose	WOCS	88.3	Sa2pSu6p	Oscoda	WCMB	95.7	Sa6pSu12p
Lexington	WUKY	91.3	Sa6pSu2p	Rogers City	WVXA	96.7	Sa6pSu12p
Louisville	WFPL	89.3	Sa6pSu12p	S. Ste. Marie	WCMZ	98.3	Sa6pSu12p
Murray	WKMS	91.3	Sa5pSu11a	Saginaw	WUCX	90.1	Sa6pSu12p
LA				Standish	WWCM	96.9	Sa6pSu12p
Alexandria	KLSA	90.7	Sa5pSu12p	**MN**			
Baton Rouge	WRKF	89.3	Sa5pSu10a	Appleton	KNCM	88.5	Sa5pSu11a
Monroe	KEDM	90.3	Sa5pSu12p	Appleton	KRSU	91.3	Sa5p
New Orleans	WWNO	89.9	Sa5pSu12p	Austin	KNSE	90.1	Sa5pSu11a
Shreveport	KDAQ	89.9	Sa5pSu12p	Bemidji	KCRB	88.5	Sa5pSu11a
Thibodeaux	KTLN	90.5	Sa5pSu2p	Bemidji	KNBJ	91.3	Sa5pSu11a
MA				Brainerd	KBPN	88.3	Sa5pSu11a
Amherst	WFCR	88.5	Sa6pSu12p	Brainerd	KBPR	90.7	Sa5p
Boston	WGBH	89.7	Sa6pSu12p	Buhl	WIRR	90.9	Sa5p
Gt Barrington	WAMQ	105.1	Sa6pSu3p	Buhl	WIRN	92.5	Sa5pSu11a
Nantucket	WNAN	91.1	Sa6pSu10a	Collegeville	KNSR	88.9	Sa5pSu11a
Woods Hole	WCAI	90.1	Sa6pSu10a	Collegeville	KSJR	90.1	Sa5p
MD				Fergus Fall	KNWF	91.5	Sa5pSu11a
Baltimore	WYPR	88.1	Sa6pSu10a	Fergus Falls	KCMF	89.7	Sa5p
Hagerstown	WETH	89.1	Sa6pSu2p	Grand Marais	WMLS	88.7	Sa5p
Salisbury	WSCL	89.5	Sa6pSu12p	Grand Marais	WLSN	89.7	Sa5pSu11a
ME				LaCrosse	WLSU	88.9	Sa5p
Augusta	WMEW	91.3	Sa6pSu12p	LaCrosse	WHLA	90.3	Sa5p
Bangor	WMEH	90.9	Sa6pSu12p	Mankato	KGAC	90.5	Sa5pSu11a
Calais	WMED	89.7	Sa6pSu12p	Mankato	KNGA	91.5	Sa5pSu11a
Camden	WMEP	90.5	Sa6pSu12p	Minneapolis	KNOW	91.1	Sa5pSu11a
Ft. Kent	WMEF	106.5	Sa6pSu12p	Moorhead	KCCD	90.3	Sa5pSu11a
Portland	WMEA	90.1	Sa6pSu12p	Moorhead	KCCM	91.1	Sa5p
Presque Isle	WMEM	106.1	Sa6pSu12p	Moorhead	KDSU	91.9	Sa5pSu2p
				Rochester	KZSE	90.7	Sa5pSu11a
				Rochester	KLSE	91.7	Sa5pSu11a
				St. Paul	KSJN	99.5	Sa5pSu11a

Minnesota	Sta.	Freq	Airtime
Thief Riv Falls	KQMN	91.5	Sa5pSu11a
Thief Riv Falls	KNTN	102.7	Sa5pSu11a
Worthington	KNSW	89.3	Sa5pSu11a
Worthington	KRSW	91.7	Sa5p
MO			
Chillicothe	KRNW	88.9	Sa5pSu3p
Columbia	KKTR	89.7	Sa5pSu12p
Columbia	KBIA	91.3	Sa5pSu12p
Maryville	KXCV	90.5	Sa5pSu3p
Point Lookout	KSMS	90.5	Sa5pSu4p
Rolla	KUMR	88.5	Sa5pSu12p
Springfield	KSMU	91.1	Sa5pSu4p
St. Louis	KWMU	90.7	Sa5pSu1p
Warrensburg	KTBG	90.9	Sa5pSu10a
West Plains	KSMW	90.9	Sa5pSu4p
MS			
Biloxi	WMAH	90.3	Sa5p
Booneville	WMAE	89.5	Sa5p
Bude	WMAU	88.9	Sa5p
Greenwood	WMAO	90.9	Sa5p
Jackson	WMPN	91.3	Sa5p
Meridian	WMAW	88.1	Sa5p
Miss. State	WMAB	89.9	Sa5p
Oxford	WMAV	90.3	Sa5p
MT			
Billings	KEMC	91.7	Sa4pSu1p
Bozeman	KBMC	102.1	Sa4pSu1p
Butte	KAPC	91.3	Sa6p
Great Falls	KGPR	89.9	Sa6p
Hamilton	KUFN	91.1	Sa6p
Helena	KUHM	91.7	Sa6p
Kalispell	KUKL	89.9	Sa6p
Miles City	KECC	90.7	Sa4pSu1p
Missoula	KUFM	89.1	Sa6p
NC			
Asheville	WCQS	88.1	Sa6pSu10a
Buxton	WBUX	90.5	Sa6pSu1p
Charlotte	WFAE	90.7	Sa6pSu12p
Elizabeth City	WUND	88.9	Sa6pSu1p
Elizabeth City	WURI	90.9	Sa6pSu1p
Franklin	WFQS	91.3	Sa6pSu1p
New Bern	WZNB	88.5	Sa6p
Greenville-	WTEB	89.3	Sa6p
New Bern	WKNS	90.3	Sa6pSu12p
Greenville-	WBJD	91.5	Sa6p
Hickory	WFHE	90.3	Sa6pSu12p
Raleigh-Dur	WUNC	91.5	Sa6pSu1p
Roanoke Rap	WZRU	88.5	Sa6pSu10a
Roanoke Rap	WZRN	90.5	Sa6pSu10a
Rocky Mount	WRQM	90.9	Sa6pSu1p
Wilmington	WHQR	91.3	Sa6pSu12p
Winston Salem	WFDD	88.5	Sa6pSu12p
ND			
Bismarck	KCND	90.5	Sa5pSu2p
Dickinson	KDPR	89.9	Sa5pSu2p
Grand Forks	KUND	89.3	Sa5pSu2p

North Dakota	Sta.	Freq	Airtime
Jamestown	KPRJ	91.5	Sa5pSu2p
Minot	KMPR	88.9	Sa5pSu2p
Williston	KPPR	89.5	Sa5pSu2p
NE			
Alliance	KTNE	91.1	Sa5pSu11a
Bassett	KMNE	90.3	Sa5pSu11a
Chadron	KCNE	91.9	Sa5pSu11a
Hastings	KHNE	89.1	Sa5pSu11a
Lexington	KLNE	88.7	Sa5pSu11a
Lincoln	KUCV	91.1	Sa5pSu11a
Merriman	KRNE	91.5	Sa5pSu11a
Norfolk	KXNE	89.3	Sa5pSu11a
North Platte	KPNE	91.7	Sa5pSu11a
NH			
Berlin	WEVC	107.1	Sa6pSu11a
Concord	WEVO	89.1	Sa6pSu11a
Hanover	WEVH	91.3	Sa6pSu11a
Jackson	WEVJ	99.5	Sa6pSu11a
Keene	WEVN	90.7	Sa6pSu11a
NJ			
Atlantic City	WNJN	89.7	Sa6pSu12p
Berlin	WNJS	88.1	Sa6pSu12p
Bridgeton	WNJB	89.3	Sa6pSu12p
Cape May	WNJZ	90.3	Sa6pSu12p
Manahawkin	WNJM	89.9	Sa6pSu12p
Sussex	WNJP	88.5	Sa6pSu12p
Trenton	WNJT	88.1	Sa6pSu12p
NM			
Albuquerque	KANW	89.1	Sa4pSu7p
Gallup	KGLP	91.7	Sa4pSu12p
Las Cruces	KRWG	90.7	Sa6pSu12p
Maljamar	KMTH	98.7	Sa4p
Portales	KENW	89.5	Sa4p
NV			
Elko	KNCC	91.5	Sa6pSu12p
Las Vegas	KNPR	89.5	Sa6pSu11a
Lund/Ely	KWPR	88.7	Sa6pSu11a
Panaca	KLNR	91.7	Sa6pSu11a
Reno	KUNR	88.7	Sa6pSu12p
Tonopah	KTPH	91.7	Sa6pSu11a
NY			
Albany	WAMC	1400	Sa6pSu3p
Albany-Schtdy.	WAMC	90.3	Sa6pSu3p
Binghamton	WSKG	89.3	Sa6Su2p
Blue Mtn Lake	WXLH	91.3	Sa6p
Brookville, NY	WCWP	88.1	Sa6p
Buffalo-Niag	WEOS	89.7	Sa6p
Buffalo-Niag	WNED	94.5	Sa6p
Brookville, NY	WCWP	88.1	Sa6p
Buffalo-Niag	WEOS	89.7	Sa6p
Buffalo-Niag	WNED	94.5	Sa6p
Buffalo-Niag	WNED	970	Su12p
Canajoharie	WCAN	93.3	Sa6pSu3p
Cortland	WSUC	90.5	Sa6pSu12p
Greenport	WSUF	89.9	Sa6p

Location	Sta.	Freq	Airtime	Location	Sta.	Freq	Airtime
NY				**OK**			
Hornell	WSQA	88.7	Sa6Su2p	Altus	KOCU	90.1	Sa5pSu1p
Ithaca	WSQG	90.9	Sa6Su2p	Ardmore	KLCU	90.3	Sa5pSu1p
Jamestown	WNJA	89.7	Sa6pSu3p	Clinton	KYCU	89.1	Sa5pSu1p
Kingston	WAMK	90.9	Sa6pSu3p	Lawton	KCCU	89.3	Sa5pSu1p
Malone	WSLO	90.9	Sa6pSu3p	Stillwater	KOSU	91.7	Sa5pSu1p
Middletown	WOSR	91.7	Sa6pSu3p	Tulsa	KWGG	89.5	Sa5pSu1p
New York	WNYC	93.9	Sa6p	**OR**			
New York	WNYC	820	Sa6pSu11a	Bend	KOAB	91.3	Sa3pSu12p
North Creek	WXLG	89.9	Sa6p	Eugene	KRVM	1280	Sa3pSu12p
Oneonta	WSQC	91.7	Sa6pSu2p	Grants Pass	KAGI	930	Sa3pSu12p
Oswego	WRVO	89.9	Sa6pSu12p	LaGrande	KTVR	89.9	Sa3pSu12p
Peru	WXLU	88.3	Sa6p	Lakeview	KOAP	88.7	Sa3pSu12p
Plattsburgh	WCEL	91.9	Sa6pSu3p	Pendleton	KRBM	90.9	Sa3pSu12p
Rochester	WXXI	91.5	Sa6p	Portland	KOPB	91.5	Sa3pSu12p
Rochester	WXXI	1370	Su12p	Portland	KOAC	550	Sa3pSu12p
Saranac Lake	WSLL	90.5	Sa6p	Roseberg	KTBR	950	Sa3pSu12p
Southampton	WLIU	88.3	Sa6p	Talent	KSJK	1230	Sa3pSu12p
Southampton	WRLI	91.3	Sa6pSu6p	**PA**			
Syracuse	WRVD	90.3	Sa6pSu12p	Erie	WQLN	91.3	Sa3pSu12p
Ticonderoga	WANC	103.9	Sa6pSu3p	Harrisburg	WITF	89.5	Sa6pSu1p
Utica-Rome	WRVN	91.9	Sa6pSu12p	Johnstown	WQEJ	89.7	Sa6p
Watertown	WSLJ	88.9	Sa6p	Kane	WPSB	90.1	Sa6pSu2p
Watertown	WRVJ	91.7	Sa6pSu12p	Philadelphia	WHYY	90.9	Sa6pSu2p
OH				Pittsburgh	WQED	89.3	Sa6p
Akron	WKSU	89.7	Sa6pSu10a	State College	WPSU	91.5	Sa6pSu2p
Akron	WNRK	90.7	Sa6pSu10a	Wilkes Barre-	WVIA	89.9	Sa6pSu8p
Athens	WOUB	91.3	Sa6p	Williamsport	WVYA	89.7	Sa6pSu8p
Bryan	WGBE	90.9	Sa6pSu12p	**SC**			
Cambridge	WOUC	89.1	Sa6p	Aiken	WLJK	89.1	Su10aSu12p
Canton	WSLU	89.5	Sa6p	Aiken	WLJK	89.1	Sa6p
Chillicothe	WVXC	89.3	Sa6pSu12p	Beaufort	WJWJ	89.9	Sa6pSu12
Chillicothe	WOUH	91.9	Sa6p	Charleston	WSCI	89.3	Sa6pSu10a
Cincinnati	WVXU	91.7	Sa6pSu12p	Columbia	WLTR	91.3	Sa6pSu10a
Cleveland	WCPN	90.3	Sa6p	Conway	WHMC	90.1	Sa6pSu10a
Columbus	WCBE	90.5	Sa6pSu10a	Greenville-	WEPR	90.1	Sa6pSu10a
Defiance	WGDE	91.9	Sa6pSu2p	Sumter	WRJA	88.1	Sa6pSu12p
Dover	WKRJ	91.5	Sa6pSu10a	**So. Dakota**			
Ironton	WOUL	89.1	Sa6pSu2p	Aberdeen	KDSD	90.9	Sa5pSu12p
Lima	WGLE	90.7	Sa6pSu2p	Brookings	KESD	88.3	Sa5pSu12p
Mt. Gilead	WVXG	95.1	Sa6pSu12p	Faith	KPSD	97.1	Sa5pSu12p
Oxford	WMUB	88.5	Sa6pSu10a	Lowry	KQSD	91.9	Sa5pSu12p
Thompson	WKSV	89.1	Sa6pSu10a	Martin	KZSD	102.5	Sa5pSu12p
Toledo	WGTE	91.3	Sa6pSu2p	Pierre	KTSD	91.1	Sa5pSu12p
West Union	WVXW	89.5	Sa6pSu12p	Rapid City	KBHE	89.3	Sa5pSu12p
Wooster	WKRW	89.3	Sa6pSu10a	Sioux Falls	KRSD	88.1	Sa5pSu12p
Youngstown	WYSU	88.5	Sa6p	Sioux Falls	KCSD	90.9	Sa5pSu12p
Zanesville	WOUZ	90.1	Sa6p	Vermillion	KUSD	89.7	Sa5pSu12p

Location	Sta.	Freq	Airtime	Location	Sta.	Freq	Airtime
TN				VT			
Chattanooga	WUTC	88.1	Sa6p	Bennington	WBTN	94.3	Sa6pSu12p
Cookeville	WHRS	91.7	Sa5pSu2p	Burlington	WVPS	107.9	Sa6pSu12p
Jackson	WKNP	90.1	Sa5pSu12p	Rutland	WRVT	88.7	Sa6pSu12p
Johnson City	WETS	89.5	Sa6pSu2p	St. Johnsbury	WVPA	88.5	Sa6pSu12p
Knoxville	WUOT	91.9	Sa6pSu2p	WA			
Memphis	WKNO	91.1	Sa5pSu12p	Bellingham	KZAZ	91.3	Sa5pSu12p
Nashville	WPLN	90.3	Sa5pSu2p	Clarkston	KNWV	90.5	Sa5pSu12p
Tullahoma	WTML	91.5	Sa5pSu2p	Ellensburg	KNWR	90.7	Sa5pSu12p
TX				Moses Lake	KLWS	91.5	Sa3pSu10a
Abilene	KACU	89.7	Sa5pSu2p	Mt. Vernon	KMWS	90.1	Sa3pSu10a
Amarillo	KJJP	94.9	Sa5pSu1p	Omak/Okanogan	KQWS	90.1	Sa3pSu10a
Austin	KUT	90.5	Sa5pSu12p	Port Angeles	KNWP	90.1	Sa5pSu12p
Beaumont	KVLU	91.3	Sa5pSu9a	Pullman	KWSU	1250	Sa3pSu10a
Bryan-College Sta	KAMU	90.9	Sa5pSu12p	Seattle-Tacoma	KUOW	94.9	Sa3pSu10a
Bushland	KTXP	91.5	Sa5pSu1p	Spokane	KPBX	91.1	Sa3pSu6p
Commerce	KETR	88.9	Sa5pSu4p	Tri-Cities	KFAE	89.1	Sa5pSu12p
Corpus Christi	KEDT	90.3	Sa5pSu4p	Walla Walla	KWWS	89.7	Sa3pSu10a
Dallas-Ft. Worth	KERA	90.1	Sa5pSu10a	Yakima	KNWY	90.3	Sa5pSu12p
El Paso	KTEP	88.5	Sa4p	WV			
Houston-Galveston	KUHF	88.7	Sa5pSu12p	Beckley	WVPB	91.7	Sa6pSu1p
Ingram	KTXI	90.1	Sa5pSu2p	Buckhannon	WVPW	88.9	Sa6pSu1p
Killeen-Temple	KNCT	91.3	Sa5p	Charleston	WVPN	88.5	Sa6pSu1p
Lubbock	KOHM	89.1	Sa5p	Martinsburg	WVEP	88.9	Sa6pSu1p
Lufkin	KLDN	88.9	Sa5pSu12p	Morgantown	WVPM	90.9	Sa6pSu1p
Odessa	KOCV	91.3	Sa5pSu2p	WI			
San Angelo	KUTX	90.1	Sa5pSu12p	Auburndale	WLBL	930	Sa5p
San Antonio	KSTX	89.1	Sa5pSu2p	Brule	WHSA	89.9	Sa5p
Spearman	KTOT	89.5	Sa5pSu1p	Eau Claire	WUEC	89.7	Sa5p
Victoria	KVRT	90.7	Sa5pSu4p	Green Bay	WHID	88.1	Sa5p
Waco	KWBU	103.3	Sa5pSu10a	Green Bay	WPNE	89.3	Sa5p
Wichita Falls	KMCU	88.7	Sa5pSu1p	Highland	WHHI	91.3	Sa5p
UT				Kenosha	WGTD	91.1	Sa5p
Logan	KUSR	89.5	Sa6pSu12p	LaCrosse	KXLC	91.1	Sa5pSu11a
Logan	KUSU	91.5	Sa6pSu12p	Madison	WERN	88.7	Sa5p
Salt Lake City	KUER	90.1	Sa5pSu2p	Madison	WHA	970	Sa5p
VA				Menomonie	WHWC	88.3	Sa5p
Charlottesville	WVTW	88.5	Sa6pSu12p	Menomonie	WVSS	90.7	Sa5p
Charlottesville	WVTU	89.3	Sa6pSu12p	Milwaukee	WUWM	89.7	Sa5pSu12p
Charlottesville	WMRY	103.5	Sa6pSu11a	Milwaukee	WHAD	90.7	Sa5p
Farmville	WMLU	91.3	Sa6pSu11a	Oshkosh	WRST	90.3	Sa5p
Harrisonburg	WMRA	90.7	Sa6pSu11a	Park Falls	WHBM	90.3	Sa5p
Lexington	WMRL	89.9	Sa6pSu11a	Rhinelander	WXPR	91.7	Sa5p
Marion	WVTR	91.9	Sa6pSu12p	River Falls	WRFW	88.7	Sa5p
Norfolk-Va Bch	WHRV	89.5	Sa6pSu8a	Sheboygan	WSHS	91.7	Sa5p
Richmond	WCVE	88.9	Sa6pSu11a	Sister Bay	WHND	89.7	Sa5p
Roanoke-Lynchb'rg	WVTF	89.1	Sa6pSu12p	Sister Bay	WHDI	91.9	Sa5p
Wise	WISE	90.5	Sa6pSu12p	Superior	WSCD	92.9	Sa5p
Windsor	WVPR	89.5	Sa6pSu12p	Superior	WSCN	100.5	Sa5pSu11a
				Wausau	WHRM	90.9	Sa5p
				Wausau	WLBL	91.9	Sa5p

AK	Station	Freq	Airtime
Chevak	KCUK	88.1	Sa10A
Fairbanks	KIYU	104.7	Sa9a
Galena	KIYU	910	Sa9a
Glenallen	KXGA	90.5	Sa9a
Haines	KHNS	102.3	Sa 9a
Homer	KBBI	890	Sa11
Juneau	KTOO	104.3	Sa10A
Kenai	KDLL	91.9	Sa 11a
Ketchikan	KRBD	105.9	Su n W2a
Mc Carthy	KXKM	89.7	Sa 9a
Petersburg	KFSK	91.1	Sa 9a
Sitka	KCAW	104.7	Sa 11a
St. Paul	KUHB	540	Sat 11a
Talkeetna	KTNA	88.5	Sat 11a
Valdez	KCHU	770	Sat 9a
Wrangell		101.7	Sat 9a
AL			
Birmingham	WBHM	90.3	Sa9, Su 5
Dothan	WRWA	88.7	Sa 9
Gadsen	WSGN	91.5	Sa9, Su 5
Huntsville	WLRH	89.3	Sa9, Su11
Jacksonville	WLJS	91.9	Sat 9 am
Muscle Shls	WQPR	88.7	Sat 10 am
Selma	WAPR	88.3	Sa 10
Troy	WTSU	89.9	Sa 9
Tuscaloosa	WUAL	91.5	Sa 10
AR			
Asu	KASU	91.9	Sa9
El Dorado	KBSA	90.9	Sa9aSu2
Eldorado	KBSA	90.9	Sa9aSu2p
Little Rock	KUAR	89.1	Sa9Su3
AZ			
Flagstaff	KNAQ	89.3	Su4p
Flagstaff	KNAU	88.7	Sa10a
Grand Canyon	KNAG	90.3	Su4p
Page	KNAD	91.7	Su4p
Prescott	KNAQ	89.3	Su4p
Prescott, AZ	KNAQ	89.3	Sun 4 pm
Showlow	KNAA	90.7	Su4p
Tempe	KJZZ	91.5	Sa10aSu12
Tuba City	KGHT	91.5	Su4p
Tucson	KUAT	1550	Sa10aSu5p
Tucson	KUAZ	89.1	Sa10aSu5p
Yuma	KAWC	1320	Sa4pTh11a
CA			
Bakersfield	KPRX	89.1	Sa10a
Burney	KNCA	89.7	Sa11a
Chico	KCHO	91.7	Sa4Su10a
Fresno	KUBO	88.7	Su8a
Fresno	KVPR	89.3	SA10a

CA	Station	Freq	Airtime
Mt Shasta	KNSQ	88.1	Sa11a
Pacific Grove	KAZU	90.3	Sa10a
Pasadena	KPCC	89.3	Sa10Su10
Philo	KZYX	90.7	Sa12
Redding	KFPR	88.9	Sa4Su10a
Rohnert Park	KRCB	91.1	Sa9a
Sacramento	KXJZ	88.9	Sa10Su10
San Bernard'o	KVCR	91.9	Sa10Su10
San Diego	KPBS	89.5	Sa10Su10
San Francisco	KALW	91.7	Sa9W1P
San Francisco	KQED	88.5	Sa10Su10
S. Luis Obispo	KCBX	90.1	Sa10a
Santa Cruz	KUSP	88.9	Su9a
Stockton	KUOP	91.3	Sa10Su10
Tahoe City	KKTO	88.9	Sa10Su10
1000 Oaks	KCLU	88.3	Sa10aSu12
Yreka	KNYR	91.3	Su11a
Yreka	KSRG	88.3	Sa11a
CO			
Alamosa	KRZA	88.7	Sa4pTh11a
Aspen	RPTR	100.1	Su10aW11p
Basalt	KDNK	90.5	Su10aW11p
Carbondale	KRCC	91.5	Sa10aSu4p
Colorado Sprgs	KBUT	90.3	Sa1p
Crested Butte	KCFR	1340	Sa10aSu12
Denver	KUWR	88.9	Su10aW11p
Glenwood Sps	KPRN	89.5	Sa10aSu12
Grd Junction	KUNC	91.5	SA10Su10
Greeley	KSUT	91.3	Sa11a
Ignacio	KSJD	91.3	Su8a
Mancos	KPRH	88.3	Sa10aSu12
Montrose	KPRH	88.3	Sa10aSu12
Montrose	KVNF	90.9	Sa10a
Paonia	KCFP	91.9	Sa10aSu12
Pueblo	KPRH	91.9	Sa10aSu12
Pueblo	RPTR	90.9	Su10aW11p
Snowmass	KOTO	91.7	F 12
Telluride	KPRE	89.9	Sa10aSu12
Vail	WSHU	91.1	Sa6,7,10a
CT			
Fairfield	WPKT	89.1	Sa10Su10
Hartford	WNPR	89.1	Sa10Su10
Norwich	WEDW	88.5	Sa10Su10
Stamford	WMMM	1260	Sa6,7,10a
Westport	WAMU	88.5	Sa8a
Washington	N/A	N/A	N/A
FL	WGCU	90.1	Sa10a
FT MEYERS	WQCS	88.9	Sa10a
Ft Pierce	WUFT	89.1	Sa10a
Gainesvulle	WJCT	89.9	Sa10Su10a
Jacksonville	WFIT	89.5	Sa10aSu1p
Melbourne	WLRN	91.3	Sa10a
Miami	WMFE	90.7	Sa10aSu12p
Orlando	WMFE	90.7	Sa10aSu12p

FL	Station	Freq	Airtime	IN	Station	Freq	Airtime
Panama City	WFSW	89.1	Sa9aM12p	Anderson	WBSB	89.5	Sa10Su10
Panama City	WKGC	90.7	Sa9aSa5p	Bloomington	WFIU	103.7	Sa10a
Pensacola	WUWF	88.1	Sa9aSu5a	Elkhart	WVPE	88.1	Sa10aSu12p
Tallahassee	WFSU	88.9	Sa10aM1p	Evansville	WNIN	88.3	Sa9aSu9a
Tampa	WUSF	89.7	Sa10a	Ft. Wayne	WBNI	89.1	Sa10a
W. Palm Bch	WXEL	90.7	SA10a	Indianapolis	WFYI	90.1	Sa11a
GA				Marion	WBSW	90.9	Sa10Su10a
Albany	WUNV	91.7	Sa10aSu6p	Muncie	WBST	92.1	Sa10Su10a
Athens	WUGA	91.7	Sa10a	Newcastle	WBSH	91.1	Sa10aSu10a
Atlanta	WABE	90.1	Sa10Su6p	Portland	WBSJ	91.7	Sa10Su10a
Atlanta	WJSP	88.1	Sa10aSu6p	W. Lafayette	WBAA	920	S10aSu1p
Augusta	WACG	90.7	Sa10aSu6p	W. Lafayette	WBAA	101.3	S10aSu1p
Brunswick	WWIO	89.1	Sa10aSu6p	**KS**			
Carrollton	WWGC	90.7	Sa10aSu6p	Garden City	KANZ	91.3	Sa9a
Columbus	WTJB	91.7	Sa9aM1p	Great Bend	KHCD	89.5	Sa9aSu12p
Macon	WDCO	89.7	Sa10aSu6p	Hill City	KZNA	90.5	Sa9a
Tifton	WABR	91.1	Sa10aSu6p	Hutchinson	KHCC	90.1	Sa9aSu12p
Waycross	WXVS	90.1	Sa10aSu6p	Lawrence	KANU	91.5	Sa9a
HI				Pittsburg	KRPS	89.9	Sa9aSu6p
Hilo	KKNO	91.1	Sa5p Su10a	Salina	KHCT	90.9	Sa9aSu12p
Honolulu	KHPR	88.1	Sa10a	Wichita	KMUW	89.1	Sa9aSu5p
Honolulu	KIPO	89.3	Sa10a	**KY**			
Pearl City	KIFO	1380	Sa10a	Bowling Green	WKYU	88.9	Sa9aSu5p
Wailuku	KIPO	89.3	Sa5p Su10a	Elizabethtown	WKUE	90.9	Sa10aSu5p
IA				Hazard	WEKH	90.9	Sa 10a
Ames	WOI	90.1	Sa9aSu12p	Henderson	WKPB	89.5	Sa9aSu5p
Carroll	WOI	90.7	Sa9aSu12p	Lexington	WUKY	91.3	Su1p
Cedar Falls	KUNI	90.9	Sa9aSu4p	Louisville	WFPL	89.3	Sa10a
Des Moines	KUNI	101.7	Sa9aSu4p	Morehead	WMKY	90.3	Su1p
Desorah	KLNI	88.7	Sa11Su10a	Murray	WKMS	91.3	Sa9aSu5p
Ft Dodge	KTPR	91.1	Sa9aSu12p	Richmond	WEKU	88.9	Sa10a
Iowa City	WSUI	910	Sa12pSu4p	Somerset	WDCL	89.7	Sa10a
Mason City	KUNY	91.5	Sa9Su1p	**LA**			
ID				Alexandria	KLSA	90.7	Sa9aSu2p
Boise	KBSX	91.5	F11aSa8aSu4	Baton Rouge	WRKF	89.3	Sa10aSu12p
Moscow	KRFA	91.7	Sa10aSu4p	Grambling	KDAQ	90.7	Sa9aSu2p
Rexburg	KBYI	100.5	Sa8a	Houma	KTLN	90.5	Sa9aSu5p
Twin Falls	KBSW	91.7	F11aSa8aSu4	Lafayette	KRVS	88.7	Sa5p
IL				Monroe	KEDM	89.9	Sa9aSu11a
Carbondale	WSIU	91.9	Sa9aSu12p	New Orleans	WWNO	89.9	Sa9aSu5p
Chicago	WBEZ	91.5	Sa9a	Shreveport	KDAQ	89.9	Sa9aSu2p
Dekalb	WNIJ	89.5	Sa9aSu12p	**MA**			
Macomb	WIUM	91.3	Sa9aSu1p	AMHERST	WFCR	88.5	Sa11aSu4p
Normal	WGLT	89.1	Sa9Su11a	Boston	WBUR	90.9	Sa11aSu6p
Olney	WUSI	90.3	Sa9aSu12p	Gr. Barrington	WAMQ	105.1	Sa10aSu1p
Peoria	WCBU	89.9	Sa9a	Harwich	WCCT	90.3	Sa11Su6p
Pittsfield	WIPA	89.3	Sa9aSu3p	Sandwich	WSDH	91.5	Sa11Su6p
Quincy	WQUB	90.3	Sa9a	Woods Hole	WCAI	90.1	Sa10aSu12p
Rock Island	WVIK	90.3	Sa9a	Woods Hole	WNAN	91.1	Sa10aSu12p
Springfield	WUIS	91.9	SA9aSu3p	**MD**			
Urbana	WILL	580	S9aSu11a	Baltimore	WPYR	88.1	Sa10aSa6p
Warsaw	WIUW	89.5	Sa9Su1p	Hagerstown	WETH	89.1	Sa10aSu1p
				Salisbury	WSCL	89.5	Sa10a

ME	Station	Freq	Airtime	MS	Station	Freq	Airtime
Bangor	MPR	90.9	Sa10aSu4p	Biloxi	WMAH	90.3	Sa9a
Calais	WMED	89.7	Sa10aSu4p	Booneville	WMAE	89.5	Sa9a
Camden	MPR	90.5	Sa10aSu4p	Bude	WMAU	88.9	Sa9a
Ft. Kent	WMEA	106.5	Sa10aSu4p	Greenwood	WMAO	90.9	Sa9a
Portland	WMEA	90.1	Sa10aSu4p	Jackson	WMPN	91.3	Sa9a
Presque Isle	WMEM	106.1	Sa10aSu4p	Meridian	WMAW	88.1	Sa9a
Waterville	WMEW	91.3	Sa10aSu4p	Miss. State	WMAB	89.9	Sa9a
MI				Oxford	WMAV	90.3	Sa9a
Alpena	WCML	91.7	Sa4p	Senatobia	WKNA	88.9	Sa9aSu2p
Ann Arbor	WUOM	91.7	Sa10aSa4p	**MT**			
Bay City	WUCX	90.1	Sa10a	Billings	KEMC	91.7	Sa10a
E. Jordan	WIZY	100.9	Sa10aSu2p	Bozeman	KBMC	102.1	Sa10a
E. Lansing	WKAR	870	Sa1p	Great Falls	KGPR	89.9	Su5p
Flint	WFUM	91.1	Sa10aSa4p	Havre Hills	KNMC	90.1	Sa10a
Grand Rapids	MIR	88.5	Sa10a	Miles City	KECC	90.7	Sa10a
Grand Rapids	WGVU	1480	Sa10aSu10a	Missoula	KUFM	89.1	Su5p
Grand Rapids	WVGR	104.1	Sa10aSa4p	**NC**			
Harbor Sprgs	WCMW	103.9	Sa4p	Ashville	WCQS	88.1	Sa10a
Interlochen	WIAA	88.7	Sa10aSu2p	Chapel Hill	WUNC	91.5	Sa11aSu3p
Interlochen	WICA	91.5	Sa10aSu2p	Charlotte	WFAE	90.7	Sa10aSu11a
Marquette	WNMU	102.3	Sa10a	Elizabeth City	WRVS	89.9	M6p
Mt. Pleasant	WCMU	89.5	Sa4p	Fayetteville	WFSS	91.9	Sa10a
S. Ste. Marie	WCMZ	98.3	Sa4p	Franklin	WFQS	91.3	Sa10a
Ypsilanti	WEMU	89.1	Sa10a	Hickory	WFHE	90.3	Sa10aSu11a
MN				New Bern	WTEB	89.3	Sa10a
Appleton	KNCM	91.3	Sa11Su10a	Roanoke Rap.	WZRU	88.5	S12p
Bemidji	KAXE	105.3	Sa9a	Wilmington	WHQR	91.3	Sa10a
Bemidji	KNBJ	91.3	Sa11Su10a	Winston-Salem	WFDD	88.5	Sa10a
Brainerd	KAXE	89.5	Sa9a	**ND**			
Brainerd	KBPR	90.7	Sa11Su10a	Ashville	WCQS	88.1	Sa10a
Collegeville	KNSR	88.9	Sa11Su10a	Chapel Hill	WUNC	91.5	Sa11aSu3p
Duluth	WSCN	100.5	Sa11Su10a	Charlotte	WFAE	90.7	Sa10aSu11a
Grand Rapids	KAXE	91.7	Sa9a	Elizabeth City	WRVS	89.9	M6p
Moorehead	KCCD	90.3	Sa11aSu10a	Fayetteville	WFSS	91.9	Sa10a
Rochester	KZSE	90.7	Sa11aSu10a	Franklin	WFQS	91.3	Sa10a
St. Cloud	KNSR	88.9	Sa11Su10a	Hickory	WFHE	90.3	Sa10aSu11a
St. Paul	KNOW	91.1	Sa11aSu10a	New Bern	WTEB	89.3	Sa10a
St. Paul	MPR	91.1	Sa11aSu10a	**NE**			
St. Peter	KNGA	91.5	Sa11aSu10a	Alliance	KTNE	91.1	Sa9aSu10a
Thief Riv. Falls	KNTN	102.7	Sa11aSu10a	Basset	KMNE	90.3	Sa9aSu10a
Virginia	WIRN	92.5	Sa11aSu10a	Chadron	KCNE	91.9	Sa9aSu10a
MO				Columbius	KUCV	90.3	Sa9aSu10a
Cape Gir'deau	KRCU	90.9	Sa9aSu5p	Hastings	KHNE	89.1	Sa9aSu10a
Chillicothe	KRNW	88.9	Sa9aSu1p	Lexington	KLNE	88.7	Sa9aSu10a
Columbia	KBIA	91.3	Sa9Su11a	Lincoln	KUCV	90.9	Sa9aSu10a
Kansas City	KCUR	89.3	Sa9a	Merriman	KRNE	91.5	Sa9aSu10a
Maryville	KXCV	90.5	Sa9aSu1p	Norfolk	KXNE	89.3	Sa9aSu10a
Osage Beach	KTBG	104.9	Sa9aSu12p	North Platte	KPNE	91.7	Sa9aSu10a
Pt. Lookout	KSMS	90.5	Sa12pSu6p	Omaha	KIOS	91.5	Sa9aSu5p
Rolla	KUMR	88.5	Sa9aSu11a	**NH**			
Springfield	KSMU	91.1	Sa12pSu6p	Berlin	WEVC	107.1	Sa11aSu3p
St. Louis	KWMU	90.7	Sa9aSu3p	Concord	NHPR	89.1	Sa10aSu3p
Warrensburg	KTBG	90.9	Sa9aSu12p	Dover	NHPR	90.5	Sa10aSu3p
				Hanover	WEVH	91.3	Sa11aSu3p
				Keene	WEVN	90.7	Sa11aSu3p
				Littleton	NHPR	104.3	Sa10aSu3p
				Nashua	NHPR	90.3	Sa10aSu3p

NJ	Station	Freq	Airtime
Lincroft	WBJB	90.5	Sa10a
NM			
Albuquerque	KANW	89.1	Sa3pSu6p
Deming	NMPR	93.5	Sa10aSu2p
Dulce	KCIE	90.5	Su3p
Gallup	KGLP	91.7	Sa11a
Las Cruces	KRWG	90.7	Sa10aSu2p
Maljamar	KMTH	98.7	Sa8a
Pine Hill	KTDB	89.7	Sa7a
Portales	KENW	89.5	Sa8a
Silver City	NMPR	91.3	Sa10aSu2p
Truth/Conseq.	NMPR	91.9	Sa10aSu2p
NV			
Elko	KNCC	91.7	SA9a
Las Vegas	KNPR	89.5	Sa10aSu4p
Panaca	KLNR	91.5	Sa10aSu4p
Reno	KUNR	88.7	Sa9a
Tonopah	KTPH	91.7	Sa10aSu4p
NY			
Albany	WAMC	90.3	Sa10aM1p
Binghamton	WSKG	89.3	Su1p
Blue Mountain	WXLH	91.3	Sa12pW7p
Buffalo	WBFO	88.7	Sa6aSa10a
Canajoharie	WCAN	93.3	Sa10aSu1p
Canton	WSLU	89.5	Sa12pW7p
Corning	WSQE	91.1	Su1p
Geneva	WEOS	89.7	Sa10a
Ithaca	WSQG	90.9	Su1p
Jamestown	WUBJ	88.1	Sa6aSa10a
Jeffersonville	WJFF	90.5	Sa10a
Kingston	WAMK	90.9	Sa10aSu1p
Malone	WSLO	90.9	Sa12pW7p
Middletown	WOSR	91.7	Sa10aSu1p
New York	WNYC	93.9	Sa11a
New York	WNYC	820	Sa12p
North Creek	WXLG	89.9	Sa12pW7p
Olean	WOLN	91.3	Sa6aSa10a
Oneonta	WSQC	91.7	Su1p
Oswego	WRVO	89.9	Sa10aSu6p
Peru	WXLU	88.3	Sa12pW7p
Plattsburg	WCEL	91.9	Sa10aSu2p
Rochester	WXXI	1370	Sa10aF11p
Saranac Lake	WSLL	90.5	Sa12pW7p
Southhampton	WPBX	91.3	Sa11a
Ticonderoga	WANC	103.9	Sa10aSu1p
Utica	WRVN	91.9	Sa10aSu6p
Watertown	WRVJ	91.7	Sa10aSu6p
Watertown	WSLJ	88.5	Sa12pW7p
OH			
Athens	WOUB	91.3	Sa10a
Bryan	WGBE	90.9	Sa10aSu6p
Cambridge	WOUC	89.1	Sa10a
Chillicothe	WOUH	91.9	Sa10a
Cincinnati	WGUC	90.9	Sa10a
Cleveland	WCPN	90.3	Sa10a
Columbus	WCBE	90.5	Sa10a

OH	Station	Freq	Airtime
Columbus	WOSU	820	Sa10aSu11a
Defiance	WGDE	91.9	Sa10aSu6p
Dover	WKRJ	91.5	Sa10aSa5p
Ironton	WOUL	89.1	Sa10a
Kent	WKSU	89.7	Sa10aSa5p
Lima	WGLE	90.7	Sa10aSa5p
Oxford	WMUB	88.5	Sa10aSu12p
Thompson	WKSV	89.1	Sa10Su12p
Toledo	WGTE	91.3	Sa10aSu6p
Wooster	WKRW	89.3	SA10aSa5p
Yellow Sprg	WYSO	91.3	Sa10a
Youngstown	WYSU	88.5	Sa10a
Zanesville	WOUZ	89.1	Sa10a
OK			
Lawton	KCCU	89.3	Sa9aSu12p
Lawton	KCCU	89.9	Sa9aSu12p
Norman	KGOU	106.3	Sa9a
Oklahoma City	KROU	105.7	Sa9a
Stillwater	KOSU	91.7	Sa9a
Tulsa	KWGS	89.5	Sa9a
OR			
Ashland	KSMF	89.1	Sa11a
Ashland	KSOR	90.1	Sa11a
Ashland	KSRG	88.3	Su11a
Astoria	KMUN	91.9	Sa10a
Bend	KOAB	91.3	Sa10Su11a
Coos Bay	KSBA	88.5	Sa11a
Corvallis	KOAC	550	Sa10aSu11a
Klamath Falls	KSKF	90.9	Sa11a
Pendleton	KRBM	90.9	Sa10Su11p
Portland	KOPB	91.5	Sa10Su11a
Roseburg	KSRS	91.5	Su11a
PA			
Bethlehem	WDIY	88.1	Sa10a
Dubois	WPSU	92.1	Sa10aSu4p
Erie	WQLN	91.3	Sa10aSu2p
Harrisburg	WITF	89.5	Sa10aSu10a
Huntingdon	WKVR	92.3	Sa10aSu4p
Kane	WPSB	90.1	Sa10aSu4p
Philadelphia	WHYY	90.9	Sa10a
Pittsburgh	WDUQ	90.5	Sa10a
Pittston	WVIA	89.9	Sa10aSu6p
University Pk	WPSU	91.5	Sa10aSu4p
RI			
Providence	WRNI	1290	Sa11aSu6p
Providence	WRNI	1230	Sa11aSu6p
SC			
Aiken	WLJK	89.1	Sa10a
Beaufort	WJWJ	89.9	Sa10a
Charleston	WSCI	89.3	Sa10a
Columbia	WLTR	91.3	Sa10a
Conway	WHMC	90.1	Sa10a
Greenville	WEPR	90.1	Sa10a
Rock Hill	WNSC	88.9	Sa10a
Sumter	WRJA	88.1	Sa10a

SD	Station	Freq	Airtime		Station	Freq	Airtime
Brookings	KESD	88.3	Sa9a	**VA**			
Faith	KPSB	97.1	Sa9a	Richmond	WCVE	88.9	Sa10aSu6p
Lowry	KQSD	91.9	Sa9a	Roanoke Rap.	WVTF	89.1	Sa10aSu11
Martin	KZSD	102.5	Sa9a	Wise	WISE	90.5	Sa10aSu4p
Pierpont	KDSD	90.9	Sa9a	**VT**			
Rapid City	KBHE	89.3	Sa9a	Bennington	WVPR	94.3	Sa10a
Reliance	KTSD	91.1	Sa9a	Colchester	VPR	107.9	Sa10a
Sioux Falls	KCSD	90.9	Sa9a	Colchester	WVPS	88.9	Sa10a
Vermillion	KUSD	89.7	Sa9a	Rutland	WRVT	88.7	Sa10a
Worthington	KNSW	91.7	Sa11Su10a	St. Johnsbury	WVPR	88.5	Sa10a
TN				Windsor	WVPR	89.5	Sa10a
Chattanooga	WUTC	88.1	Sa10aSu6p	**WA**			
Dyersburg	WKNQ	90.7	Sa9aSu2p	Bellingham	KZAZ	91.7	Sa10Su4p
Hickory	WETS	91.5	Sa10a	Clarkston	KNWY	90.5	Sa10Su4p
Jackson	WKNP	90.1	Sa9aSu2p	Cottonwood	KNOW	90.1	Sa10Su4p
Johnson City	WETS	89.5	Sa10a	Ellensburg	KNWR	90.7	Sa10Su4p
Knoxville	WUOT	91.9	Sa10a	Moses Lake	KLWS	91.5	Sa10Su10
Memphis	WKNO	91.1	Sa9aSu2p	Pullman	KWSU	1250	Sa10Su10a
Nashville	WPLN	1430	Sa9aSu5p	Pullman	KWSU	91.7	Sa10aSu4p
Nashville	WPLN	90.3	Sa9aSu5p	Seattle	KUOW	94.9	Sa9a
TX				Spokane	KPBX	91.1	Sa6pSu9a
Abilene	KACU	89.7	Sa9aSu4p	Tacoma	KPLU	88.5	Sa10aSu2p
Austin	KUT	90.5	Sa9a	Tri-Cities	KFAE	89.1	Sa10aSu4p
Beaumont	KVLU	91.3	Sa9aF6p	Walla Walla	KWFS	89.7	Sa10Su10a
Beaumont	KVLU	91.9	Sa9aF6p	Yakima	KNWY	90.3	Sa10aSu4p
Big Sandy	KBAU	90.7	Sa9aSu11a	**WI**			
College Stn	KAMU	90.9	Sa9a	Appleton	WLFM	91.1	Sa9a
Corpus Christi	KEDT	90.3	Sa9a	Ashland	WPR	102.9	Sa9a
Dallas	KERA	90.1	Sa10aSu12p	Auburndale	WLBL	930	Sa9aSa4p
El Paso	KTEP	88.5	F12pSa10a	Delafield	WHAD	90.7	Sa9a
Harlingen	KMBH	88.9	Sa9Su1p	Green Bay	WGBW	91.5	Sa9aSa4p
Houston	KUHF	88.7	Sa9aSa7p	Highland	WHHI	91.3	Sa9aSa4p
Lubbock	KOHM	89.1	Sa9a	La Crosse	WHLA	90.3	Sa9a
Lufkin	KLDN	88.9	Sa9aSu2p	La Crosse	WLSU	88.9	Sa9a
Mcallen	KHID	88.1	Sa9aSu1p	Madison	WHAD	970	Sa9a
Odessa	KOCV	91.3	Sa9aSu4p	Menomonie	WHWC	88.3	Sa9a
Redland	KLDN	88.9	Sa9aSa5p	Milwaukee	WUWM	89.7	Sa9aSu11a
San Angelo	KUTX	90.1	Sa9a	Oshkosh	WRST	90.3	Sa9aSa4p
San Antonio	KSTX	89.1	Sa9aSu1p	Park Falls	WHBM	90.3	Sa9a
Texarkana	KTXK	91.5	Sa9aSu5p	Rhinelander	WXPR	91.7	Sa11a
Victoria	KVRT	90.7	Sa9a	Superior	KUWS	91.3	Sa9aSa4p
Waco	KWBU	103.3	Sa9aSu3p	W. Madison	WPR	90.9	Sa9a
UT				Wausau	WLBL	91.9	Sa9aSa4p
Logan	KUSU	91.5	Sa10Su5pM4p	Wausau	WXPW	91.9	Sa11a
Park City	KCPW	88.3	Sa9aSu10a	**WV**			
Salt Lake City	KCPW	105.3	Sa9aSa5pSu10	Beckley	WVBP	91.7	Sa10a
Salt Lake City	KUER	90.1	Sa10aSu12p	Buckhannon	WVPW	88.9	Sa10a
VA				Charleston	WVPN	88.5	Sa10a
Arlington	WETA	90.9	Sa10aSu2p	Huntington	WVWV	89.5	Sa10a
Charlottesville	WVTU	89.3	Sa10aSu11a	Martinsburg	WVEP	88.9	Sa10a
Crozet	WMRY	91.3	Sa10aSu4p	Morgantown	WVPM	90.9	Sa10a
Harrisonburg	WMRA	90.7	Sa10aSu4p	Parkersburg	WVPG	90.3	Sa10a
Lexington	WMRL	89.9	Sa10aSu4p	Wheeling	WVNP	89.9	Sa10a
Marion	WVTR	91.9	Sa10aSu11a	**WY**			
Norfolk	WHRV	89.5	Sa10a	Jackson	KUWJ	90.3	Sa9aSu7p
Richmond	WCVE	88.9	Sa10aSu6p	Laramie	KUWR	91.9	Sa9aSu7p
				Rock Springs	KUWZ	90.5	Sa9aSu7p

Angels				Athletics			
				CA	Redding	KMCA	1450
CA	Anaheim	KIKF	94.3	CA	Sacramento	KHTK	1140
CA	Bakersfield	KBID	1350	CA	San Francisco	KABL	960
CA	El Centro	KAMP	1430	CA	Sonora	KVML	1450
CA	Los Angeles	KLAC	570	NV	Reno	KCBN	1230
CA	Palm Desert	KESO	1400	OR	Medford	KCMX	580
HI	Honolulu	KGU	760	Blue Jays			
ND	Jamestown	KSJB	600	MAN	Winkler	CKMW	1570
Astros				MAN	Winnipeg	CKY	580
LA	Baton Rouge	WIBR	1300	MI	S. Ste Marie	WKNW	1400
LA	Lafayette	KVOL	1330	NS	Digby	CKDY	1420
LA	Lake Charles	KLCL	1470	NS	Kentville	CKEN	1490
LA	New Orleans	WBYU	1450	NS	Middleton	CKAD	1750
LA	Shreveport	KFLO	1300	NS	Weymouth	CKDY	103.3
TX	Alpine	KVLF	1240	NS	Windsor	CFAB	1450
TX	Austin	KFON	1490	ON	Toronto	CHUM	1050
TX	Beaumont	KLVI	560	ON	Algoma	CKNR	94.1
TX	Brenham	KWHI	1280	ON	Bancroft	CJNH	1240
TX	College Station	WTAW	1150	ON	Collingwood	CKCB	95.1
TX	Center	KORI	104.7	ON	Hamilton	CHAM	820
TX	Columbus	KULM	98.3	ON	Kingston	CGTO	960
TX	Corpus Christi	KSIX	1230	ON	Kitchener	CKGL	570
TX	Crockett	KIVY	92.7	ON	Leamington	CHYR	96.7
TX	El Campo	KULP	1390	ON	London	CJBK	1290
TX	Fredricksburg	KNAF	910	ON	Niagra Falls	CJRN	710
TX	Gonzales	KCTI	106.3	ON	Ottawa	OSR	1200
TX	Gonzales	KCTI	1450	ON	Peterborough	CRUZ	980
TX	Houston	KTRH	740	ON	Sarnia	CHOK	1070
TX	Houston	KXYZ	1320	ON	Smith Falls	CJET	630
TX	Jacksonville	KEBE	1400	ON	Stratford	CJCS	1240
TX	Lampasas	KCYL	1450	ON	Trenton	SJTN	1270
TX	Lufkin	KRBA	1340	SK	Estevan	CJSL	1280
TX	Madisonville	KMVL	1220	SK	Kindersley	CFYM	1210
TX	Navasota	KWBC	1550	SK	Regina	CJME	1300
TX	Raymondville	KSOX	1240	SK	Rosetown	CJYM	1330
TX	Rockdale	KRXT	98.5	SK	Saskatoon	CKOM	650
TX	San Antonio	KENS	1160	SK	Swift Current	CIMG	94.1
TX	Temple	KTEM	1400	SK	Weyburn	CFSL	1190
TX	Tyler	KGLD	1330	Braves			
TX	Tyler	KYZS	1490	AL	Anniston	WDNG	1450
TX	Victoria	KAMG	1340	AL	Auburn	WAUD	1230
TX	Waco	KBBW	1010	AL	Birmingham	WAPI	1070
TX	Woodville	KVLL	1490	AL	Birmingham	WJOX	690
Athletics				AL	Brewton	WKNU	106.3
CA	San Francisco	KABL	960	AL	Centre	WEIS	990
CA	Auburn	KAHI	950	AL	Clanton	WEZZ	97.7
CA	Concord	KATD	990	AL	Clanton	WKLF	980
CA	Eureka	KINS	980	AL	Cullman	WKUL	92.1
CA	Mendocino	KPMO	1300	AL	Dothan	WAGF	101.3
CA	Modesto	KANM	970	AL	Foley	WHEP	1310
CA	Napa	KVON	1440	AL	Gadsden	WGAD	1350
CA	Oakland	KNEW	910	AL	Guntersville	WTWX	95.9
				AL	Guntersville	WGSV	1270

249

Braves					Braves			
AL	Jasper	WZPQ	1360		GA	Greensboro	WDDK	103.9
AL	Madison	WUMP	730		GA	Griffin	WKEU	1450
AL	Mobile	WABB	1480		GA	Hawkinsville	WQSY	103.9
AL	Montgomery	WNZZ	950		GA	Hawkinsville	WCEH	610
AL	Oneonta	WCRL	1570		GA	Helen	WHEL	105.5
AL	Piedmont	WPID	1280		GA	Jesup	WLOP	1370
AL	Scottsboro	WWIC	1050		GA	Jesup	WIFO	105.5
AL	Selma	WMRK	1340		GA	Lagrange	WMXY	720
AL	Sylacauga	WFEB	1340		GA	Lagrange	WZLG	98.1
AL	Tallassee	WACQ	1130		GA	Louisville	WPEH	1420
AL	Tallassee	WACQ	99.9		GA	Louisville	WPEH	92.1
AL	Thomasville	WJDB	630		GA	Lyons	WBBT	1340
AL	Thomasville	WJDB	95.5		GA	Lyons	WLYU	100.9
AL	Tuscumbia	WVNA	1590		GA	Macon	WMAC	940
AL	Tuscumbia	WVNA	100.3		GA	Mcrae	WYIS	1410
AL	Valley	WRLD	95.3		GA	Mcrae	WYSC	102.7
FL	Crestview	WAAZ	104.7		GA	Milledgeville	WMVG	1450
FL	Crestview	WJSB	1050		GA	Milledgeville	WKZR	102.3
FL	Gainesville	WRUF	850		GA	Monroe	WKUN	1580
FL	Jacksonville	WNZS	930		GA	Moultrie	WMGA	580
FL	Marianna	WTYS	1340		GA	Newnan	WCOH	1400
FL	Marianna	WBNF	94.1		GA	Newnan	WMKJ	96.7
FL	Orlando	WQTM	540		GA	Quitman	WSFB	1490
FL	Panama City	WLTG	1430		GA	Ringold	WSGC	101.9
FL	Pensacola	WCOA	1370		GA	Rome	WLAQ	1410
GA	Albany	WGPC	1450		GA	Rome	WATG	95.7
GA	Albany	WGPC	104.5		GA	Sandersville	WSNT	1490
GA	Americus	WISK	1390		GA	Sandersville	WSNT	99.9
GA	Americus	WISK	98.7		GA	Savannah	WBMQ	630
GA	Athens	WGAU	1340		GA	Savannah	WEAS	900
GA	Atlanta	WKLS	96.1		GA	Statesboro	WMCD	100.1
GA	Atlanta	WGST	640		GA	Statesboro	WWNS	1240
GA	Augusta	WGAC	580		GA	Swainsboro	WJAT	800
GA	Blakely	WBBK	1260		GA	Swainsboro	WJAT	98.1
FL	Pensacola	WCOA	1370		GA	Thomaston	WTGA	1590
GA	Blakely	WBBK	93.1		GA	Thomaston	WTGA	101.1
GA	Brunswick	WSFN	790		GA	Thomson	WTWA	1240
GA	Calhoun	WJTH	900		GA	Thomson	WTHO	101.7
GA	Carrollton	WBTR	1330		GA	Tifton	WTIF	1340
GA	Carrollton	WBTR	92.1		GA	Tifton	WTIF	107.5
GA	Cartersville	WBHF	1450		GA	Trenton	WKWN	1420
GA	Claxton	WCLA	1470		GA	Valdosta	WSTI	105.3
GA	Claxton	WCLA	107.3		KY	Madisonville	WTTL	1310
GA	Columbus	WRCG	1420		MS	Jackson	WSLI	930
GA	Cornelia	WCON	1450		NC	Charlotte	WFNZ	610
GA	Cornelia	WCON	99.3		NC	Charlotte	WGIV	1600
GA	Dalton	WBLJ	1230		NC	Cherryville	WCSL	1590
GA	Dalton	WYYU	104.5		NC	Concord	WEGO	1410
GA	Donalsonville	WGMK	106.3		NC	Fayetteville	WAZZ	1490
GA	Douglas	WDMG	860		NC	Gastonia	WGNC	1450
GA	Douglas	WDMG	99.5		GA	Greensboro	WDDK	103.9
GA	Dublin	WMLT	1330		GA	Griffin	WKEU	1450
GA	Elberton	WEHR	105.1		GA	Hawkinsville	WQSY	103.9
GA	Gainesville	WDUN	550		GA	Hawkinsville	WCEH	610
GA	Clarksville	WMJE	102.9		GA	Helen	WHEL	105.5
					GA	Jesup	WLOP	1370
					GA	Jesup	WIFO	105.5

Braves					Braves			
GA	Lagrange	WMXY	720		SC	Camden	WPUB	94.3
GA	Lagrange	WZLG	98.1		SC	Charleston	WQSC	1340
GA	Louisville	WPEH	1420		SC	Clinton	WPCC	1410
GA	Louisville	WPEH	92.1		SC	Columbia	WCOS	1400
GA	Lyons	WBBT	1340		SC	Fountain Inn	WFIS	1600
GA	Lyons	WLYU	100.9		SC	Greenville	WPEK	98.1
GA	Macon	WMAC	940		SC	Greenwood	WCRS	1450
GA	McRae	WYIS	1410		SC	Kingstree	WDKD	1310
GA	McRae	WYSC	102.7		SC	Kingstree	WWKT	99.3
GA	Milledgeville	WMVG	1450		SC	Myrtle Beach	WKZQ	1450
GA	Milledgeville	WKZR	102.3		SC	Spartanburg	WSPA	950
GA	Monroe	WKUN	1580		SC	Union	WBCU	1460
GA	Moultrie	WMGA	580		SC	Walterboro	WALI	93.7
GA	Newnan	WCOH	1400		TN	Clarksville	WJZM	1400
GA	Newnan	WMKJ	96.7		TN	Cookeville	WATX	1590
GA	Quitman	WSFB	1490		TN	Dresden	WCDZ	95.1
GA	Ringold	WSGC	101.9		TN	Dresden	WCMT	1410
GA	Rome	WLAQ	1410		TN	Etowah	WCPH	1220
GA	Rome	WATG	95.7		TN	Johnson City	WJCW	910
GA	Sandersville	WSNT	1490		TN	Johnson City	WKIN	1320
GA	Sandersville	WSNT	99.9		TN	Lawrenceburg	WDXE	1370
GA	Savannah	WBMQ	630		TN	Lawrenceburg	WDXE	95.9
GA	Savannah	WEAS	900		TN	Lenoir City	WLIL	730
GA	Statesboro	WMCD	100.1		TN	Lenoir City	WLIL	93.5
GA	Statesboro	WWNS	1240		TN	Manchester	WMSR	1320
GA	Swainsboro	WJAT	800		TN	Nashville	WGNS	1450
GA	Swainsboro	WJAT	98.1		TN	Newport	WLIK	1270
GA	Thomaston	WTGA	1590		WV	Logan	WVOW	1290
GA	Thomaston	WTGA	101.1		WV	Logan	WVOW	101.9
GA	Thomson	WTWA	1240		Brewers			
GA	Thomson	WTHO	101.7		MI	Escanaba	WDBC	680
GA	Tifton	WTIF	1340		MI	Iron River	WIKB	1230
GA	Tifton	WTIF	107.5		MI	Iron River	WIKB	99.1
GA	Trenton	WKWN	1420		MI	Ironwood	WJMS	590
GA	Valdosta	WSTI	105.3		MN	Duluth	WDSM	710
KY	Madisonville	WTTL	1310		WI	Antigo	WRLO	105.3
MS	Jackson	WSLI	930		WI	Appleton	WHBY	1150
NC	Charlotte	WFNZ	610		WI	Ashland	WATW	1400
NC	Charlotte	WGIV	1600		WI	Beaver Dam	WBEV	1430
NC	Cherryville	WCSL	1590		WI	Beloit	WGEZ	1490
NC	Concord	WEGO	1410		WI	Berlin	WISS	1090
NC	Fayetteville	WAZZ	1490		WI	Berlin	WBJZ	102.3
NC	Gastonia	WGNC	1450		WI	Black Riv Falls	WWIS	1090
NC	Greensboro	WTCK	1320		WI	Black Riv Falls	WWIS	99.7
NC	Greensboro	WWBG	1470		WI	Clintonville	WFCL	1380
NC	Lincolnton	WLON	1050		WI	Durand	WRDN	1430
NC	Murphy	WCVP	600		WI	Durand	WRDN	95.9
NC	Murphy	WCNG	102.7		WI	Eau Claire	WBIZ	1400
NC	Rutherfordton	WCAB	590		WI	Fond du Lac	KFIZ	1450
NC	Salisbury	WSTP	1490		WI	Fort Atkinson	WFAW	940
NC	Shelby	WOHS	730		WI	Green Bay	WDUZ	1400
NC	Sylva	WRGC	680		WI	Hayward	WRLS	92.3
NC	Valdese	WSVM	1490		WI	Ishpeming	WJPD	92.3
NC	Winston-Salem	WTOB	1380		WI	Janesville	WCLO	1230
SC	Abbeville	WABV	1590		WI	La Crosse	WKTY	580
SC	Barnwell	WBAW	740					
SC	Barnwell	WBAW	99.1					

Brewers					Cardinals			
WI	Madison	WIBA	1310		IL	Decatur	WSOY	1340
WI	Manitowoc	WOMT	1240		IL	Dixon	WIXN	1460
WI	Marshfield	WDLB	1450		IL	Effingham	WCRC	95.7
WI	Medford	WIGM	1490		IL	Galesburg	WGIL	1400
WI	Medford	WIGM	99.3		IL	Harrisburg	WEBQ	102.3
WI	Milwaukee	WTMJ	620		IL	Herrin	WJPF	1340
WI	Monroe	WEKZ	1260		IL	Jacksonville	WEAI	107.1
WI	Park Falls	WCQM	980		IL	Kewanee	WKEI	1450
WI	Park Falls	WNBI	98.3		IL	Lasalle	WLPO	1220
WI	Platteville	WPVL	1590		IL	Lincoln	WVAX	1370
WI	Portage	WPDR	1350		IL	Loves Park	WLUV	1520
WI	Prairie Du Chein	WPRE	980		IL	Loves Park	WLUV	96.7
WI	Prairie Du Chein	WQPC	94.3		IL	Mt Carmel	WSJD	100.5
WI	Reedsburg	WRDB	1400		IL	Mt Vernon	WMIX	940
WI	Rhinelander	WOBT	1240		IL	Murphysboro	WINI	1420
WI	Rice Lake	WJMC	1240		IL	Olney	WLVN	740
WI	Richland Ctr	WRCO	1450		IL	Olney	WSEI	92.9
WI	River Falls	WEVR	1550		IL	Peoria	WMBD	1470
WI	River Falls	WEVR	106.3		IL	Quincy	WGEM	105.1
WI	Shawano	WTCH	960		IL	Robinson	WTAY	1570
WI	Sheboygan	WHBL	1330		IL	Shelbyville	WRAN	98.3
WI	Stevens Point	WSPT	1010		IL	Sparta	WHCO	1230
WI	Sturgeon Bay	WDOR	93.9		IL	Springfield	WTAX	1240
WI	Waupaca	WDUX	92.7		IL	Taylorville	WTIM	97.3
WI	Wausau	WSAU	550		IL	Virginia	WVIL	101.3
WI	Wisconsin Rap	WFHR	1320		IN	Evansville	WGBF	103.1
Cardinals					IN	Vincennes	WFML	96.7
AR	Batesville	KBTA	1340		IN	Washington	WAMW	107.9
AR	Benton	KEWI	690		KY	Benton	WCBL	102.3
AR	Conway	KASR	92.7		KY	Hartford	WKHB	106.3
AR	Crossett	KAGH	800		KY	Henderson	WSON	860
AR	Crossett	KAGH	104.9		KY	Mayfield	WYMC	1430
AR	Fayetteville	KREB	1390		KY	Morganfield	WMSK	1550
AR	Fayetteville	KREB	99.5		KY	Morganfield	WMSK	95.3
AR	Fort Smith	KFPW	1230		KY	Murray	WSJP	1130
AR	Helena	KFFA	1360		KY	Paducah	WDXR	1450
AR	Hot Springs	KXOW	1420		KY	Wickliffe	WGKY	95.9
AR	Jonesboro	KJBR	93.7		MO	St. Louis	KMOX	1120
AR	Morrilton	KVOM	800		MO	Cpe Giradeau	KZIM	960
AR	Paragould	KDRS	1490		MO	Columbia	KFRU	1400
AR	Paragould	KLQZ	107.1		MO	Doniphan	KOEA	97.5
AR	Salem	KSAR	95.9		MO	Fulton	KKCA	100.5
IA	Atlantic	KJAN	1220		MO	Hannibal	KHMO	1070
IA	Indianola	KJJC	106.9		MO	Jeferson City	KWOS	1240
IA	Keokuk	KOKX	95.3		MO	Joplin	WMBH	1450
IA	Maquoketa	KMAQ	95.3		MO	Kennett	KTMO	105.5
IA	Newton	KCOB	95.9		MO	Kirksville	KIRX	1450
IA	Ottumwa	KBIZ	1240		MO	Lebanon	KJEL	103.7
IL	Aledo	WRMJ	102.3		MO	Memphis	KMEM	100.5
IL	Anna	WIBH	1440		MO	Mexico	KXEO	1340
IL	Bloomington	WJBC	1230		MO	Moberly	KRES	104.7
IL	Cairo	WKRO	1490		MO	Monett	KKBL	95.9
IL	Carmi	WROY	1460		MO	Park Hills	KFMO	1240
IL	Champaign	WDWS	1400		MO	Piedmont	KPWB	104.9
IL	Danville	WDAN	1490		MO	Poplar Bluff	KWOC	94.9

Cardinals

State	City	Call	Freq
MO	Rolla	KTTR	1490
MO	Salem	KSMO	1340
MO	Sedalia	KSIS	1050
MO	Sikeston	KBXB	97.9
MO	Southwest City	KLTK	1140
MO	Southwest City	KWMQ	100.3
MO	Springfield	KTXR	101.3
MO	St. Joseph	KSFT	1550
MO	Sullivan	KTUI	100.9
MO	Thayer	KALM	1290
MO	Waynesville	KJPW	102.3
MO	West Plains	KWPM	1450
MO	Willow Springs	KUKU	100.3
MS	Tupelo	WTUP	1490
OK	Bartlesville	KWON	1400
OK	Enid	KCRC	1390
OK	Norman	KNOR	1400
OK	Ponca City	WBBZ	1230
OK	Taft	KHJM	100.3
TN	Covington	WKBL	93.5
TN	Dyersburg	WTRO	1330
TN	Lexington	WZLT	99.3
TN	Memphis	WHBQ	560
TN	Milan	WTKB	93.7
TN	Ripley	WTRB	1570
TN	Union City	WENK	1240

Cubs

State	City	Call	Freq
IA	Boone	KWBG	1590
IA	Burlington	KCPS	1150
IA	Carroll	KCIM	1380
IA	Davenport	KJOC	1170
IA	Des Moines	KRNT	1350
IA	Elkader	KADR	1400
IA	Iowa City	KXIC	800
IA	Keokuk	KOXX	1310
IA	Ottumwa	KLEE	1480
IA	Sheldon	KIWA	1550
IA	Sheldon	KIWA	105.3
IL	Carbondale	WCIL	1020
IL	Carlinville	WCNL	95.9
IL	Champaign	WDWS	1400
IL	Chicago	WGN	720
IL	Decatur	WDZ	1340
IL	Highland	WINU	880
IL	Lawrenceville	WAKO	910
IL	Lawrenceville	WAKO	103.1
IL	Metropolis	WMOK	920
IL	Monmouth	WRAM	1330
IL	Monmouth	WMOI	97.7
IL	Peoria	WIRL	1290
IL	Quincy	WGEM	1440
IL	Springfield	WFMB	1450

Cubs (continued)

State	City	Call	Freq
IN	Anderson	WHBU	1240
IN	Auburn	WIFF	1570
IN	Boonville	WBNL	1540
IN	Boonville	WBNL	107.1
IN	Goshen	WKAM	1460
IN	Marion	WBAT	1400
IN	Mt Vernon	WPCO	1590
IN	Rockville	WAXI	104.3
MI	Kalamazoo	WQSN	1470
NV	Las Vegas	KRLV	1340
OK	Tulsa	KTRT	1270
SD	Vermillion	KOSZ	1570

Devil Rays

State	City	Call	Freq
FL	Brandon	WLCC	760
FL	Brooksville	WWJB	1450
FL	Chat'hoochee	WCTL	1580
FL	Dade City	WDCF	1350
FL	Englewood	WENG	1530
FL	Gainesville	WGTT	1230
FL	Jacksonville	WJGR	1320
FL	Lakeland	WONN	1230
FL	Ocala	WMOP	900
FL	Orlando	WHOO	990
FL	Sarasota	WSPB	1450
FL	Sebring	WWTK	730
FL	Tallahassee	WNLS	1270
FL	Tampa	WFLA	970
FL	Venice	WAMR	1320

Diamondbacks

State	City	Call	Freq
AZ	Flagstaff	KVNA	600
AZ	Globe	KIKO	1340
AZ	Holbrook	KZUA	92.1
AZ	Kingman	KAAA	1230
AZ	Lake Havasu	KBBC	980
AZ	Phoenix	KTAR	620
AZ	Phoenix	KPHX	1480
AZ	Prescott Vly	KQNA	1130
AZ	Prescott Vly	KPPV	106.7
AZ	Safford	KATO	1230
AZ	Safford	KPPV	106.7
AZ	Sedona	KAZM	780
AZ	Sierra Vista	KTAN	1420
AZ	Tucson	KFFN	1490
AZ	Williams	KYET	1180
AZ	Yuma	KJOK	1400
FL	Century	WPFL	105.1
MEX	Hermosillo	XEOH	1540
MEX	Nogales	XENY	760
NM	Albuquerque	KNML	1050
NM	Artesia	KSVP	990
NM	Farmington	KCQL	1340
NM	Silver City	KNFT	950
NV	Las Vegas	KSFN	1140
TX	El Paso	KAMA	750
TX	El Paso	KHEY	690

Major League Baseball

Diamondbacks

State	City	Call	Freq
AZ	Tucson	KCUB	1290
CA	Bakersfield	KNZR	1560
CA	Banning	KMET	1490
CA	Bishop	KBOV	1230
CA	Brawley	KROP	1300
CA	Calexico	KICO	1490
CA	Fresno	KCBL	1340
CA	Lancaster	KAVL	610
CA	Los Angeles	KWKW	1330
CA	Los Angeles	KWKW	1330
CA	Los Angeles	KFWB	980
CA	Palm Springs	KPSI	920
CA	Riverside	KCKC	1350
CA	S Luis Obispo	KKAL	99.7
CA	Santa Clarita	KHTS	1220
CA	Santa Maria	KSMA	1240
CA	1000 Oaks	KBET	850
CA	Ventura	KVEN	1450
CA	Victorville	KWRN	1550
CA	Visalia	KVBL	1400
FL	Vero Beach	WTTB	1370
HI	Kahalui, Maui	KAOI	1110
MEX	Ensenada	XESD	920
NM	Albuquerque	KDEF	1150
NM	Gallup	KTHR	1230
NV	Las Vegas	KSFN	1140
NV	Las Vegas	KRLV	1340
VI	St. Thomas	WSTA	1340

Expos

State	City	Call	Freq
NS	Halifax (E)	CJCH	920
ON	London (E)	CJBK	800
ON	Ottawa (E)	CFRA	580
QC	Alma (F)	CFGT	1270
QC	Amos (F)	CHAD	1340
QC	Amqui (F)	CFVM	1220
QC	Chapais (F)	CFED	1340
QC	Chibougamou	CJMD	1240
QC	Chicoutimi (F)	CKRS	590
QC	Dolbeau (F)	CHVD	1230
QC	Gaspe' (F)	CHGM	1150
QC	Gatineau (F)	CJRC	1150
QC	La Tuque (F)	CFLM	1240
QC	Matane (F)	CHRM	1290
QC	Montreal (F)	CKAC	730
QC	Montreal (E)	CIQC	600
QC	New Carlisle	CHNC	610
QC	Rimouski	CFLP	1000
QC	Riviere	CJFP	103.7
QC	Roberval	n/a	910
QC	Rouyn-Nor'da	CKRN	1400
QC	Sept-Iles	CKCN	560
QC	Sherbrooke	CHLT	630
QC	Ste-Foy	CHRC	800
QC	Trois-Rivieres	CHLN	550

Giants

State	City	Call	Freq
CA	S Francisco	KNBR	680
CA	Atwater	KTFN	1580
CA	Eureka	KATA	1340
CA	Fort Bragg	KLLK	96.7
CA	Fresno	KYNO	1330
CA	King City	KRKC	1490
CA	Lakeport	KXBX	1270
CA	Mendocino	KMFB	92.7
CA	Mt. Shasta	KMJC	620
CA	Redding	KNRO	600
CA	San Jose	KZSF	1370
CA	S Luis Obispo	KKJL	1400
CA	Tulare	KJUG	1270
CA	Willows	KIQS	1560
CA	Yreka	KMJC	620
CA	Yreka	KSYC	1490
HI	Hilo	KPUA	670
HI	Honolulu	KGU	760
HI	Lihue	KQNG	570
HI	Wailuki	KAOI	1110
NV	Reno	KPLY	1270
OR	Hillsboro	KUIK	1360
OR	Medford	KTMT	580

Indians

State	City	Call	Freq
NY	Dunkirk	WDOE	1410
NY	Olean	WMNS	1360
NY	Springville	WSPQ	1330
OH	Akron	WAKR	1590
OH	Ashtabula	WFUN	970
OH	Barnesville	WBNV	93.5
OH	Bellefontaine	WBLL	1390
OH	Bryan	WQCT	1520
OH	Bryan	WBNO	100.9
OH	Bucyrus	WQEL	92.7
OH	Canton	WHBC	1480
OH	Cleveland	WTAM	1100
OH	Columbus	WMNI	920
OH	Delaware	WDLR	1550
OH	Delphos	WDOH	107.1
OH	Dover	WJER	1450
OH	Elyria	WEOL	930
OH	Findlay	WFIN	1330
OH	Fostoria	WFOB	1430
OH	Lancaster	WLOH	1320
OH	Mansfield	WMAN	1400
OH	Marion	WMRN	1490
OH	Marysville	WURO	1270
OH	McConnelsv'le	WJAW	100.9
OH	Mt Vernon	WMVO	1300
OH	Newark	WCLT	1430
OH	Norwalk	WLKR	95.3
OH	Painesville	WBKC	1460
OH	Sandusky	WLEC	1450
OH	Toledo	WSPD	1370

	City	Call	Freq	#		City	Call	Freq
QC	VAL D'or	CKVD	900		OH	Urichsville	WBTC	1540
QC	Ville-Marie	CKVM	710		OH	Warren	WANR	1570
Indians					**Orioles**			
OH	Wooster	WQKT	104.5		DC	Washington	WTOP	1500
OH	Youngstown	WRTK	1390		DE	Newark	WNRK	1260
OH	Zanesville	WYBZ	107.3		MD	Aberdeen	WAMD	870
PA	Corry	WWCB	1370		MD	Baltimore	WBAL	1090
PA	Erie	WRIE	1260		MD	Cambridge	WCEM	106.3
PA	Sharon	WPIC	790		MD	Cumberland	WTBO	1450
Mariners					MD	Frederick	WFMD	930
AK	Anchorage	KAXX	1020		MD	Hagerstown	WARK	1490
AK	Juneau	KINY	800		MD	Prince Fredr.	WMJS	92.7
AK	Naknek	KAKN	100.9		MD	Salisbury	WTGM	960
AK	Petersburg	KRSA	580		MD	Westminster	WTTR	1470
BC	Abbotsford	CKMA	850		NC	Salisbury	WSAT	1280
ID	Coeur d'alene	KVNI	1080		PA	Chambersburg	WCHA	800
ID	Lewiston	KRLC	1350		PA	Hanover	WHVR	1280
ID	Orofino	KLER	1300		PA	York	WSBA	910
MT	Missoula	KGRZ	1450		VA	Winchester	WINC	1400
OR	Albany/Corval.	KIHR	990		WV	Martinsburg	WEPM	1340
OR	Hood River	KIHR	1340		**Padres**			
OR	Portland	KFXX	1520		CA	San Diego	KOGO	600
OR	Salem	KSLM	1390		CA	San Diego	KURS	1200
OR	Seaside	KSWB	840		**Phillies**			
WA	**Seattle**	KIRO	710		DE	Wilmington	WDEL	1150
WA	Aberdeen	KXRO	1320		NJ	Atlantic City	WGYM	1490
WA	Bellingham	KPUG	1170		NJ	Wildwood	WCMC	1230
WA	Centralia	KELA	1470		PA	**Philadelphia**	WPHT	1210
WA	Colfax	KMAX	840		PA	Allentown	WAEB	790
WA	Colville	KCVL	1240		PA	Gettysburg	WGET	1320
WA	Ellensburg	KXLE	1240		PA	Harrisburg	WWKL	1460
WA	Forks	KVAC	1490		PA	Hazeltown	WAZL	1490
WA	Longview/Kelso	KEDO	1400		PA	Lancaster	WLPA	1490
WA	Moses Lake	KWIQ	100.3		PA	Lebanon	WLBR	1270
WA	Mt. Vernon	KBRC	1430		PA	Levittown	WBCB	1490
WA	Olympia	KGY	1240		PA	Lock Haven	WBPZ	1230
WA	Omak	KOMW	680		PA	Pottsville	WPPA	1360
WA	Pasco	KFLD	870		PA	Reading	WEEU	850
WA	Port Angeles	KONP	1450		PA	Scranton	WARM	590
WA	Prosser	KZXR	101.7		PA	Selinsgrove	WYGL	1240
WA	Shelton	KMAS	1030		PA	Williamsport	WRAK	1400
WA	Spokane	KXLY	920		PA	Yeagertown	WCHX	105.5
WA	Twisp	KZLR	100.3		PA	York	WQXA	1250
WA	Vancouver	KVAN	1550	1	**Pirates**			
WA	Walla Walla	KGDC	1320	2	FL	Bradenton	WWPR	1490
WA	Wenatchee	KKRT	900	3	MD	Cumberland	WCBC	1270
WA	Yakima	KIT	1280	4	OH	Byesville	WILE	97.7
Marlins				5	OH	Cambridge	WILE	1270
FL	Belle Glade	WBGF	93.5	6	OH	E. Liverpool	WOHI	1490
FL	Daytona	WELE	1380	7	OH	Warren	WRBP	1440
FL	Ft. Myers	WWCN	770	8	PA	**Pittsburgh**	KDKA	1020
FL	Key West	WKWF	1600	9	PA	Altoona	WVAM	1430
FL	Melbourne	WMEL	920	0	PA	Ambridge	WMBA	1460
FL	Miami	WQAM	560	1	PA	Beaver Falls	WBVP	1230
FL	Miami	WQBA	1140	2	PA	Bradford	WESB	1490
FL	West Palm Bch	WJNO	1040	3	PA	Charleroi	WESA	940
Mets				4	PA	Connellsville	WCVI	1340

NY	New York	WFAN	660	5				
NY (Span)	New York	WADO	1280					
Pirates					**Rangers**			
PA	Coudersport	WFRM	600		TX	Nacodoches	KSFA	860
PA	Dubois	WOWQ	102.1		TX	Orange/Beau	KOGT	1600
PA	Erie	WFLP	1330		TX	Paris	KPLT	1490
PA	Franklin	WFRA	1450		TX	Pecos	KPTX	98.3
PA	Indiana	WCCS	1160		TX	Plainview	KKYN	1090
PA	Johnstown	WNTJ	1490		TX	Quanah	KIXC	100.9
WV	Clarksburg	WHAR	1340		TX	San Angelo	KKSA	1260
WV	Fairmont	WTCS	1490		TX	San Antonio	KTKR	760
WV	Morgantown	WAJR	1440		TX	Seymour	KSEY	94.3
WV	New Martins.	WETZ	1330		TX	Stamford	KVRP	1400
WV	Weirton	WEIR	1430		TX	Stephenville	KCUB	98.3
WV	Wheeling	WBBD	1400		TX	Sulphur Spgs	KSST	1230
Rangers					TX	Texarkana	KCMC	740
AR	Fort Smith	KFDF	1580		TX	Tyler	KTTB	600
LA	Shreveport	KFLO	1300		TX	Weatherford	KZEE	1220
OK	Ardmore	KVSO	1240		TX	Wichita Falls	KTUB	990
OK	Duncan	KKEN	1380		TX	Nacodoches	KSFA	860
OK	Lawton	KXCA	1350		**Red Sox**			
OK	Shawnee	KGFF	1450		CT	Hartford	WTIC	1080
TX	Arlington	KRLD	1080		CT	Norwalk	WNCR	1350
TX	Abilene	KGMM	1280		CT	Putnam	WINY	1350
TX	Alpine	KVLF	1240		CT	Stamford	WSTC	1400
TX	Amarillo	KGNC	710		CT	Willimantic	WILI	1400
TX	Austin	KVET	1300		MA	Boston	WEEI	850
TX	Big Spring	KBST	1490		MA	Boston	WRCA	1330
TX	Breckenridge	KROO	1430		MA	Cape Cod	WWKJ	101.1
TX	Brownfield	KKUB	1300		MA	Fall River	WSAR	1480
TX	Brownwood	KBWD	1380		MA	Fitchburg	WEIM	1280
TX	College Station	KTAM	1240		MA	Greenfield	WHAI	1240
TX	Carthage	KGAS	1590		MA	Greenfield	WHAI	98.3
TX	Carthage	KGAS	104.3		MA	Lawrence	WHAV	1490
TX	Childress	KCTX	1510		MA	Milford	WMRC	1490
TX	Clifton	KWOW	103.3		MA	New Bedford	WBSM	1420
TX	Colorado City	KVMC	1320		MA	North Adams	WNAW	1230
TX	Colorado City	KAUM	1320		MA	Northampton	WHMP	1400
TX	Comanche	KCOM	1550		MA	Pittsfield	WBEC	1420
TX	Corpus Christi	KCCT	1150		MA	Springfield	WHYN	560
TX	Daingerfield	KEGG	1560		MA	Springfield	WACE	730
TX	Eastland	KEAS	97.7		MA	Ware	WARE	1250
TX	El Campo	KULF	1390		MA	Worcester	WTAG	580
TX	Gainesville	KGAF	1580		MA	Worcester	WCRN	830
TX	Gatesville	KRYL	103.1		ME	Bangor	WZON	620
TX	Graham	KSWA	1330		ME	Brunswick	WJJB	900
TX	Houston	KSEV	760		ME	Calais	WQDY	1230
TX	Jackonville	KLJT	102.3		ME	Calais	WQDY	102.5
TX	Lamesa	KPET	690		ME	Camden	WQSS	102.5
TX	Lampasas	KACQ	101.9		ME	Dover	WDME	103.1
TX	Longview	KEES	1430		ME	Ellsworth	WDEA	1370
TX	Lubock	KKAM	1340		ME	Farmington	WKTJ	99.3
TX	Madisonville	KMVL	1220		ME	Houlton	WHOU	100.1
TX	Marshall	KCUL	1410		ME	Kennebunkp't	WQEZ	104.7
TX	Marshall	KCUL	92.3		ME	Machias	WALZ	95.3
TX	Monahans	KLBO	1330		ME	Mexico	WTBM	100.7

Red Sox				**Reds**				
ME	Monticello	WREM	710	OH	Marietta	WJAW	100.9
ME	Norway	WOXO	92.7	OH	Middleport	WMPO	92.1
ME	Portland	WJAB	1440	OH	Portsmouth	WNXT	99.3
ME	Presque Isle	WEGP	1390	OH	Zanesville	WHIZ	1240
ME	Skowhegan	WSKW	1160	WV	Charleston	WCHS	580
ME	Skowhegan	WHQO	107.9	WV	Huntington	WRVC	930
NY	Port Henry	WMNT	92.1	WV	Martinsburg	WRNR	740
RI	Providence	WPRO	630	WV	Oak Hill/Beck	WAXS	94.1
RI	Providence	WRIB	1220	WV	Parkersburg	WLTP	1450
RI	Woosocket	WNRI	1380	WV	Williamson	WBTH	1400
VT	Brattleboro	WKVT	1490	**Rockies**			
VT	Burlington	WJOY	1230	AR	Conway	KASR	1330
VT	Middlebury	WFAD	1490	CO	Aspen	KNFO	106.1
VT	Rutland	WSYB	1380	CO	Colorado Spr	KTWK	740
VT	St. Johnsbury	WSTJ	1340	CO	Denver	KOA	850
VT	Waterbury	WDEV	550	CO	Durango	KIQX	101.3
VT	Waterbury	WDEV	96.1	CO	Ft. Collins	KIIX	600
Reds				CO	Ft. Morgan	KSIR	1010
FL	Sarasota	QOSA		FL	CO	Grand Junct	KBKL	107.9
IN	Batesville	WRBI		IN	CO	Greeley	KFKA	1310
IN	Indianapolis	WNDE		IN	CO	Lamar	KLMR	920
IN	Muncie	WLBC		IN	CO	Pueblo	KCSJ	590
IN	Muncie	WLBC		IN	CO	Rifle	KRGS	690
IN	North Vernon	WJCP		IN	CO	Steamboat Sp	KBCR	1230
IN	Portland	WPGW		IN	CO	Sterling	KNNG	104.7
IN	Portland	WPGW		IN	CO	Walsenburg	KSPK	102.3
IN	Richmond	WHON		IN	CO	Yuma	KNEC	100.9
IN	Shelbyville	WOOO		IN	ID	Twin Falls	KTFI	1270
IN	Sullivan	WNDI		IN	KS	Garden City	KSKL	94.5
KY	Ashland	WCMI		KY	KS	Goodland	KLOE	730
KY	Bowling Green	WBGN		KY	MT	Billings	KGHL	790
KY	Burnside	WKEQ		KY	MT	Bozeman	KMMS	1450
KY	Corbin	WCTT		KY	MT	Dillon	KDBM	1490
KY	Danville	WHIR		KY	MT	Glendive	KXGN	1400
KY	Eminence	WKSF		KY	MT	Kalispell	KOFI	1180
KY	Frankfort	WFKY		KY	MT	Lewistown	KXLO	1230
KY	Grayson	WUGO		KY	MT	Malta	KMMR	100.1
KY	Lexington	WLAP		KY	MT	Miles City	KMTA	1050
KY	Louisville	WAVG	970	MT	Missoula	KYLT	1340
KY	Martin	WMDJ	1440	MT	Wolf Point	KVCK	1450
KY	Maysville	WFTM	96.9	ND	Bowman	KPOK	1340
KY	Maysville	WFTM	96.9	ND	Dickenson	KDIX	1230
KY	Somerset	WSFC	1240	NE	Alliance	KCOW	1400
KY	Whitley City	WHAY	105.9	NE	Chadron	KCSR	610
OH	Athens	WATH	970	NE	Kearney	KKPR	1460
OH	Celina	WKKI	943	NE	Lincoln	KLMS	1480
OH	Chillicothe	WBEX	1490	NE	Scottsbluff	KOAK	690
OH	Cincinnati	WLW	700	SD	Belle Fourche	KBFS	1450
OH	Columbus	WTVN	610	SD	Lemmon	KBJM	1400
OH	Dayton	WHIO	1290	SD	Rapid City	KIMM	1150
OH	Findlay	WBVI	97.7	WY	Casper	KTWO	1030
OH	Ironton	WIRO	1230	WY	Cheyenne	KGAB	650
OH	Lima	WIMA	1150	WY	Douglas	KKTY	1470

OH	Marietta	WMOA	1490		WY	Green River	KUGR	1490
Rockies					**Royals**			
WY	Kemmerer	KMER	950		MO	Butler	KMAM	1530
WY	Laramie	KOWB	1290		MO	Butler	KMOE	92.1
WY	Powell	KPOW	1260		MO	Clinton	KDKD	95.3
WY	Rawlins	KRAL	1240		MO	Columbia	KTGR	1580
WY	Riverton	KVOW	1450		MO	Joplin	KQYX	1560
WY	Sheridan	KROE	103.1		MO	Joplin	KQYX	1560
WY	Sundance	KYDT	930		MO	Kansas City	KMBZ	980
WY	Thermopolis	KTHE	1240		MO	Lebanon	KLWT	1230
WY	Wheatland	KYCN	1340		MO	Marshall	KMMO	1300
WY	Worland	KWOR	1340		MO	Marshall	KMMO	102.9
Royals					MO	Maryville	KNIM	1500
AR	Conway	KTOD	1330		MO	Maryville	KNIM	95.3
IA	Cherokee	KCHE	1440		MO	Monett	KRMO	990
IA	Cherokee	KCHE	92.1		MO	Neosho	KBTN	1420
IA	Des Moines	KWKY	1150		MO	Nevada	KNEM	1240
IA	Shenandoah	KMA	960		MO	Nevada	KNMO	97.7
KS	Beloit	KVSV	1190		MO	Osage Beach	KRMS	1150
KS	Chanute	KKOY	1460		MO	Rolla	KTTR	1490
KS	Chanute	KKOY	105.5		MO	Sedalia	KDRO	1490
KS	Coffeyville	KGGF	690		MO	St. James	KKTR	99.7
KS	Colby	KXXX	790		MO	St. Joseph	KFEQ	680
KS	Dodge City	KGNO	1370		MO	Warrensburg	KOKO	1450
KS	Emporia	KVOE	1400		NE	Fairbury	KGMT	1310
KS	Fort Scott	KMDO	1600		NE	Fairbury	KUTT	99.3
KS	Fort Scott	KOMB	103.9		NE	Falls City	KTNC	1230
KS	Garden City	KIUL	1240		NE	Lexington	KRVN	880
KS	Hays	KAYS	1400		NE	Lincoln	KFOR	1250
KS	Hiawatha	KNZA	103.9		NE	Nebraska Cit	KNCY	1600
KS	Hutchinson	KWBW	1450		NE	Omaha	KBBX	1420
KS	Independence	KIND	101.7		NE	Ord	KNLV	1060
KS	Junction City	KJCK	1420		NE	Ord	KNLV	103.9
KS	Larned	KNNS	1510		NE	Scottsbluff	KNEB	960
KS	Lawrence	KLWN	1320		NM	Clayton	KLMX	1450
KS	Liberal	KYUU	1470		OK	Tulsa	KAKC	1300
KS	Manhattan	KMAN	1350		SD	Belle Fourche	KBFS	1450
KS	Marysville	KNDY	103.1		SD	Yankton	KYNT	1450
KS	Mcpherson	KNGL	1540		**Tigers**			
KS	Mcpherson	KBBE	96.7		MI	Alma	WFYC	1280
KS	Parsons	KLKC	1540		MI	Battle Creek	WRCC	1400
KS	Parsons	KLKC	93.5		MI	Big Rapids	WBRN	1460
KS	Pittsburg	KKOW	860		MI	Cadillac	WATT	1240
KS	Pratt	KWLS	1290		MI	Charlevoix	WMKT	1270
KS	Russell	KRSL	990		MI	Cheboygan	WCBY	1240
KS	Russell	KCAY	95.9		MI	Detroit	WJR	760
KS	Salina	KSAL	580		MI	Escanaba	WCHT	600
KS	Topeka	WIBW	580		MI	Fremont	WSHN	100.1
KS	Wellington	KWME	93.5		MI	Grand Rapids	WBBL	1340
KS	Winfield	KKLE	1550		MI	Greenville	WPLB	1380
MO	Atchison	KAIR	1470		MI	Hillsdale	WCSR	1340
MO	Ava	KKOZ	1430		MI	Holland	WHTC	1450
MO	Ava	KKOZ	95.9		MI	Houghton	WCCY	1400
MO	Bethany/Albany	KAAN	95.9		MI	Jackson	WIBM	1450
MO	Bolivar	KYOO	1200		MI	Kalamazoo	WKZO	590

MO	Bolivar	KYOO	106.3	MI	Marquette	WDMJ	1320
Tigers				**Twins**			
MI	Midland	WMPX	1490	ND	Bismarck	KLXX	1270
MI	Mt. Pleasant	WMMI	830	ND	Carrington	KDAK	1600
MI	Mt. Pleasant	WCZY	104.3	ND	Devil's Lake	KDLR	1240
MI	Newberry	WNBY	93.7	ND	Dickinson	KLTC	1460
MI	Petoskey	WMBN	1340	ND	Fargo	KFGO	790
MI	Port Huron	WPHM	1380	ND	Grafton	KXPO	1340
MI	Saginaw	WSGW	790	ND	Harvey	KHND	1470
MI	S.Ste. Marie	WSOO	1230	ND	Jamestown	KQDJ	1400
MI	St. Ignace	WIDG	940	ND	Lisbon	KQLX	106.1
MI	Tawas/Oscoda	WKJC	104.7	ND	Mayville	KMAV	1520
OH	Toledo	WLQR	1470	ND	Minot	KRRZ	1390
ON	London	CFPL	980	ND	Oakes	KDDR	1220
Twins				ND	Valley City	KOVC	1490
IA	Mason City	KGLO	1300	ND	Wahpeton	KBMW	1450
MN	Ada	KRJB	106.3	ND	Williston	KEYZ	600
MN	Aitkin	KKIN	930	ND	Yankton	WNAX	570
MN	Albert Lea	KATE	1450	SD	Aberdeen	KSDN	930
MN	Alexandria	KSTQ	99.3	SD	Huron	KOKK	1210
MN	Austin	KAUS	1480	SD	Pierre	KGFX	1060
MN	Bemidji	KBUN	1450	SD	Rapid City	KTOQ	1340
MN	Brainerd	WWWI	1270	SD	Sioux Falls	KSOO	1140
MN	Crookston	KROX	1260	SD	Watertown	KWAT	950
MN	Duluth	KDAL	610	SD	Yankton	WNAX	570
MN	Faribault	KDHL	920	**White Sox**			
MN	Fergus Falls	KBRF	1250	IA	Davenport	KJOC	1170
MN	Glenwood	KMGK	107.1	IL	Aurora	WBIG	1280
MN	Grand Rapids	KOZY	1320	IL	Canton	WBYS	107.9
MN	Intern'al Falls	KGHS	1230	IL	Champaign	WBCP	1580
MN	Little Falls	WYRQ	92.1	IL	Chicago	WMVP	1000
MN	Long Prairie	KEYL	1400	IL	Decatur	WDZ	1050
MN	Mankato	KTOE	1420	IL	Freeport	WFRL	1570
MN	Marshall	KMHL	1400	IL	Galesburg	WAIK	1590
MN	Minne/St. Paul	WCCO	830	IL	Galva	WHHK	102.5
MN	Morris	KMRE	1230	IL	Genesco	WGEN	1500
MN	New Ulm	KMUJ	860	IL	Harrisburg	WEBQ	1240
MN	Olivia	KOLV	100.1	IL	Harrisburg	WEBQ	102.3
MN	Owatonna	KRFO	1390	IL	Highland	WINU	880
MN	Preston	KFIL	103.1	IL	Kankakee	WKLZ	1470
MN	Rochester	KROC	1340	IL	Lincoln	WVAX	1370
MN	Sauk Centre	KMSR	103.1	IL	Mattoon	WLBH	1170
MN	St. Cloud	WJON	1240	IL	Mattoon	WLBH	96.9
MN	Staples	KNSP	1430	IL	Metropolis	WMOK	920
MN	Thief Riv. Falls	KTRF	1230	IL	Peoria	WOAM	1350
MN	Virginia	WHLB	1400	IL	Princeton	WZOE	1490
MN	Wabasha	KMFX	1190	IL	Quincy	WGEM	1440
MN	Wadena	KWAD	920	IL	Rochelle	WRHL	1060
MN	Waseca	KRUE	92.1	IL	Rockford	WROK	1440
MN	Willmar	KWLM	1340	IL	Sparta	WHCO	1230
MN	Windom	KDOM	1380	IL	Springfield	WTAX	1240
MN	Windom	KDOM	94.3	IL	Streator	WIZZ	1250
MN	Winona	KWNO	1230	IN	Goshen	WKAM	1460
MN	Worthington	KWOA	730	IN	Monticello	WMRS	107.7
ND	Beulah	KHOL	1410	IN	Oxford	WIBN	98.1

					IN	Rensselaer	WLQI	97.7
Yankees								
CT	Bridgeport	WICC	600					
CT	Danbury	WLAD	800					
CT	Hartford	WPOP	1410					
CT	Waterbury	WWCO	1240					
FL	Belle Glade	WSWN	900					
FL	Tampa	WZTM	820					
MA	Pittsfield	WBRK	1340					
MA	Springfield	WNNZ	640					
MA	Worcester	WWTM	1440					
NH	Dover	WTSN	1270					
NM	Albuquerque	KDEF	1150					
NY	New York	WABC	770					
NY	Albany	WSRD	104.9					
NY	Binghampton	WENE	1430					
NY	Elmira	WELM	1410					
NY	Gloversville	WENT	1340					
NY	Ithaca	WPIE	1160					
NY	Jamestown	WJTN	1240					
NY	Kingston	WKNY	1490					
NY	Newark	WACK	1420					
NY	Plattsburgh	WIRY	1340					
NY	Rochester	WNNR	103.5					
NY	Syracuse	WHEN	620					
NY	Utica	WIBX	950					
NY	Watertown	WTNY	790					
PA	Easton	WEEX	1320					
PA	Pottsville	WPAM	1450					
PA	Scranton	WEJL	630					
PA	Wilkes-Barre	WBAX	1240					
RI	Hope Valley	WJJF	1180					
RI	Providence	WLKW	550					
VT	Burlington	WKDR	1390					

Bears

IA	Burlington, IA	KCPS	1150
IA	Clinton, IA	KCLN	1390
IA	Davenport, IA	KJOC	1170
IA	Des Moines, IA	KWKY	1150
IL	Aurora, IL	WBIG	1280
IL	Chicago, IL	WBBM	780
IL	Decatur, IL	WSOY	1340
IL	DeKalb, IL	WLBK	1360
IL	Galesburg, IL	WAIK	1590
IL	Gibson City, IL	WGCY	106.3
IL	Kankakee, IL	WVLI	95.1
IL	Macomb, IL	WLMD	104.7
IL	Marion, IL	WDDD	810
IL	Peoria, IL	WWFS	1290
IL	Pontiac, IL	WTRX	93.7
IL	Quincy, IL	WGEM	105.1
IL	Rochelle, IL	WRHL	1060
IL	Rochelle, IL	WRHL	102.3
IL	Rockford, IL	WROK	1440
IL	Sparta, IL	WHCO	1230
IL	Springfield, IL	WFMB	1450
IL	Sterling, IL	WZZT	102.7
IL	Watseka, IL	WIVR	101.7
IN	Michigan City,	WIMS	1420
IN	Rensselaer, IN	WLQI	97.7
IN	South Bend, IN	WSBT	960
MI	Kalamazoo, MI	WKLZ	1470

Bengals

KY	Burnside	WKEQ	910
KY	Lexington	WLAP	630
KY	Louisville	WXXA	790
KY	Paintsville	WKYH	600
OH	Athens	WATH	970
OH	Canton	WTIG	990
OH	Celina	WCSM	96.7
OH	Chillicothe	WBEX	1490
OH	Cincinnati	WCKY	1360
OH	Cincinnati	WOFX	92.5
OH	Cincinnati	WLW	700
OH	Columbus	WTVN	610
OH	Dayton	WTUE	104.7
OH	Hillsboro	WSRW	1590
OH	Ironton	WIRO	1230
OH	Lima	WIMA	1150
OH	Logan	WLGN	1510
OH	London	WFTG	1400
OH	Marietta	WMOA	1490
OH	Middleport	WMPO	1390
OH	Portsmouth	WIOI	1010
OH	Wash Ct Hse	WKSI	1250
WV	Huntington	WVHU	800

Bills

NY	Albany	WTMM	1300
NY	Alfred	WZKZ	101.9
NY	Auburn	WAUB	1590
NY	Bath	WABH	1380
NY	Binghamton	WNBF	1290
NY	Bradford	WBRR	100.1
NY	Buffalo	WGRF	96.9
NY	Buffalo	WEDG	103.3
NY	Dunkirk	WDOE	1410
NY	Elmira	WELM	1410
NY	Fredonia	WBKX	96.5
NY	Geneva	WGVA	1240
NY	Ithaca	WKRT	920
NY	Ithaca	WPIE	1160
NY	Jamestown	WJTN	1240
NY	Jamestown	WWSE	99.3
NY	Newark	WACK	1420
NY	Olean	WPIG	95.7
NY	Remsen	WADR	1480
NY	Rochester	WCMF	96.5
NY	Springville	WSPQ	1330
NY	Syracuse	WNSS	1260
NY	Utica	WUTQ	1550
ON	St. Catherines	CKTB	610
ON	St. Catherines	CKTB	610
PA	Erie	WQHZ	TBA

Broncos

CO	Alamosa	KALQ	93.5
CO	Aspen	KNFO	106.1
CO	Colorado Spr	KVOR	1300
CO	Craig	KRAI	550
CO	Denver	KOA	850
CO	Durango	KRSJ	100.5
CO	Ft. Collins	KCOL	600
CO	Ft. Morgan	KSIR	1010
CO	Glenwood Spr	KMTS	99.1
CO	Granby	KKHI	930
CO	Grand Jct	KSTR	96.1
CO	Greeley	KFKA	1310
CO	Gunnison	KPKE	1490
CO	Montrose	KUBC	580
CO	Pagosa Spr	KWUF	1400
CO	Pueblo	KGHF	1350
CO	Salida	KVRH	1340
CO	Steamboat	KBCR	1230
CO	Trinidad	KCRT	92.5
CO	Walsenburg	KSPK	102.3
KS	Garden City	KYBD	94.5
KS	Goodland	KLOE	730
KS	Hays	KAYS	1400

Broncos

State	City	Call	Freq
KS	Liberal	KSCB	1270
MT	Billings	KGHL	790
MT	Bozeman	KMMS	1450
MT	Colstrip	KOAL	106.1
MT	Glendive	KDZN	96.5
MT	Kalispell	KOFI	1180
MT	Lewistown	KXTL	1230
MT	Malta	KMMA	100.1
ND	Bismarck	KFYR	550
ND	Bowman	KPOK	1340
NE	Alliance	KCOW	1400
NE	Chadron	KCSR	610
NE	North Platte	KOOQ	1410
NE	Scottsbluff	KNEB	960
NM	Albuquerque	KABQ	1350
NM	Artesia	KSVP	990
NM	Farmington	KDAG	96.9
NV	Las Vegas	KSHP	1400
SD	Belle F'rche	KBFS	1450
SD	Lemmon	KBJM	1400
SD	Rapid City	KOTA	1380
WY	Buffalo	KBBS	1450
WY	Casper	KTWO	1330
WY	Cheyenne	KGAB	1240
WY	Cody	KODI	1400
WY	Douglas	KKTY	1470
WY	Gillette	KIML	1270
WY	Green River	KUGR	1490
WY	Kemmerer	KMER	950
WY	Lander	KOVE	1330
WY	Laramie	KOWB	1290
WY	Newcastle	KASL	1240
WY	Powell	KLZY	92.5
WY	Rawlins	KRAL	1240
WY	Riverton	KVOW	1450
WY	Sheridan	KWYO	1410
WY	Sundance	KYDT	103.1
WY	Thermopolis	KTHE	1240
WY	Torrington	KGOS	1490
WY	Wheatland	KYCN	1340
WY	Worland	KWOR	1340

Browns

State	City	Call	Freq
OH	Akron	WAKR	1590
OH	Ashland	WNCO	1340
OH	Ashtabula	WFUN	970
OH	Athens	WATH	970
OH	Bucyrus	WQEL	92.7
OH	Cadiz	WCDK	106.3
OH	Canton	WHBC	1480
OH	Celina	WKKI	94.3
OH	Chillicothe	WCHI	1350
OH	Cleveland	WMMS	100.7

Browns

State	City	Call	Freq
OH	CLEVELAND	WTAM	1100
OH	Columbus	WFJX	105.7
OH	Dayton	WONE	980
OH	Defiance	WDFM	98.1
OH	Dover	WJER	1450
OH	Elyria	WEOL	930
OH	Findlay	WQTL	106.3
OH	Fostoria	WFOB	1430
OH	Lima	WBUK	107.5
OH	Mansfield	WMAN	1400
OH	Marietta	WJAW	630
OH	Marion	WMRN	1490
OH	Mt.Vernon	WMVO	1300
OH	Newark	WHTH	790
OH	Painesville	WBKC	1460
OH	Portsmouth	WNXT	99.3
OH	Portsmouth	WNXT	1260
OH	Sandusky	WMJK	100.9
OH	Sandusky	WLEC	1450
OH	Toledo	WIOT	104.
OH	Wooster	WQKT	104.
OH	Youngstown	WNCD	93.
OH	Youngstown	WKBN	570
OH	Zanesville	WHIZ	1240

Buccaneers

Not Available

Cardinals

State	City	Call	Freq
AZ	Flagstaff	KWMX	96.
AZ	Holbrook	KZUA	92.
AZ	Lk Havasu	KNTR	980
AZ	Miami	KIKO	134
AZ	Phoenix	ESPN	86
AZ	Phoenix	KTAR	62
AZ	Phoenix	KMIA*	71
AZ	Prescott Vly	KPPV	106.
AZ	Prescott Vly	KQNA	113
AZ	Sedona	KAZM	78
AZ	Showlow	KSNX	93.
AZ	Sierra Vista	KTAN	142
AZ	Thatcher	KATO	123
AZ	Tuba City	KTBA	105
AZ	Tucson	KTKT	99
AZ	Wilcox	KHIL	125
AZ	Window Rk	KHAC	88
AZ	Yuma	KJOK	140
NM	Albuquerque	KDFF	115
NM	Lordsburg	KPSA	97
NM	Los Alamos	KRSN	149

Chargers

State	City	Call	Freq
CA	Imperial Vly	KXO	123
CA	Los Angeles	KMPC	154
CA	Palm Sprgs	KXPS	101
CA	Riverside	KTMQ	103
CA	San Diego	KIOZ	105

Chiefs

State	City	Call	Freq		State	City	Call	Freq
AR	Mammoth Spr	KAMS	95.1		**Chiefs**			
A	Atlantic	KJAN	1220		NE	Falls City	KTNC	1230
A	Burlington	KBKB	1360		NE	Grand Island	KRGI	1430
A	Des Moines	KXNO	1460		NE	Lincoln	KFOR	1240
A	Shenandoah	KKBZ	99.3		NE	Omaha	KOMJ	590
A	Sioux City	KSFT	107.1		OK	Bartlesville	KWON	1400
A	Burlington	KSNP	97.7		**Colts**			
KS	Chanute	KKOY	105.5		IL	Danville	WDAN	1490
KS	Clay Center	KCLY	100.9		IN	Indianapolis	WFBQ	94.7
KS	Coffeyville	KGGF	690		IN	Bedford	WBIW	1340
KS	Colby	KXXX	790		IN	Columbus	WYGB	102.9
KS	Colby	KQLS	100.3		IN	Crawfordsville	WIMC	103.9
KS	Emporia	KVOE	101.7		IN	Elkhart	WTRC	1340
KS	Ft. SCOTT	KMDO	1600		IN	Evansville	WYNG	94.9
KS	Ft. SCOTT	KOMB	103.9		IN	Ft. Wayne	WOWO	1190
KS	Garden City	KKJQ	97.3		IN	Greencastle	WREB	94.3
KS	Great Bend	KVGB	104.3		IN	Hardinsburg	WKLO	96.9
KS	Hays	KFIX	96.9		IN	Lafayette	WSHP	95.7
KS	Hutchinson	KWBW	1450		IN	Madison	WORX	96.7
KS	Junction City	KJCK	1420		IN	Marion	WGOM	860
KS	Larned	KNNS	1520		IN	Michigan City	WEFM	95.9
KS	Lawrence	KLWN	1320		IN	Monticello	WMRS	107.7
KS	Manhattan	KMAN	1350		IN	Muncie	WXFN	1340
KS	McPherson	KNGL	1540		IN	North Vernon	WJCP	92.7
KS	McPherson	KBBE	96.7		IN	Richmond	WHON	930
KS	Parsons	KLKC	1540		IN	Rochester	WROI	92.1
KS	Parsons	KLKC	93.5		IN	Santa Claus	WAXL	103.3
KS	Pittsburg	KKOW	860		IN	South Bend	WDND	1580
KS	Pittsburg	KKOW	96.9		IN	Terre Haute	WWVR	105.5
KS	Salina	KILS	92.7		IN	Vincennes	WAOV	1450
KS	Topeka	WIBW	580		IN	Wabash	WJOT	105.9
KS	Wellington	KLEY	1130		IN	Washington	WWBL	106.5
KS	Wichita	KTHR	107.3		**Cowboys**			
KS	Winfield	KKLE	1550		AR	Little Rock	KASR	92.7
MO	Bethany	KAAN	870		LA	Minden	KASO	1240
MO	Bethany	KAAN	95.5		LA	Shreveport	KSYB	560
MO	Chillicothe	KULH	105.9		NM	Albuquerque	KKOB	770
MO	Clinton	KDKD	95.3		NM	Artesia	KBIM	94.9
MO	Columbia	KFRU	1400		NM	Farmington	KTRA	102.1
MO	Joplin	KKOW	860		NM	Hobbs	KBIM	94.9
MO	Joplin	KKOW	96.9		TX	Dallas	KLUV	98.7
MO	Kansas City	KCFX	101.1		TX	Abilene	KWKC	1340
MO	Kirksville	KRLX	94.5		TX	Amarillo	KGNC	710
MO	Lebanon	KCLQ	107.9		TX	Austin	KZNX	1530
MO	Marshall	KMMO	1300		TX	Beaumont	KLVI	560
MO	Marshall	KMMO	102.9		TX	Belton	KTON	940
MO	Monett	KRMO	990		TX	Big Spring	KBST	95.9
MO	Neosho	KBNT	1420		TX	Brady	KNEL	95.3
MO	Osage Beach	KRMS	93.5		TX	Brownwood	KBWD	1380
MO	Sedalia	KSDL	92.3		TX	College Stn	KZNE	1150
MO	St. Joseph	KFEQ	680		TX	Carthage	KGAS	104.3
MO	St. Joseph	KSJQ	92.7		TX	Comanche	KCOM	1550
MO	Warrensburg	KOKO	1450		TX	Corpus Christi	KEYS	1440
MO	Waynesville	KFBD	97.9		TX	Edinburg	KURV	710
					TX	El Paso	KROD	600

Cowboys					Dolphins			
TX	Fredericksb'rg	KNAF	910		FL	Miami/Ft Laud	WAXY	79(
TX	Gonzales	KCTI	1450		FL	Miami/Ft Laud	WMXJ	102.7
TX	Gonzales	KCTI	106.3		FL	W. Palm Bch	WEFL	76(
TX	Haskell	KVRP	1400		FL	Vero Beach	WGNX	99.7
TX	Haskell	KVRP	97.1		FL	Orlando	WOCL	105.9
TX	Hereford	KPAN	860		FL	Ft. Meyers	WWCN	77(
TX	Hereford	KPAN	106.3		FL	Sarasota	WSRQ	145(
TX	Horseshoe Bay	KBEY	92.5		FL	Daytona Bch	WOCL	105.9
TX	Junction	KMBL	1450		FL	Melbourne	WGNX	99.7
TX	Kerrville	KERV	1230		FL	Key West	WKWF	160(
TX	Killeen	KTON	940		Eagles			
TX	Lamesa	KPET	690		PA	Allentown	WCTO	96.1
TX	Laredo	KHOY	88.1		PA	Hershey	WQXA	105.7
TX	Llano	KBEY	92.5		PA	Levittown	WBCB	149(
TX	Longview	KEYS	1430		PA	Pottsville	WPPA	136(
TX	Lubbock	KJTV	950		PA	Reading	WEEU	83(
TX	Lufkin	KTBQ	107.7		PA	Scranton/W-B	WBSX	97.9
TX	Marble Falls	KBEY	92.5		PA	Sunbury	WEGH	107.3
TX	Marshall	KMHT	103.9		PA	Williamsport	WBZD	93.3
TX	Mexia	KYCX	104.9		Falcons			
TX	Midland	KHKX	99.1		AL	Anniston	WHMA	139(
TX	Nacodoches	KTBQ	107.7		GA	Albany	WGBC	145(
TX	Odessa	KHKX	99.1		GA	Athens	WRFC	96(
TX	Ozona	KYXX	94.3		**GA**	**Atlanta**	**WZGC**	**92.9**
TX	Palestine	KNET	1450		GA	Brunswick	WSFN	79(
TX	Paris	KITX	95.5		GA	Columbus	WDAK	54(
TX	Pecos	KPTX	98.3		GA	Dahlonega	WKHC	104.3
TX	San Angelo	KGKL	960		GA	Gainesville	WGGA	124(
TX	San Angelo	KGKL	97.5		GA	Hogansville	WVCC	72(
TX	San Antonio	KTSA	550		GA	Jesup	WLOP	137(
TX	Sonora	KHOS	92.1		GA	Lagrange	WMGP	98.
TX	Stamford	KPAN	860		GA	Louisville	WPEH	142(
TX	Stamford	KPAN	106.3		GA	Louisville	WPEH	92.
TX	Stephenville	KSTV	1510		GA	Macon	WMAC	94(
TX	Stephenville	KSTV	93.1		GA	Milledgeville	WMVG	145(
TX	Sulphur Spr	KSST	1230		GA	Rome	WKCX	97.
TX	Sweetwater	KXOX	1240		GA	Savannah	WTKS	129(
TX	Temple	KTON	940		GA	Statesboro	WWNS	124(
TX	Texarkana	KCMC	740		GA	Swainsboro	WXRS	159(
TX	Tyler	KTBB	600		GA	Talking Rock	WNSY	100.
TX	Uvalde	KVOU	1400		GA	Thomaston	WTGA	159(
TX	Waco	KRZX	1660		GA	Thomaston	WTGA	101.
TX	Waco	KRZI	1580		GA	Tuscaloosa	WACT	142(
TX	Wichita Falls	KWFS	1290		GA	Vidalia	WVOP	97(
TX	Wichita Falls	KWFS	102.3		GA	Waycross	WFNS	135(
Dolphins					MS	Meridian	WFFX	145
FL	Bradenton	WSRQ	1450		TN	Chattanooga	WDOD	96.
FL	Panama City	WASG	550		VA	Bedford	WBWR	106.
FL	Marathon	WFFG	1300		VA	Blacksburg	WBRW	105.
FL	Belle Glade	WBGF	93.5		VA	Roanoke	WBWR	97.
FL	Ocala	WOCA	1370					
FL	Pensacola	WVTJ	610					

FortyNiners

ST	City	Call	Freq
CA	Bay Area	KNBR	680
CA	Bay Area	KNBR	1050
CA	Bay Area	KSAN	107.7
CA	Chico	KTHU	100.7
CA	Eureka	KATA	1340
CA	Fresno	KFIG	580
CA	Grass Valley	KNCO	830
CA	Hillsboro	KUIK	1360
CA	King City	KRKC	1490
CA	Mendocino	KMFB	96.7
CA	Modesto	KESP	970
CA	Mount Shasta	KNTK	102.3
CA	Paso Robles	KPRL	1230
CA	Red Bluff	KBLF	1490
CA	Redding	KNRO	1670
CA	Sacramento	KFBK	1530
CA	S Luis Obispo	KKJL	1400
CA	Sonora	KZSQ	92.7
CA	Susanville	KJDX	93.3
CA	Visalia	KJUG	1270
HI	Hilo	KPUA	670
HI	Honolulu	KKEA	1420
NV	Reno	KJFK	1230
OR	Eugene	KNND	1400
OR	Quincy	KPCO	1370

Giants

ST	City	Call	Freq
CT	Danbury	WLAD	800
NH	Dover	WTSN	1270
NY	Albany	WPYX	106.5
NY	Albany	WOFX	980
NY	Binghamton	WENE	1430
NY	Binghamton	WINR	680
NY	East Hampton	WHBE	96.7
NY	Johnstown	WIZR	930
NY	Kingston	WKNY	1490
NY	New York	WFAN	660
NY	Plattsburgh	WIRY	1340
NY	Rochester	WHAM	1180
NY	Syracuse	WHEN	620
PA	Easton	WEEX	1230
PA	Leighton	WYNS	1160
PA	Stroudsburg	WVPO	920
VT	Burlington	WKOL	105.1

Jaguars

ST	City	Call	Freq
FL	Altamont Spr	WHOO	1080
FL	Amelia Island	WOKV	690
FL	Crescent Bch	WFOY	1240
FL	Daytona Beach	WELE	1380
FL	Ft Pierce	WPSL	1590
FL	Gainesville	WRUF	850
FL	Jacksonville	WOKV	690
FL	Jacksonville	WKQL	96.9
FL	Madison	WSTI	105.3
FL	Melbourne	WIXC	1060
FL	Altamont Spr	WHOO	1080
FL	Amelia Island	WOKV	690
FL	Orlando	WHOO	1080
FL	Palatka	WIYD	1260
FL	Quitman	WSTI	105.3
FL	Tallahassee	WUTL	106.1
FL	Waycross	WWSN	103.3
GA	Brunswick	WBGA	107.7
GA	Cuthbert	WCUG	850
GA	Jesup	WIFO	105.5

Jets

ST	City	Call	Freq
NY	New Tork	WADO*	1280
NY	New York	ESPN	1050
NY	New York	WABC	770

Lions

ST	City	Call	Freq
MI	Ann Arbor	WAAM	1600
MI	Battle Creek	WBCK	930
MI	Battle Creek	WRCC	1400
MI	Big Rapids	WBRN	1460
MI	Coldwater	WTVB	1590
MI	Detroit	WKRK	97.1
MI	Detroit	WXYT	1270
MI	Escanaba	WYKX	104.7
MI	Flint	WTRX	1330
MI	Grand Rap's	WKLQ	107.3
MI	Hillsdale	WCSR	1340
MI	Hillsdale	WCSR	92.1
MI	Holland	WHTC	1450
MI	Houghton	WCCY	1400
MI	Iron Mtn	WMIQ	1450
MI	Jackson	WIBM	1450
MI	Kalamazoo	WKZO	590
MI	Lansing	WQTX	92.7
MI	Lapeer	WLST	1530
MI	Lapeer	WQUS	103.1
MI	Marquette	WDMJ	1320
MI	Mt. Pleasant	WMMI	830
MI	Mt. Pleasant	WCZY	104.3
MI	Newbury	WNBY	1450
MI	Petoskey	WJML	1110
MI	Petoskey	WWKK	750
MI	Pinconning	WYLZ	100.9
MI	Port Huron	WPHM	1380
MI	S.Ste. Marie	WSOO	1230
MI	Saginaw	WILZ	104.5
MI	South Haven	WGMY	940
MI	St. Ignace	WIDG	940
MI	St. Joseph	WSJM	1400
MI	Tawas City	WKJC	104.7
MI	Traverse City	WTCM	580
MI	Whitehall	WEFG	97.5
OH	Toledo	WLQR	1470

Packers					Packers			
IA	Des Moines	KPSZ	940		WI	Sheboygan	WHBL	133•
MI	Escanaba	WGLQ	97.1		WI	Siren	WXCX	105.
MI	Iron Mtn	WJNR	101.5		WI	Sparta	WCOW	97.
MI	Iron River	WIKB	1230		WI	Superior	WDSM	71•
MI	Iron River	WIKB	99.1		WI	Waupaca	WDUX	92.
MI	Ironwood	WIMI	99.7		WI	Wausau	WSAU	55•
MN	Stillwater	WMGT	1220		WI	Wautoma	WAUH	102.
ND	Bismarck	KFYR	550		WI	Whitehall	WHTL	102.
SD	Aberdeen	KGIM	1420		WI	Whitewater	WKCH	106.•
SD	Sioux Falls	KSOO	1140		WI	Wisconsin Rds	WFHR	132•
WI	Algoma	WRLU	104.1		**Rams**			
WI	Antigo	WRLO	105.3		CA	Los Angeles	KLAC	57•
WI	Appleton	WHBY	1150		CA	San Diego	XTRA	69•
WI	Ashland	WATW	1400		IL	Carthage	WCAZ	99•
WI	Ashland	WBSZ	93.3		IL	Jerseyville	WJBM	148•
WI	Beaver Dam	WBEV	1430		IL	Mt. Vernon	WMIX	94•
WI	Beloit	WTJK	1380		IL	Pittsfield	WBBA	97.•
WI	Clintonville	WFCL	92.3		IL	Quincy	WTAD	93•
WI	Clintonville	WJMQ	92.3		IL	Salem	WJBD	100.
WI	Eau Claire	WBIZ	1400		IL	Springfield	WCVA	96.
WI	Eau Claire	WBIZ	100.7		IN	Evansville	WGBE	128•
WI	Fond du Lac	KFIZ	1450		MO	Cape Girard'u	KGIE	122•
WI	Fond du Lac	WFON	107.1		MO	Columbia	KTGR	158•
WI	Ft Atkinson	WFAW	940		MO	Dexter	KDEX	159•
WI	Green Bay	**WNFL**	1440		MO	Farmngton	KREI	80•
WI	Green Bay	**WIXX**	101.1		MO	Fulton	KKCA	100.•
WI	Hayward	WRLS	92.3		MO	Jefferson City	KWOS	95•
WI	Janesville	WCLO	1230		MO	Monett	KKBL	95.•
WI	La Crosse	WKTY	580		MO	Mountain View	KUPH	96.•
WI	Lancaster	WGLR	97.7		MO	Osage Beach	KRMS	115•
WI	Madison	WIBA	1310		MO	Poplar Bluff	KWOC	93•
WI	Madison	WIBA	101.5		MO	Rolla	KZNN	105.•
WI	Manitowoc	WOMT	1240		MO	Sedalia	KPOW	97.•
WI	Marinette	WMAM	570		MO	Sikeston	KSIM	140•
WI	Marinette	WLST	95.1		MO	Springfield	KGMY	140•
WI	Marshfield	WDLB	1450		**MO**	**St. Louis**	**KLOU**	**103.**
WI	Medford	WKEB	99.3		**MO**	**St. Louis**	**KTRS**	**55•**
WI	Menomonie	WMEQ	92.1		MO	Washington	KSLQ	104.
WI	Milwaukee	WTMJ	620		NV	Las Vegas	KSHP	140•
WI	Minocqua	WMQA	95.9		OK	Ada	KADA	123•
WI	Monroe	WEKZ	1260		OK	Durant	KSEO	75•
WI	Monroe	WEKZ	93.7		OK	Norman	KREF	140•
WI	Park Falls	WCQM	98.3		**Ravens**			
WI	Prairie/ Chien	WQPC	94.3		DE	Dover	WWTX	129•
WI	Reedsburg	WNFM	104.9		DE	Dover	WDOV	141•
WI	Rhinelander	WOBT	1240		**MD**	**Baltimore**	**ESPN**	**130•**
WI	Rice Lake	WJMC	1240		**MD**	**Baltimore**	**WJAK**	**102.**
WI	Rice Lake	WJMC	96.3		MD	Cambridge	WCEM	124•
WI	Richland Ctr	WRCO	100.9		MD	Cambridge	WCEM	106.•
WI	River Falls	WEVR	1550		MD	Cumberland	WSBC	107.•
WI	River Falls	WEVR	106.3		MD	La Plata	WKIK	102.•
WI	Shawano	WTCH	960		MD	Ocean City	WRXS	106.•
WI	Shawano	WOWN	99.3					

Ravens

State	City	Call	Freq
MD	Salisbury	WICO	1320
MD	Westminster	WTTR	1470
PA	Chambersburg	WHAG	1410
PA	Hanover	WHVR	1280
PA	Lancaster	WLAN	1390
PA	York	WSBA	910
WV	Harrisonburg	WHBG	1360
WV	Martinsburg	WEPM	1340
WV	Winchester	WINC	1400

Redskins

State	City	Call	Freq
MD	Annapolis	WNAV	1430
MD	Cumberland	WCBC	1270
MD	Frederick	WFMD	930
MD	Hagerstown	WARX	1490
MD	Mechan'sville	WSMD	98.5
MD	Salisbury	WOSC	95.9
NC	Farmville	WGBH	1250
NC	Greensboro	WSJS	600
NC	Raleigh	WPTF	680
NC	Salisbury	WSTP	1490
NC	Shelby	WOHS	730
PA	Gettysburg	WGET	1320
VA	Blacksburg	WPIN	810
VA	Charlottesville	WCHV	1260
VA	Fredricksburg	WBQB	101.5
VA	Gloucester	WXGM	99.1
VA	Gloucester	WXGM	1420
VA	Harrisonburg	WSVA	550
VA	Leesburg	WAGE	1200
VA	Lynchburg	WBRG	1050
VA	Norfolk	WNIS	790
VA	Norfolk	WTAR	850
VA	Richmond	WJMO	105.7
VA	Roanoke	WFIR	960
VA	Staunton	WSVO	93.1
VA	Tappahan'ck	WRAR	105.5
WV	Martinsburg	WRNR	740

Saints

State	City	Call	Freq
AL	Mobile	WNSP	105.5
FL	Destin	WBZR	1120
FL	Valporaiso	WFSH	1340
LA	Alexandria	KEZP	104.3
LA	Baton Rouge	WJBO	1150
LA	Bogalusa	WBOX	920
LA	Bogalusa	WBOX	92.2
LA	Houma	KCIL	107.5
LA	Jennings	KJEF	1290
LA	JonesviLLE	KTGV	105.1
LA	Lafayette	KMDL	97.3
LA	Lake Charles	KLCL	1400
LA	Monroe	KNOE	540
LA	Moreauville	KLIL	92.1
LA	Morgan City	KQKI	95.3
LA	New Iberia	KANE	1240
LA	New Orleans	WWL	870
LA	New Orleans	WFND*	830
LA	Shreveport	KWKH	1130
LA	Thibodaux	KVPI	92.5
LA	Thibodaux	KVPI	1520
LA	Ville Platte	KTIB	640
MS	Byam	WJIX	1450
MS	Columbia	WCJU	104.6
MS	Hattiesburg	WKNZ	107.1
MS	Jackson	WDJX	680
MS	McComb	WHNY	1250
MS	Meridian	WZKR	103.3
MS	Natchez	WKSO	93.7
MS	Prentis	WJDR	98.3
MS	Richton	WXHB	98.5
MS	Tupelo	WTUP	1490
MS	Vicksburg	WQBC	1420

Seahawks

State	City	Call	Freq
AK	Anchorage	KTZN	550
AK	Cordova	KLAM	1450
AK	Haines	KRSA	94.9
AK	Juneau	KINY	800
AK	Petersburg	KRSA	580
AK	Sitka	KRSA	94.9
AK	Wrangell	KRSA	94.9
BC	Vancouver	MOJO	730
ID	Coeur d'Alene	KVNI	1080
ID	Lewiston	KCLK	1430
ID	Payette	KIOV	1450
ID	Pocatello	KSEI	930
ID	Rexburg	KRXK	1230
MT	Kalispell	KJJR	880
MT	Missoula	KGRZ	1450
OR	Astoria	KKEE	1230
OR	Bend	KWLZ	96.5
OR	Coos Bay	KWRO	1230
OR	Eugene	KPNW	1120
OR	Klamath Falls	KFLS	1450
OR	Lebanon	KGAL	1580
OR	Pendleton	KTIX	1240
OR	Portland	KXL	750
OR	Roseburg	KQEN	1250
OR	The Dalles	KODL	1440
OR	Tillamook	KMBD	1590
WA	Aberdeen	KWOK	1490
WA	Bellingham	KPUG	1170
WA	Centralia	KMNT	102.9
WA	Colfax	KMAX	840
WA	Colville	KCVL	1240
WA	Ellensburg	KXLE	1240
WA	Forks	KVAC	1490
WA	Grand Coulee	KEYG	98.5
WA	Longview	KEDO	1400

Seahawks

WA	Moses Lake	KBRC	1470
WA	Mount Vernon	KBRC	1430
WA	Olympia	KGY	1240
WA	Omak	KNCW	92.7
WA	Port Angeles	KONP	1450
WA	Prosser	KZXR	1310
WA	**Seattle**	**KIRO**	**710**
WA	Shelton	KMAS	1030
WA	Spokane	KXLY	920
WA	Tri-Cities	KONA	610
WA	Walla Walla	KUJ	1420
WA	Wenatchee	KPQ	560
WA	Yakima	KIT	1280

Steelers

OH	St. Mary's	WJAW	630
OH	Youngstown	WBBG	106.1
PA	Altoona	WFBG	1290
PA	Beaver Falls	WBVP	1230
PA	Beaver Falls	WMBA	1460
PA	Bedford	WAYC	100.9
PA	Bradford	WESB	1490
PA	Butler	WISR	680
PA	Chambersb'g	WCHA	800
PA	Clarion	WCCR	92.7
PA	Cooperstown	WUUZ	94.3
PA	Erie	WFNN	1330
PA	Grove City	WWGY	1400
PA	Indiana	WQMU	92.5
PA	Johnstown	WWGE	1400
PA	Latrobe	WCNS	1480
PA	New Castle	WKST	1200
PA	**Pittsburgh**	**WDVE**	**102.5**
PA	**Pittsburgh**	**WBGG**	**970**
PA	Portage	WUZI	105.7
PA	Punxsutawny	WPXZ	104.1
PA	Seagertown	WHUZ	94.3
PA	Somerset	WUZY	97.7
PÁ	State College	WMAJ	1450
PA	Uniontown	WMBS	590
PA	Warren	WKNB	104.3
PA	Washington	WJPA	1450
PA	Washington	WJPA	100.9
PA	Williamsport	WCXR	103.7
PA	Williamsport	WZXR	99.3
WV	Clarksburg	WPDX	104.9
WV	Fairmont	WMMN	920
WV	Fairmont	WRLF	94.3
WV	Morgantown	WZST	100.9
WV	Weirton	WEIR	1430
WV	Wheeling	WWVA	1170

Texans

LA	Baton Rouge	WIBR	1300
LA	Lake Charles	KEZM	1310
TX	Alpine	KVLF	1240
TX	Amarillo	KPUR	1440
TX	Austin	KICE	1370
TX	Beaumont	KIKR	1450
TX	Big Spring	KTBS	94.3
TX	Brenham	KWHI	1280
TX	Bryan	KZNE	1150
TX	Carthage	KGAS	104.3
TX	Centerville	KQBB	92.5
TX	Columbus	KLUM	98.3
TX	Corpu Christi	KNCN	101.3
TX	Crockett	KBHT	93.5
TX	Edinburgh	KSOX	1240
TX	Fredericksburg	KNAF	910
TX	**Houston**	**KILT**	**610**
TX	**Houston**	**KILT**	**100.3**
TX	Killeen	KTEM	1400
TX	Kingsville	KNCN	101.3
TX	Liberty	KSHN	99.9
TX	Longview	KFRO	95.3
TX	Lufkin	KUEZ	100.1
TX	Madisonville	KBHT	93.5
TX	Marble Falls	KHLB	1340
TX	Mcallen	KSOX	1240
TX	Midland/Odessa	KMCM	96.9
TX	Navasota	KZNE	1600
TX	New Braunfels	KGNB	1420
TX	Orange	KOGT	1600
TX	Palestine	KBHT	93.5
TX	Pittsburgh	KDVE	103.
TX	Port Arthur	KSHN	99.9
TX	San Angelo	KGKL	960
TX	San Antonio	KTKR	760
TX	San Marcos	KGNB	1420
TX	Silsbee	KSET	1300
TX	Tatum	KXAL	100.1
TX	Temple	KTEM	1400
TX	Texarkana	KTFS	940
TX	Tyler	KFRO	95.3
TX	Victoria	KAJI	94.

Titans

AL	Arab	WAFN	92.
AL	Birmingham	WJOX	690
AL	Cullman	WFMH	1340
AL	Florence	WQLT	107.
AL	Gadsden	WJBY	930
AL	Huntsville	WRTT	95.
AL	Scottsboro	WWIC	105
AL	Sylacauga	WFEB	1340
AR	Stuttgart	KWAK	1240
AR	Stuttgart	KWAK	105.
KY	Bowling Green	WPTQ	103.

Titans				Titans			
KY	Cadiz	WKDZ	106.5	TN	Parsons	WKJQ	97.3
KY	Calvert City	WCCK	95.7	TN	Pulaski	WKSR	98.3
KY	Campbellsville	WTCO	1450	TN	Sevierville	WSEV	105.5
KY	Elizabethtown	WIEL	1400	TN	Sevierville	WSEV	930
KY	Elkton	WEKT	1070	TN	Shelbyville	WZNG	1400
KY	Henderson	WSON	860	TN	S. Pittsburg	WEPG	910
KY	Leitchfield	WMTL	870	TN	Tri Cities	WGOC	640
KY	Leitchfield	WKHG	104.9	TN	Union City	WYVY	104.9
KY	Madisonville	WHRZ	97.7	TN	Winchester	WCDT	1340
KY	Mayfield	WNGO	1320				
KY	Murray	WNBS	1340	**Vikings**			
KY	Murray	WRKY	1130	IA	Ames	KASI	1430
KY	Owensboro	WOMI	1490	IA	Cedar Rapids	WMTL	600
KY	Paducah	WKYX	570	IA	Cedar Rapids	KMJM	1360
KY	Paducah	WPAD	1560	IA	Mason City	KGLO	1300
MS	Greenville	WGVM	1260	IA	Spencer	KICD	1240
MS	Greenville	WDMS	100.7	MN	Alexandria	KIKV	100.7
MS	Holly Springs	WKRA	1110	MN	Atkin	KKIN	94.3
MS	Jackson	WPBQ	1240	MN	Austin	KNFX	970
TN	Camden	WFWL	1220	MN	Bemidji	KBHP	101.1
TN	Centerville	WNKX	96.7	MN	Bemidji	KBUN	1450
TN	Chattanooga	WGOW	1150	MN	Benson	KSCR	93.5
TN	Chattanooga	WGOW	102.3	MN	Blue Earth	KBEW	98.1
TN	Clarksville	WZZP	97.5	MN	Brainerd	KBLB	93.3
TN	Cleveland	WCLE	104.1	MN	Crookston	KROX	1260
TN	Cleveland	WCLE	1570	MN	Detroit Lakes	KDLM	1340
TN	Columbia	WMCP	1280	MN	Detroit Lakes	KBOT	104.1
TN	Cookeville	WBXE	93.7	MN	Duluth	KKCB	105.5
TN	Covington	WKBQ	93.5	MN	Duluth	WEBC	560
TN	Crossville	WPBX	99.3	MN	Fergus Falls	KBRF	1250
TN	Dickson	WDKN	1260	MN	Grand Rapid	KMFY	96.9
TN	Dyersburg	WTRO	1450	MN	HIBBING	WNMT	650
TN	Fayetteville	WYTM	105.5	MN	Hutchinson	KARP	106.9
TN	Franklin	WAKM	950	MN	Internat'l Falls	KSDM	104.1
TN	Harriman	WBZH	92.7	MN	Jackson	KRAQ	105.7
TN	Harrogate	WRWB	740	MN	Mankato	KYSM	1230
TN	Hartsville	WTNK	1090	MN	Mankato	KXLP	93.1
TN	Jackson	WYNU	92.3	MN	Moose Lake	WMOZ	106.9
TN	Knoxville	WOKI	100.3	MN	Morristown	KKOK	95.7
TN	La Follette	WQLA	104.9	MN	NEWCASTLE	KNUJ	107.3
TN	Lawrenceburg	WDXE	1370	MN	Owatonna	KRFO	1390
TN	Lebanon	WANT	98.9	MN	Park Rapids	KPRM	870
TN	Lebanon	WCOR	900	MN	Park Rapids	KDKK	97.5
TN	Lewisburg	WAXO	1220	MN	Pine City	WCMP	100.9
TN	Manchester	WMSR	1320	MN	Rochester	KRCH	101.7
TN	Martin	WCMT	1410	MN	Rochester	KWEB	1270
TN	Martin	WCMT	101.3	MN	Rosseau	KCAJ	102.1
TN	McKenzie	WHDM	1440	MN	Spring Grove	KQYB	98.3
TN	McMinnville	WTRZ	103.9	MN	St. Cloud	WJON	1240
TN	Memphis	WGKX	105.9	MN	Thief Riv. Falls	KTRF	1230
TN	Morristown	WCRK	1150	MN	Twin Cities	KQQL	107.9
TN	Nashville	WKDF	103.3	MN	Twin Cities	KFAN	1130
TN	Paris	WMUF	1000	MN	Wabasha	KMFW	1190
TN	Paris	WMUF	104.7				

Vikings

MN	Wadena	KSKK	94.7					
MN	Walker	KQKK	101.9					
MN	Waseca	KOWZ	1170					
MN	Willmar	KWLM	1340					
MN	Winona	KWNO	1230					
MT	Miles City	KKRY	92.5					
ND	Bismarck	KXMR	710					
ND	Carrington	KDAK	1600					
ND	Devils Lake	KDLR	1240					
ND	Dickinson	KLTC	1460					
ND	Fargo	KFGO	790					
ND	Grand Forks	KKXL	1440					
ND	Jamestown	KQDJ	1400					
ND	Minot	KRRZ	1390					
ND	Oakes	KDDR	1220					
ND	Valley City	KOVC	1490					
ND	Williston	KEYZ	600					
SD	Aberdeen	KSDN	930					
SD	Millbank	KMSD	1510					
SD	Mobridge	KMLO	100.7					
SD	Pierre	KPLO	94.5					
SD	Sioux Falls	KWSF	107.9					
SD	Sisseton	KBWS	102.9					
SD	Watertown	KWAT	950					
WI	Balsam Lake	WLMX	104.9					
WI	Hayward	WHSM	101.1					
WI	Menomonie	WMEQ	880					
WI	Sparta	WKLJ	1290					

St	City	Station	Freq		St	City	Station	Freq
AL					AK			
	Birmingham	WBHM	90.3			Petersburg	KFSK	100.9
	Dothan	WRWA	88.7			Point Baker	KFSK	88.1
	Gadsden	WSGN	91.5			Point Hope	KBRW	101.5
	Huntsville	WLRH	89.3			Point Lay	KBRW	88.1
	Huntsville/Normal	WJAB	90.9			Port Alexander	KCAW	91.9
	Jacksonville	WLJS	91.9			Port Protection	KFSK	91.1
	Mobile	WHIL	91.3			Prudhoe Bay	KSKA	91.1
	Montgomery	WVAS	90.7			Scow Bay	KFSK	103.1
	Montgomery-Troy	WTSU	89.9			Seward	KSKA	88.1
	Muscle Shoals	WQPR	88.7			Sitka	KCAW	90.1
	Phenix City	WTJB	91.7			SITKA	KCAW	104.7
	Selma	WAPR	88.3			St. Paul	KUSB	91.9
	Tuscaloosa	WUAL	91.5			Talkeetna	KTNA	88.5
AK						Tenakee Springs	KCAW	91.9
	Anaktuvuk Pass	KBRW	101.5			Unalaska	KIAL	1450
	Anchorage	KNBA	90.3			Valdez	KCHU	770
	Anchorage	KSKA	91.1			Women's Bay	KMXT	100.1
	Angoon	KCAW	105.5			Wrangell	KSTK	101.7
	Barrow	KBRW	91.9			Yakutat	KCAW	90.1
	Barrow	KBRW	680		AR			
	Bethel	KYUK	640			El Dorado	KBSA	90.9
	Chevak	KCUK	88.1			Fayetteville	KUAF	91.3
	Chiniak	KMXT	100.1			Jonesboro	KASU	91.9
	Dillingham	KDLG	670			Little Rock	KUAR	89.1
	Excursion Inlet	KTOO	89.9			Little Rock	KLRE	90.5
	Fairbanks	KUAC	89.9		AZ			
	Fort Yukon	KZPA	900			Flagstaff	KNAU	88.7
	Galena	KIYU	910			Flagstaff	KNAQ	91.7
	Girdwood	KSKA	91.9			Page	KNAD	91.7
	Glennallen	KXGA	90.5			Phoenix	KBAQ	89.5
	Gustavus	KTOO	88.1			Phoenix	KJZZ	91.5
	Haines	KHNS	102.3			Prescott	KPUB	89.3
	Homer	KBBI	890			Safford	KUAT	89.5
	Hoonah	KTOO	91.9			Show Low	KNAA	90.7
	Juneau	KTOO	104.3			Sierra Vista	KUAT	89.7
	Kake	KCAW	107.1			Tuba City	KGHR	91.5
	Kaktovik	KBRW	88.1			Tucson	KUAZ	89.1
	Kenai	KDLL	91.9			Tucson	KUAT	90.5
	Ketchikan	KRBD	105.9			Tucson	KUAT	1550
	Kodiak	KMXT	100.1			Yuma	KAWC	88.9
	Lemon/Switzer Crk	KTOO	101.7			Yuma	KAWC	1320
	McCarthy	KXKM	89.7					
	McGrath	KSKO	870					
	Mendenhall Valley	KTOO	103.4					
	Nuiqsut	KBRW	88.1					
	Ouzinkie	KMXT	100.1					
	Pasagshak	KMXT	100.1					
	Pelican	KCAW	91.7					

CA				CO		
Arcata	KHSU	90.5		Alamosa	KRZA	88.7
Bakersfield	KPRX	89.1		Aspen	KAJX	91.5
Bishop	KUNR	90.9		Bayfield	KSUT	90.1
Burney	KNCA	89.7		Boulder	KGNU	88.5
Chico	KCHO	91.7		Boulder	KCFC	1490
Crescent City	KHSR	91.9		Breckenridge	KUNC	88.3
Fresno	KVPR	89.3		Buena Vista	KUNC	89.9
Groveland	KXSR	91.7		Buena Vista	KRCC	95.7
Indio	KCRY	89.3		Canon City	KRCC	105.7
Long Beach	KLON	88.1		Carbondale	KDNK	90.5
Los Angeles	KUSC	91.5		Colorado Springs	KRCC	91.5
Mount Shasta	KNSQ	88.1		Cortez	KSUT	89.5
North Hollywood	KPFK	90.7		Cortez	KSJD	91.5
Northridge	KCSN	88.5		Craig	KPRN	90.1
Oxnard	KCRU	89.1		Crested Butte	KBUT	90.3
Palm Springs	KPSC	88.5		Delta	KPRU	103.3
Pasadena	KPCC	89.3		Denver	KUVO	89.3
Philo	KZYX	90.7		Denver	KUVO	89.3
Redding	KFPR	88.9		Denver	KVOD	90.1
Redding	KSOR	90.9		Denver	KCFR	1340
Ridgecrest	KVCR	88.7		Dolores	KSUT	91.9
Sacramento	KXJZ	88.9		Dove Creek	KPRU	88.7
Sacramento	KXPR	90.9		Durango	KSUT	89.5
San Bernardino	KVCR	91.9		Eagle	KUWR	89.9
San Diego	KPBS	89.5		Estes Park	KUNC	90.9
San Francisco	KQED	88.5		Front Range	KUNC	91.5
San Francisco	KALW	91.7		Ft. Collins	KGNU	99.9
San Luis Obispo	KCBX	90.1		Glenwood Springs	KUWR	88.9
San Mateo	KCSM	91.1		Grand Junction	KPRN*	89.5
Santa Barbara	KFAC	88.7		Greeley	KUNC	91.5
Santa Cruz	KUSP	88.9		Gunnison	KVOD	89.1
Santa Monica	KCRW	89.9		Haxton/Holyoke	KUNC	89.9
Santa Rosa	KRCB	91.1		Ignacio	KSUT	91.3
Stockton	KUOP	91.3		Julesburg/Ovid	KUNC	89.1
Susanville	KUNR	91.9		La Junta	KRLJ	89.1
Thousand Oaks	KCLU	88.3		Leadville	KVOD	103.9
Thousand Oaks	KCPB	91.1		Limon	KRCC	89.9
Truckee	KUNR	88.1		Manitou Springs	KRCC	90.1
Tulelake	KSOR	91.9		Meeker	KPRN	91.1
Victorville	KVCR	88.1		Montrose	KPRH	88.3
Weed	KSOR	89.5		Montrose	KVMT	89.1
Willits	KZYZ	91.5		Ouray	KPRN	91.5
Yreka	KNYR	91.3		Pagosa Springs	KSUT	105.3
				Paonia	KVNF	90.9
				Parachute	KPRN	88.3
				Pueblo	KCFP	91.9
				Pueblo	KKPC	1230
				Rangley	KPRN	91.1
				Rifle	KPRN	88.3

CO					FL			
	Rio Blanco Cty	KPRN	88.7			Sarasota	WMNF	88.5
	Salida	KUNC	89.9			Sarasota	WUSF	89.7
	Silverton	KVOD	91.5			Tallahassee	WFSU	88.9
	Steamboat Springs	KUNC	88.5			Tallahassee	WFSQ	91.5
	Sterling	KUNC	90.5			Tampa/St Pet'b'rg	WMNF	88.5
	Summit County	KPRN	89.3			Tampa/St Pet'b'rg	WUSF	89.7
	Summit County	KUNC	107.1			West Palm Beach	WXEL	90.7
	Telluride	KOTO	91.7		GA			
	Trinidad	KRCC	91.1			Albany	WUNV	91.7
	Vail	KPRE	89.9			Athens	WUGA	91.7
	Vail/Eagle Valley	KUNC	99.7			Atlanta	WRFG	89.3
	Ward	KGNU	93.7			Atlanta	WABE	90.1
	Westcliffe/Gardner	KRCC	88.5			Atlanta	WCLK	91.9
	Wray	KUNC	93.5			Augusta	WACG	90.7
	Yuma	KUNC	88.7			Brunswick	WWIO	89.1
CT						Carrollton	WWGC	90.7
	Fairfield	WSHU	91.1			Columbus	WTJB	88.1
	Meriden/Hartford	WPKT	90.5			Dahlonega	WNGU	89.5
	New Haven	WSHU	91.1			Demorest	WPPR	88.3
	Northfield	WSUF	93.3			Fort Gaines	WJWV	90.9
	Norwich	WNPR	89.1			Macon	WDCO	89.7
	Stamford	WEDW	88.5			Savannah	WSVH	91.1
	Stamford	WSHU	90.1			Tifton	WABR	91.1
	Westport	WMMM	1260			Valdosta	WWET	91.7
	Willimantic	WECS	90.1			Warm Springs	WJSP	88.1
	Fairfield	WSHU	91.1			Waycross	WXVS	90.1
DE					HI			
	Dover	WRTX	91.7			Honolulu	KHPR	88.1
DC						Honolulu	KIPO	89.3
	Washington	WAMU	88.5			Oahu North Shore	KIPO	88.7
	Washington	WETA	90.9			Pearl City	KIFO	1380
FL						Wailuku (Maui)	KKUA	90.7
	Fort Myers/Naples	WGCU	90.1		ID			
	Fort Pierce	WQCS	88.9			Bellvue	KBSW	89.3
	Gainesville	WUFT	89.1			Boise	KBSU	90.3
	Inverness	WJUF	90.1			Boise	KBSX	91.5
	Jacksonville	WJCT	89.9			Boise	KBSU	730
	Marco	WMKO	91.7			Burley	KBSY	88.5
	Melbourne/Titusville	WFIT	89.5			Cascade	KBSU	89.3
	Melbourne/Titusville	WMFE	90.7			Challis	KBSW	89.7
	Miami/Ft Lauderdale	WLRN	91.3			Cottonwood	KNWO	90.1
	Orlando	WUCF	89.9			Grace	KUSU	90.5
	Orlando	WMFE	90.7			Ketchum	KBSW	93.5
	Panama City	WFSW	89.1			Lower Stanley	KBSU	91.1
	Panama City	WKGC	90.7			Mccall	KBSM	91.7
	Panama City	WKGC	1480			Moscow	KRFA	91.7
	Pensacola	WUWF	88.1			New Meadows	KBSX	93.5

ID					IN			
	Orofino	KRFA	102.3			Marion	WBSW	90.9
	Pocatello	KUER	107.7			Muncie	WBST	92.1
	Rexburg	KRIC	100.5			North Manchester	WBKE	89.5
	Salmon	KBSW	91.5			Portland	WBSJ	91.7
	Stanley	KBSU	106.3			Richmond	WVXR	89.3
	Sun Valley	KBSX	91.1			Terre Haute	WFIU	95.1
	Sun Valley	KWRV	91.9			W. Lafayette	WBAA	101.3
	Twin Falls	KBSW	91.7		IA			
IL						Ames	WTPR	88.1
	Carbondale	WSIU	91.9			Ames	WOI	90.1
	Champaign	WEFT	90.1			Ames	WOI	640
	Champaign-Urbana	WILL	90.9			Cedar Falls	KHKE	89.5
	Champaign-Urbana	WILL	580			Cedar Falls	KUNI	90.9
	Chicago	WBEZ	91.5			Cedar Rapids	KCCK	88.3
	Dekalb	WNIU	90.5			Davenport	KUNI	94.5
	Edwardsville	WSIE	88.7			Decorah	KLNI	88.7
	Freeport	WNIE	89.1			Decorah	KLCD	89.5
	La Salle	WNIW	91.5			Des Moines	KUNI	101.7
	Macomb	WIUM	91.3			Dubuque	KUNI	98.7
	Normal	WGLT	89.1			Eldridge	KUNI	102.1
	Olney	WUSI	90.3			Fort Dodge	KTPR	91.1
	Peoria	WCBU	89.9			Iowa City	WSUI	910
	Pittsfield	WIPA	89.3			Mason City	KHKE	90.7
	Quincy	WQUB	90.3			Mason City	KUNY	91.5
	Rock Island	WVIK	90.3			Mason City	KRNI	1010
	Rockford	WNIJ	89.5			Sioux City	KWIT	90.3
	Springfield	WUIS	91.9		KS			
	Sterling	WNIQ	91.5			Ashland	KANZ	92.9
	Warsaw	WIUW	89.5			Atwood	KANZ	95.3
	Rock Island	WVIK	90.3			Colby	KANZ	88.9
	Rockford	WNIJ	89.5			Dodge City	KANZ	96.3
	Springfield	WUIS	91.9			Elkhart	KANZ	98.3
	Sterling	WNIQ	91.5			Garden City	KANZ	91.1
	Warsaw	WIUW	89.5			Goodland	KANZ	89.7
IN						Great Bend	KHCT	90.9
	Anderson	WBSB	89.5			Hays	KANZ	98.3
	Bloomimgton	WFIU	103.7			Herndon	KZNA	91.3
	Columbus	WFIU	100.7			Hill City	KZNA	90.5
	Elkhart	WVPE	88.1			Hutchinson	KHCC	90.1
	Evansville	WNIN	88.3			Hutchinson	KHCC	90.1
	Fort Wayne	WBNI	89.1			Lamar	KANZ	90.7
	Goshen	WGCS	91.1			Lawrence	KANU	91.5
	Hagerstown	WBSH	91.1			Liberal	KANZ	96.3
	Indianapolis	WFYI	90.1			Ness City	KANZ	92.9
	Indianapolis	WFIU	103.7			Pittsburg	KRPS	89.9
	Kokomo	WFIU	106.1			Salina	KHCD	89.5
	Lafayette	WBAA	920			Sharon Springs	KANZ	90.7
						St. Francis	KANZ	96.3
						Tribune	KANZ	89.5
						Washburn	KANZ	91.3
						Wichita	KMUW	89.1

KY					ME			
	Bowling Grn	WKYU	88.9			Bangor	WMEH	90.9
	Elizabethtwn	WKUE	90.9			Calais	WMED	89.7
	Hazard	WEKH	90.9			Fort Kent	WMEF	106.5
	Henderson	WKPB	89.5			Portland	WMEA	90.1
	Highland Hts	WNKU	89.7			Presque Isle	WMEM	106.1
	Lexington	WUKY	91.3			Waterville	WMEW	91.3
	Louisville	WFPL	89.3		MI			
	Louisville	WFPK	91.9			Allendale	WGVU	88.5
	Morehead	WMKY	90.3			Alpena	WCML	91.7
	Murray	WKMS	91.3			Ann Arbor	WUOM	91.7
	Paducah	WKMS	92.1			Bay City	WUCX	90.1
	Richmond	WEKU	88.9			Detroit	WDET	101.9
	Somerset	WDCL	89.7			East Jordan	WIZY	100.9
LA						East Lansing	WKAR	90.5
	Alexandria	KLSA	90.7			East Lansing	WKAR	870
	Baton Rouge	WRKF	89.3			Escanaba	WNMU	107.1
	Baton Rouge	WBRH	90.3			Flint	WFUM	91.1
	Grambling	KDAQ	89.9			Grand Rapids	WBLU	88.9
	Grambling	KDAQ	90.7			Grand Rapids	WMUK	102.1
	Lafayette	KRVS	88.7			Grand Rapids	WVGR	104.1
	Lake Charles	KVLU	90.9			Harbor Springs	WCMW	103.9
	Monroe	KEDM	90.3			Houghton	WGGL	91.9
	New Orleans	WWNO	89.9			Interlochen	WIAA	88.7
	Shreveport	KDAQ	89.9			Kalamazoo	WMUK	102.1
	Thibidaux	KTLN	90.5			Kentwood	WGVU	1480
MA	Amherst	WFCR	88.5			Manistee	WVXM	97.7
	Amherst	WPNI	1430			Manistique	WNMU	91.9
	Boston	WGBH	89.7			Marquette	WNMU	90.1
	Boston	WBUR	90.9			Marquette	WNMU	102.3
	Boston	WUMB	91.9			Menominee	WNMU	91.3
	Falmouth	WFPB	91.9			Mount Pleasant	WCMU	89.5
	Great Barrington	WAMQ	105.1			Muskegon	WGVS	850
	Harwich	WCCT	90.3			Newberry	WNMU	91.1
	Nantucket	WNAN	91.1			Oscoda	WCMB	95.7
	New Bedford	WFPR	91.9			Rogers City	WXVA	96.7
	Orleans	WFPB	1170			Sault Ste. Marie	WCMZ	98.3
	Sandwich	WSDH	91.5			Standish	WWCM*	96.9
	Springfield	WFCR	88.5			Stephenson	WNMU	107.3
	West Yarmouth	WBUR	1240			Traverse City	WIAA	100.7
MD						Twin Lake	WBLV	90.3
	Baltimore	WJHU	88.1			Whitehall	WGVS	95.3
	Baltimore	WEAA	88.9			Ypsilanti	WEMU	89.1
	Frostburg	WFWM	91.9					
	Hagerstown	WETH	89.1					
	Ocean City	WSDL	90.7					
	Princess Anne	WESM	91.3					
	Salisbury	WSCL	89.5					
	Worton	WKHS	90.5					

MN	City	Call	Freq	MO/MS/MT	City	Call	Freq
				MO			
	Albert Lea	KZSE	103.9		Branson	KSMS	90.5
	Alexandria	KSJR	90.9		Cape Girardeau	KRCU	90.9
	Appleton	KNCM	88.5		Chillicothe	KRNW	88.9
	Appleton	KRSU	91.3		Columbia	KOPN	89.5
	Austin	KZSE	103.3		Columbia	KBIA	91.3
	Austin	KLSE	103.9		Joplin	KSMU	98.9
	Bemidji	KCRB	88.5		Kansas City	KCUR	89.3
	Bemidji	KNBJ	91.3		Lebanon	KUMR	96.3
	Bemidji	KAXE	94.7		Maryville	KXCV	90.5
	Brainerd	KAXE	89.5		Mountain Grove	KSMU	88.1
	Brainerd	KBPR	90.7		Neosho	KSMU	103.7
	Buhl	WIRN	92.5		Rolla	KUMR	88.5
	Collegeville	KNSR	88.9		Springfield	KSMU	91.1
	Collegeville	KSJR	90.1		St. Louis	WSIE	88.7
	Duluth	WSCD	92.9		St. Louis	KWMU	90.7
	Duluth	WSCN	100.5		Warrensburg	KCMW	90.9
	Ely	WIRR	89.5		West Plains	KSMU	101.7
	Ely	WIRN	101.7	MS			
	Fergus Falls	KCCM	89.7		Biloxi	WMAH	90.3
	Grand Rapids	KAXE	91.7		Booneville	WMAE	89.5
	Grand Rapids	KCRB	104.1		Bude	WMAU	88.9
	Grand Rapids	WSCN	107.3		Greenwood	WMAO	90.9
	International Falls	KNBJ	88.1		Holly Springs	WURC	88.1
	International Falls	KCRB	97.7		Jackson	WJSU	88.5
	International Falls	KCRB	97.7		Jackson	WMPN	91.3
	La Crescent	KXLC	91.1		Lorman	WPRL	91.7
	Moorhead	KCCD	90.3		Meridian	WMAW	88.1
	Moorhead	KCCM	91.1		Mississippi State	WMAB	89.9
	Northfield	WCAL	89.3		Oxford	WMAV	90.3
	Owatonna	KNGA	103.9		Senatobia	WKNA	88.9
	Owatonna	KGAC	105.7	MT			
	Rochester	KMSE	88.7		Big Sky	KBMC	95.9
	Rochester	KZSE	90.7		Big Timber	KEMC	90.5
	Rochester	KLSE	91.7		Billings	KEMC	91.7
	Roseau	KQMN	90.9		Bozeman	KBMC	102.1
	St. Cloud	KNSR	88.9		Bozeman	KEMC	106.7
	St. Cloud	KSJR	90.1		Butte	KAPC	91.3
	St. Paul	KNOW	91.1		Chester	KEMC	100.1
	St. Paul	KSJN	99.5		Colstrip	KEMC	88.5
	St. Peter	KGAC	90.5		Columbus	KEMC	88.5
	St. Peter	KNGA	91.5		Cut Bank	KEMC	88.9
	Thief River Falls	KDSU	88.3		Dillon	KAPC	91.7
	Thief River Falls	KNTN	102.7		Forsyth	KEMC	88.5
	Virginia/Hibbing	WIRR	90.9		Ft. Belknap Agcy	KGVA	88.1
	Virginia/Hibbing	WIRN	92.5		Glasgow	KEMC	91.9
	Winona	KLSE	101.9		Great Falls	KGPR	89.9
	Winona	KZSE	107.3		Hamilton	KUFN	91.9
	Worthington	KRSW	88.9		Hardin	KEMC	91.7
	Worthington	KRSW	89.3		Havre	KNMC	90.1
	Worthington	KNSW	91.7				
	Worthington	KNSW	91.7				

MT					NM			
	Helena	KUHM	91.7			Albuquerque	KANW	89.1
	Helena	KEMC	97.1			Albuquerque	KUNM	89.9
	Kalispell	KUKL	89.9			Arroyo Seco	KUNM	91.1
	Lewistown	KEMC	88.5			Clayton	KENW	93.5
	Livingston	KEMC	88.5			Des Moines	KENW	106.1
	Marysville	KAPC	107.1			Gallup	KGLP	91.7
	Miles City	KECC	90.7			Gallup	KSUT	91.7
	Missoula	KUFM	89.1			Las Cruces	KRWG	90.7
	No. Missoula	KAPC	91.5			Las Vegas	KUNM	91.9
	Paradise Vlly	KEMC	91.1			Las Vegas	KENW	107.1
	Plentywood	KDSU	91.9			Maljamar	KMTH	98.7
	Red Lodge	KEMC	89.1			Montoya	KENW	90.7
	Shelby	KEMC	90.3			Pine Hill	KTDB	89.7
	Swan Lake	KUFM	91.1			Portales	KENW	89.5
	Terry	KEMC	91.9			Quay	KENW	91.3
	White Sulphur Sprg	KUFM	98.3			Ramah	KTDB	89.7
	Whitefish	KUFM	91.7			Raton	KENW	104.7
NE						Roy	KENW	104.9
	Alliance	KTNE	91.1			Ruidoso	KMTH	91.3
	Bassett	KMNE	90.3			Santa Fe	KUNM	89.9
	Chadron	KCNE	91.9			Socorro	KUNM	91.9
	Columbus	KUCV	90.3			St. Augustine	KENW	102.9
	Culbertson	KUCV	92.7			Taos	KUNM	91.9
	Falls City	KUCV	91.7			Tucumcari	KENW	104.5
	Harrison	KUCV	89.5			Wagon Mound	KENW	92.1
	Hastings	KHNE	89.1		NV			
	Lexington	KLNE	88.7			Battle Mtn	KUNR	100.5
	Lincoln	KUCV	91.1			Beatty	KNPR	91.7
	Max	KUCV	93.3			Crescent Valley	KUNR	91.5
	Merriman	KRNE	91.5			Elko	KNCC	91.5
	Norfolk	KXNE	89.3			Eureka	KUNR	90.9
	North Platte	KPNE	91.7			Hawthorne	KUNR	91.5
	Omaha	KIOS	91.5			Incline Village	KUNR	89.9
NH						Las Vegas	KNPR	89.5
	Concord	WEVO	89.1			Las Vegas	KUNV	91.5
	Gorham	WEVC	107.1			Lovelock	KUNR	89.9
	Hanover	WEVH	91.3			Panaca	KLNR	91.7
	Keene	WEVN	90.7			Reno	KUNR	88.7
NJ						Searchlight	KNPR	105.1
	Atlantic City	WNJN	89.7			Tonopah	KTPH	91.7
	Berlin	WNJS	88.1			Verde	KUNR	91.7
	Bridgeton	WNJB	89.3			Winnemucca	KUNR	91.3
	Cape May	WNJZ	90.3			Yerington	KUNR	91.9
	Lincroft	WBJB	90.5					
	Manahawkin	WNJM	89.9					
	Newark	WBGO	88.3					
	Ocean City	WRTQ	91.3					
	Sussex	WNJP	88.5					
	Trenton	WNJT	88.1					

NY					NY			
	Albany	WAMC	90.3			Poughkeepsie	WRHV	88.7
	Alexandria Bay	WSLU	88.7			Ridge	WSHU	91.7
	Alfred	WALF	89.7			Rochester	WXXI	91.5
	Binghamton	WSKG	89.3			Rochester	WXXI	1370
	Binghamton	WSQX	91.5			Saranac Lake	WSLL	90.5
	Blue Mountain Lake	WXLH	91.3			Schenectady	WMHT	89.1
	Brooklyn	WNYE	91.5			Selden	WSUF	89.9
	Brookville	WCWP	88.1			Southampton	WPBX	88.3
	Buffalo/Niagara Falls	WBFO	88.7			Southhampton	WRLI	91.3
	Buffalo/Niagara Falls	WNED	94.5			Syracuse	WAER	88.3
	Buffalo/Niagara Falls	WNED	970			Syracuse	WRVD	90.3
	Canajoharie	WCAN	93.3			Syracuse	WRVD	90.3
	Canton	WSLU	89.5			Syracuse	WCNY	91.3
	Corning	WSQE	91.1			Ticonderoga	WANC	103.9
	Cortland	WSUC	90.5			Troy/Rensselaer	WAMC	93.1
	Elmira	WSKG	91.9			Utica	WUNY	89.5
	Geneva	WEOS	89.7			Utica	WRVN	91.9
	Greensport	WSUF	89.9			Vestal	WSQX	89.7
	Houghton	WJSL	90.3			Watertown	WSLJ	88.9
	Huntington Station	WSHU	91.3			Watertown	WJNY	90.9
	Ithaca	WSQG	90.9			Watertown	WRVJ	91.7
	Jamestown	WUBJ	88.1		NC			
	Jamestown	WNJA	89.7			Asheville	WCQS	88.1
	Jeffersonville	WJFF	90.5			Beech Mountain	WNCW	95.5
	Keene	WSLU	89.7			Boone	WNCW	92.9
	Kingston	WAMK	90.9			Buxton	WBUX	90.5
	Lake Placid	WSLU	91.7			Chapel Hill	WUNC	91.5
	Long Lake	WSLU	91.5			Charlotte	WFAE	90.7
	Lowville	WSLU	88.1			Charlotte	WNCW	100.7
	Malone	WSLO	90.9			Davidson	WDAV	89.9
	Middletown	WOSR	91.7			Durham	WNCU	90.7
	New York City	WFUV	90.7			Fayetteville	WFSS	91.9
	New York City	WNYC	93.9			Franklin	WFQS	91.3
	New York City	WNYC	820			Greensboro	WUNC	91.5
	Newburgh	WAMC	107.7			Greenville	WTEB	89.3
	North Creek	WXLG	89.9			Hickory	WFHE	90.3
	Noyack	WSHU	103.3			Kinston	WKNS	90.5
	Old Forge	WSLU	88.7			Manteo	WURI	91.5
	Olean	WOLN	91.3			Morehead City	WBJD	91.5
	Oneonta	WAMC	88.9			Roanoke Rapids	WZRU	88.5
	Oneonta	WSQC	91.7			Rocky Mount	WRQM	90.9
	Oswego	WRVO	89.9			Spindale	WNCW	88.7
	Peru	WXLU	88.3			Wilmington	WHQR	91.3
	Plattsburg	WCEL	91.9			Winston-Salem	WFDD	88.5

Radio on the Road

NPR

ND					OH			
	Beach	KDSU	91.9			Portsmouth	WOSP	91.5
	Belcourt	KEYA	88.5			Thompson	WSKU	89.1
	Bismarck	KCND	90.5			Toledo	WGTE	91.3
	Bowman	KDSU	91.9			West Union	WVXW	89.5
	Crary	KDSU	89.5			Wooster	WKRW	89.3
	Crosby	KDSU	91.9			Yellow Springs	WYSO	91.3
	Devils Lake	KFJM	90.7			Youngstown	WYSU	88.5
	Devils Lake	KDSU	91.5			Zanesville	WOUZ	90.1
	Dickinson	KDPR	89.9		OK			
	Fargo	KRSD	88.1			Altus	KCCU	89.9
	Fargo	KCCM	91.1			Ardmore	KLCU	90.3
	Fargo	KDSU	91.9			Lawton	KCCU	89.3
	Grand Forks	KUND	89.3			Lawton	KCCU	102.9
	Grand Forks	KFJM	90.7			Norman	KGOU	106.3
	Grand Forks	KFJM	1370			Oklahoma City	KROU	105.7
	Hettinger	KDSU	91.9			Oklahoma City	KROU	105.7
	Jamestown	KPRJ	91.5			Stillwater	KOSU	91.7
	Lakota	KFJM	91.7			Tulsa	KWGS	89.5
	Minot	KMPR	88.9		OR			
	Tioga	KDSU	91.9			Ashland	KSRG	88.3
	Williston	KPPR	89.5			Ashland	KSMF	89.1
OH						Ashland	KSOR	90.1
	Athens	WOUB	91.3			Ashland	KSJK-AM	1230
	Athens	WOUB	1340			Astoria	KMUN	91.9
	Bryan	WGBE	90.9			Baker	KOPB	91.5
	Cambridge	WOUC	89.1			Bandon	KSOR	91.7
	Chillicothe	WVXC	89.3			Bend	KLCC	88.9
	Chillicothe	WOUH	91.9			Bend	KOAB	91.3
	Cincinnati	WMUB	88.5			Big Bend	KSOR	91.3
	Cincinnati	WGUC	90.9			Brookings	KSOR	91.1
	Cincinnati	WVXU	91.7			Burns	KOPB	91.7
	Cleveland	WCPN	90.3			Callahan	KSOR	89.1
	Columbus	WOSU	89.7			Camas Valley	KSOR	88.7
	Columbus	WCBE	90.5			Canyonville	KSOR	91.9
	Columbus	WOSU	820			Chiloquin	KSOR	91.7
	Coshocton	WOSE	91.1			Coos Bay	KSBA	88.5
	Dayton	WMUB	88.5			Coos Bay	KSOR	89.1
	Defiance	WGDE	91.9			Coquille	KSOR	88.1
	Ironton	WOUL	89.1			Corvallis	KOAC	103.1
	Kent	WKSU	89.7			Corvallis	KOAC	550
	Lima	WGLE	90.7			Cottage Grove	KLCC	91.5
	Mansfield	WOSV	91.7			Eugene	KLCC	89.7
	Marion	WOSB	91.1			Florence	KLFO	88.1
	Mount Gilead	WVXG	95.1			Gasquet	KSOR	89.1
	New Paris	WVXR	89.3			Gold Beach	KSOR	91.5
	New Philadelphia	WKRJ	91.5			Grants Pass	KSOR	88.9
	Oxford	WMUB	88.5			Grants Pass	KSMF	89.1
						Grants Pass	KAGI	930

279

	OR				PA		
	John Day	KOPB	91.5		Philadelphia	WRTI	90.1
	Joseph/Enterprise	KFAE	100.9		Philadelphia	WHYY	90.9
	Klamath Falls	KSOR	90.5		Pittsburgh	WQED	89.3
	Klamath Falls	KSKF	90.9		Pittsburgh	WDUQ	90.5
	La Grande	KOPB	89.9		Pittsburgh	WYEP	91.3
	La Pine	KSOR	89.1		Pottsville	WVIA	94.9
	Lakeview	KSOR	89.5		Reading	WRTI	97.7
	Lakeview	KOPB	91.1		Scranton	WVIA	89.9
	Lincoln	KSOR	88.7		State College	WPSU	91.5
	Newport	KLCO	90.5		Stroudsburg	WVIA	94.3
	Oak Ridge	KLCC	91.5		Summerdale	WJAZ	91.7
	Pendleton	KFAE	89.1		Sunbury	WVIA	105.7
	Pendleton	KRBM	90.9		Treasure Lake	WPSB	95.1
	Pendleton	KOPB	91.5		Williamsport	WVIA	89.3
	Port Orford	KSOR	90.5	RI			
	Portland	KOPB	91.5		Providence	WRNI	1290
	Portland	KOAC	550		Westerly	WXNI	1230
	Portland	KBPS	1450	SC			
	Reedsport	KLCC	90.9		Aiken	WLJK	89.1
	Richland	KOPB	91.9		Beaufort	WJWJ	89.9
	Roseburg	KLCC	88.5		Charleston	WSCI	89.3
	Roseburg	KSRS	91.5		Clemson	WEPR	90.1
	Roseburg	KSRS	91.5		Columbia	WLTR	91.3
	Roseburg	KSOR	91.9		Conway	WHMC	90.1
	Silver Lake	KOPB	91.7		Greenville	WNCW	97.3
	Sisters	KLCC	90.3		Greenville/Sp'b'g	WEPR	90.1
	The Dalles	KOPB	91.5		Rock Hill	WNSC	88.9
	Wagontire	KOPB	90.3		Sumter	WRJA	88.1
PA	Allentown	WVIA	99.3	SD			
	Allentown/Bethlm	WDIY	88.1		Belle Fourche	KBHE	88.1
	Allentown/Bethl'm	WDIY	93.9		Brookings	KESD	88.3
	Altoona	WPSB	106.7		Edgemont	KBHE	91.1
	Bethlehem	WVIA	105.7		Faith	KPSD	97.1
	Bradford	WPSB	100.9		Hot Springs	KBHE	88.1
	Clarks Summitt	WVIA	90.3		Lowry	KQSD	91.9
	Clearfield	WPSB	104.7		Martin	KZSD	102.5
	Erie	WQLN	91.3		Mobridge	KQSD	91.9
	Harrisburg	WXPH	88.1		Pierpont	KDSD	90.9
	Harrisburg	WITF	89.5		Pringle	KBHE	91.9
	Harrisburg	WJAZ	91.7		Rapid City	KBHE	89.3
	Johnstown	WQEJ	89.7		Reliance	KTSD	91.1
	Kane	WPSB	90.1		Sioux Falls	KRSD	88.1
	Lancaster	WITF	99.7		Sioux Falls	KNCM	88.5
	Lewisburg	WVIA	100.1		Sioux Falls	KCSD	90.9
	Mt. Pocono	WRTY	91.1		Vermillion	KUSD	89.7
	North East Pa	WVIA	89.9				
	Philadelphia	WXPN	88.5				

Radio on the Road

NPR

TN					UT			
	Chattanooga	WUTC	88.1			Annabella	KUER	101.7
	Collegedale	WSMC	90.5			Bear Lake	KUSU	89.3
	Cookeville	WHRS	91.7			Beaver	KUER	90.1
	Dyersburg	WKNQ	90.7			Blanding	KUER	90.1
	Jackson	WKNP	90.1			Bonanza	KUER	91.5
	Johnson City	WETS	89.5			Brigham City	KUSU	89.5
	Knoxville	WUOT	91.9			Cedar City	KUER	88.3
	Knoxville	WNCW	96.7			Cedar City	KUSU	97.3
	Madison	WPLN	1430			Coalville	KUER	88.5
	Memphis	WKNO	91.1			Coalville	KCUA	92.5
	Murfreesboro	WMOT	89.5			Delta	KUER	88.1
	Nashville	WPLN	90.3			Delta	KUSU	89.5
	Paris	WKMS	99.5			Duchesne	KUER	104.9
	Tullahoma	WTML	91.5			Escalante	KUER	91.1
TX						Ferron	KUER	88.3
	Abilene	KACU	89.7			Fillmore	KUER	90.1
	Andrews	KENW	90.9			Green River	KUSU	90.9
	Austin	KUT	90.5			Logan	KUSU	91.5
	Beaumont	KVLU	91.3			Milford	KUER	88.5
	College Station	KEOS	89.1			Milford	KUSU	90.7
	College Station	KAMU	90.9			Moab	KUER	91.5
	Corpus Christi	KEDT	90.3			Monticello	KUER	90.1
	Dallas/Ft Worth	KERA	90.1			**Utah**	(cont'd)	
	Denton	KNTU	88.1			Myton/Duchesne	KUSU	106.3
	El Paso	KTEP	88.5			Ogden	KUSU	92.5
	Harlingen	KMBH	88.9			Panguitch	KUER	101.7
	Houston	KUHF	88.7			Park City	KPCW	91.9
	Ingram	KTXI	90.1			Parowan	KUSU	89.5
	Lubbock	KOHM	89.1			Price	KUER	89.5
	Lufkin	KLDN	88.9			Price	KUSU	91.5
	Mcallen	KHID	88.1			Provo	KUSU	88.7
	Midland	KENW	88.9			Randolph	KUER	88.3
	Midland	KMTH	88.7			Randolph	KUSU	91.1
	Odessa	KOCV	91.3			Redwash	KUSU	89.7
	Prairie View	KPVU	91.3			Richfield	KUSU	91.5
	San Angelo	KUTX	90.1			Richfield	KUER	101.7
	San Antonio	KPAC	88.3			Roosevelt	KUSU	100.1
	San Antonio	KSTX	89.1			Salt Lake City	KCPW	88.3
	Texarkana	KTXK	91.5			Salt Lake City	KRCL	90.9
	Victoria	KVRT	90.7			Salt Lake City	KUSU	96.7
	Waco	KWBU	107.1			Salt Lake City	KUER	90.1
	Wichita Falls	KMCU	88.7			St. George	KUER	90.5
						St. George	KUSU	90.9
						Vernal	KUER	90.1
						Vernal	KUSU	91.5
						Washington	KUSU	90.1
						Zion Nat'l Park	KUSU	90.3

VA					WA			
	Charlottesville	WVTW	88.5			Spokane	KSFC	91.9
	Charlottesville	WVTU	89.3			Tacoma	KPLU	88.5
	Charlottesville	WMRY	103.5			Tri-Cities	KWWS	89.7
	Christiansburg	WWVT	1260			Walla Walla	KFAE	89.1
	Harrisonburg	WMRA	90.7			Walla Walla	KWWS	89.7
	Lexington	WMRL	89.9			Wenatchee	KNWR	90.7
	Marion	WVTR	91.9			West Seattle	KPLU	88.1
	Norfolk/Virg Bch	WHRV	89.5			Yakima	KNWY	90.3
	Norfolk/Virg Bch	WHRO	90.3			Yakima	KNWY	90.3
	Norfolk/Virg Bch	WNSB	91.1		WI			
	Richmond	WCVE	88.9			Appleton	WLFM	91.1
	Roanoke	WVTF	89.1			Auburndale	WLBL	930
	Winchester	WMRA	94.5			Brule	WHSA	89.9
	Wise	WISE	90.5			Delafield	WHAD	90.7
VT						Eau Claire	WUEC	89.7
	Bennington	WBTN	94.3			Green Bay	WHID	88.1
	Burlington	WVPS	107.9			Green Bay	WPNE	89.3
	Rutland	WRVT	88.7			Highland	WHHI	91.3
	St. Johnsbury	WVPA	88.5			Kenosha	WGTD	91.1
	Windsor	WVPR	89.5			La Crosse	KLSE	88.1
WA						La Crosse	WLSU	88.9
	Aberdeen	KPLU	90.1			La Crosse	WHLA	90.3
	Bellevue	KBCS	91.3			Madison	WERN	88.7
	Bellingham	KPLU	88.7			Madison	WSHS	91.7
	Bellingham	KUOW	90.5			Madison	WHA	970
	Bellingham Area	KZAZ	91.7			Menomonie	WHWC	88.3
	Brewster	KPBX	91.9			Menomonie	WVSS	90.7
	Cashmere	KNWR	91.3			Milwaukee	WUWM	89.7
	Centralia Area	KPLU	90.1			Oshkosh	WRST	90.3
	Chelan	KNWR	91.7			Park Falls	WHBM	90.3
	Clarkston	KNWV	90.5			Rhinelander	WXPR	91.7
	Cottonwood	KNWO	90.1			River Falls	WRFW	88.7
	Ellensburg Area	KNWR	90.7			Sheboygan	WSHS	91.7
	Goldendale	KFAE	90.5			Sister Bay	WHND	89.7
	Kamiah/Kooskia	KRFA	102.5			Sister Bay	WHDI	91.1
	Moses Lake	KLWS	91.5			Superior	KUWS	91.3
	Mount Vernon	KPLU	91.1			Wausau	WHRM	90.9
	Omak	KQWS	90.1			Wausau	WLBL	91.9
	Port Angeles	KNWP	90.1			Wausau	WXPW	91.9
	Port Angeles	KPLU	91.5		WV	Beckley	WVPB	91.7
	Pullman	KRFA	91.7			Bridgeport	WVPB	107.3
	Pullman	KWSU	1250			Buckhannon	WVPW	88.9
	Raymond	KPLU	90.3			Charleston	WVPN	88.5
	Richland	KFAE	89.1			Clarksburg	WVPB	107.3
	Seattle Area	KCMU	90.3			Huntington	WWWV	89.9
	Seattle Area	KUOW	94.9			Martinsburg	WVEP	88.9
	Spokane	KPBX	91.1			Matewan	WVPB	91.7

WV								
	Morgantown	WVPM	90.9					
	Parkersburg	WVPG	90.3					
	Petersburg	WAUA	89.5					
	Union	WVNP	91.5					
	Weston	WVEP	88.9					
	Wheeling	WVNP	89.9					
WY								
	Afton	KUWA	91.3					
	Buffalo	KBUW	90.5					
	Buffalo	KEMC	91.9					
	Casper	KUWR	88.7					
	Casper	KUWC	91.3					
	Cheyenne	KUWR	91.9					
	Cody	KEMC	88.5					
	Cody	KEMC	88.5					
	Dubois	KUWR	91.3					
	Evanston	KUWR	93.5					
	Gillette	KEMC	88.7					
	Gillette	KUWG	90.9					
	Green River	KUWZ	90.5					
	Greybull/Big Horn	KEMC	91.1					
	Jackson	KUWJ	90.3					
	Lander	KUWR	89.5					
	Laramie	KUWR	91.9					
	Lovell	KEMC	91.1					
	Lyman	KUER	91.9					
	Newcastle	KUWN	90.5					
	Powell	KEMC	88.5					
	Powell	KUWP	90.1					
	Rawlins	KUWR	89.1					
	Riverton	KUWR	90.9					
	Rock Springs	KUWZ	90.5					
	Rock Springs	KUWZ	90.5					
	Sheridan	KEMC	89.9					
	Sheridan	KSUW	91.3					
	Sheridan	KEMC	104.9					
	Thermopolis	KEMC	88.9					
	Torrington	KUWR	89.9					
	Worland	KEMC	88.5					
	Yellowstone Park	KEMC	104.9